Aquaculture and Fish Farming

Aquaculture and Fish Farming

Edited by Roger Creed

SYRAWOOD
PUBLISHING HOUSE

New York

Published by Syrawood Publishing House,
750 Third Avenue, 9th Floor,
New York, NY 10017, USA
www.syrawoodpublishinghouse.com

Aquaculture and Fish Farming
Edited by Roger Creed

International Standard Book Number: 978-1-68286-377-0 (Hardback)

Cataloging-in-publication Data

Aquaculture and fish farming / edited by Roger Creed.
    p. cm.
Includes bibliographical references and index.
ISBN 978-1-68286-377-0
1. Aquaculture. 2. Fish culture. 3. Fisheries. I. Creed, Roger.
SH135 .A68 2017
639.8--dc23

Printed in the United States of America.

TABLE OF CONTENTS

PREFACE

Aquaculture also known as Aquafarming is defined as cultivation of aquatic organisms like fish and aquatic plants. A major branch of this field is fish farming which involves raising fish for commercial purposes. This book outlines the processes and applications of aquaculture and fish farming in detail. The book also discusses multiple techniques in aquaculture and fish farming. It also unfolds the innovative aspects of fish farming which will be crucial for the progress of this field in the future. This book includes some of the vital pieces of work being conducted across the world, on various topics related to this discipline. This text is an apt resource for aquaculturists, fish farmers and researchers who wish to study the unexplored aspects of this field.

Every book is initially just a concept; it takes months of research and hard work to give it the final shape in which the readers receive it. In its early stages, this book also went through rigorous reviewing. The notable contributions made by experts from across the globe were first molded into patterned chapters and then arranged in a sensibly sequential manner to bring out the best results.

It has been my immense pleasure to be a part of this project and to contribute my years of learning in such a meaningful form. I would like to take this opportunity to thank all the people who have been associated with the completion of this book at any step.

Editor

Intersex Occurrence in Rainbow Trout (*Oncorhynchus mykiss*) Male Fry Chronically Exposed to Ethynylestradiol

Sophie Depiereux[1]*, Mélanie Liagre[1], Lorraine Danis[1], Bertrand De Meulder[2], Eric Depiereux[2], Helmut Segner[3], Patrick Kestemont[1]*

1 Unit of Research in Environmental and Evolutionary Biology (URBE-NARILIS), Laboratory of Ecophysiology and Ecotoxicology, University of Namur, Namur, Belgium, 2 Unit of Research in Molecular Biology (URBM-NARILIS), University of Namur, Namur, Belgium, 3 Centre for Fish and Wildlife Health, Vetsuisse Faculty, University of Bern, Bern, Switzerland

Abstract

This study aimed to investigate the male-to-female morphological and physiological transdifferentiation process in rainbow trout (*Oncorhynchus mykiss*) exposed to exogenous estrogens. The first objective was to elucidate whether trout develop intersex gonads under exposure to low levels of estrogen. To this end, the gonads of an all-male population of fry exposed chronically (from 60 to 136 days post fertilization – dpf) to several doses (from environmentally relevant 0.01 μg/L to supra-environmental levels: 0.1, 1 and 10 μg/L) of the potent synthetic estrogen ethynylestradiol (EE2) were examined histologically. The morphological evaluations were underpinned by the analysis of gonad steroid (testosterone, estradiol and 11-ketotestosterone) levels and of brain and gonad gene expression, including estrogen-responsive genes and genes involved in sex differentiation in (gonads: cyp19a1a, ER isoforms, vtg, dmrt1, sox9a2; sdY; cyp11b; brain: cyp19a1b, ER isoforms). Intersex gonads were observed from the first concentration used (0.01 μg EE2/L) and sexual inversion could be detected from 0.1 μg EE2/L. This was accompanied by a linear decrease in 11-KT levels, whereas no effect on E2 and T levels was observed. Q-PCR results from the gonads showed downregulation of testicular markers (dmrt1, sox9a2; sdY; cyp11b) with increasing EE2 exposure concentrations, and upregulation of the female vtg gene. No evidence was found for a direct involvement of aromatase in the sex conversion process. The results from this study provide evidence that gonads of male trout respond to estrogen exposure by intersex formation and, with increasing concentration, by morphological and physiological conversion to phenotypic ovaries. However, supra-environmental estrogen concentrations are needed to induce these changes.

Editor: Patrick Prunet, Institut National de la Recherche Agronomique (INRA), France

Funding: Fonds de la Recherche Scientifique – FNRS, Belgium, grant number: FC 83738 (http://www.frs-fnrs.be/). The funders had no role in study design, data collection and analysis, decision to publish, or preparation of the manuscript.

Competing Interests: The authors have declared that no competing interests exist.

* Email: sophiedepiereux@gmail.com (SD); patrick.kestemont@unamur.be (PK)

Introduction

Gonads are unique among all organs given their bipotential nature: they can develop either into an ovary or a testis from a single primordium. The germinal epithelium is composed of germ cells which develop into sperm or eggs and somatic cells, involved in gamete maturation, which differentiate into Sertoli or follicular cells, depending on the destiny of the gonad. This leads to a plasticity of this developmental pathway. In mammals, the way by which cells choose between either fate is genetically controlled by the expression of a main male gene SRY that leads irrefutably to testis development [1]. Though these pathways are still not well understood in mammals, the situation is even more complex in fish, since they display a great variety of sex determination systems, and because sex differentiation can be subject to exogenous influences, which can override the presumptive developmental route (*i.e.* temperature in zebrafish) [2].

Sex steroids play a key role in these differentiation processes, acting as inducers of the developmental pathway and organizers of cellular differentiation [3]. Numerous studies in different fish species have demonstrated that the administration of exogenous hormones during the critical period of gonadal development can override endogenous sex determining/differentiating mechanisms in the developing fish, enabling the induction of phenotypic sex reversal. Given that one mating type can be more marketable, this practice is now commonly used in fish farming to obtain unisexual populations [4–6].

From an environmental perspective, this lability in gonad sexual differentiation leads to a sensitivity of fish to endocrine disruptors, which are molecules that mimic natural endogenous hormones or interfere with the endogenous synthesis of hormones. These molecules can reach surface waters through wastewaters if they are not properly removed by sewage treatment plants, and can locally attain harmful concentrations for wild fish [7–10]. The effective concentrations appear to be very low, being comprised within the ng to μg/L range, depending on the potency of the molecule and the duration of the exposure [10]. Nowadays there is considerable evidence that fish exposed to these chemicals display reproductive failures [11–13]. Among the endocrine disruptors, xenoestrogens such as the synthetic estrogen ethynylestradiol (EE2 - the main component of the contraceptive pill) in particular have attracted attention [14]. EE2 is widespread in surface waters with mean reported concentrations around 0.05 ng EE2/L but can attain higher levels locally, in exceptional cases up to 831 ng EE2/L [15]. Xenoestrogens can induce feminization of male fish, characterized by changes such as the production of the female egg yolk protein, vitellogenin (VTG, encoded by the gene *vtg*) by males and the appearance of ovarian-like morphological features within testes. Vitellogenin induction is recognised as a biomarker of xenoestrogenic exposure [16] and both immunochemical methods such as ELISAs for vitellogenin protein and RT-PCR methods for vitellogenin mRNA have been developed to measure it in a variety of fish species, including the rainbow trout (*Oncorhynchus mykiss*) [17–19]. Male fish displaying ovotestis (*i.e.* oocytes scattered within testis tissue) have been found in numerous fish species exposed to xenoestrogens in the field or in laboratory experiments [7,20–23].

The rainbow trout is one of the most widely used fish species in ecotoxicology and constitutes a model organism in reproductive physiology of fish. It develops as a synchronous differentiated gonochorist teleost with an annual reproductive cycle [24,25]. Sex determination is genetically controlled by a male heterogamety system (XX – XY) with a main male gene *sdY* recently found by Yano *et al.* (2012) [26]. Several studies have shown an impact of xenoestrogens and effluents on reproductive parameters in juvenile and mature rainbow trout [27,28]. The labile period, during which gonad phenotypic differentiation is sensitive to steroid hormones, has been determined to be between 44 and 51 days post fertilization (dpf) [29]. Vitellogenin induction in trout in response to xenoestrogens is very sensitive [27,30], with positive responses recorded at EE2 concentrations as low as 0.1 to 0.5 ng/L [31]. It has also been recorded in juvenile fish exposed to low concentrations of EE2 (1, 10 and 100 ng/L) [28]. While low concentrations of xenoestrogens are able to induce vitellogenin in trout, the situation is less clear with respect to ovotestis or intersex, as all studies on salmonids have failed to record morphological disturbances following exposure to xenoestrogens or effluents [27,32–37]. Moreover, in their whole-lake experiment, Palace *et al.* [38] exposed fish to 5 ng/L of EE2 (nominal concentration), which induced elevated levels of VTG in Lake trout (*Salvelinus namaycush*), but remained without effect on gonad morphology, whereas fathead minnow (*Pimephales promelas*) and pearl dace (*Margariscus margarita*) also present in the same lake developed intersex gonads. Based on these considerations and to our knowledge, no study exists establishing intersex development in rainbow trout in response to xenoestrogenic exposure. This, however, does not mean that trout are not responsive at all. Indeed, high doses of estrogens, when used for aquaculture practices, are able to induce functional sex reversal of juvenile male rainbow trout into phenotypic females [39,40].

The goal of the present study was thus to investigate the consequences of chronic exposure on gonad morphology of juvenile rainbow trout to EE2. Considering the apparently low propensity of the rainbow trout to develop ovotestis under chronic exposure to estrogen in environmental conditions (see above), we covered a large range of EE2 concentrations, from low, environmentally relevant (0.01 µg/L), to supra-environmental levels (0.1, 1 and 10 µg/L). The concentrations we used approached on the one side environmental levels (0.01 ug/L), although this concentration is still 10 to 20 times higher than what is usually measured in the environment (except in highly polluted sites, downstream of some sewage treatment plants), but it is 20 000 times lower than those used in aquaculture to reverse fish (*i.e.* 20 mg/kg of food [40] or 250 µg/L by immersion [39]). Special attention was devoted to the question as to whether xenoestrogen exposure would result in the formation of intersex gonads. To gain insight into the physiological changes associated with EE2

treatment, we also assayed gonad sex steroid levels (T, E2 and 11-KT) and expression of gonad and brain genes that are potentially regulated by xenoestrogen exposure. Another goal of our experiment was to characterize the whole transcriptome of the fish through the use of microarray assays. The results will be presented in another report. A first step was to measure the expression of key genes involved in sex differentiation and/or EE2 exposure by Q-PCR. The results of these analyses are presented here. As they are the first site of biological actions of environmental estrogens, the expression of the four isoforms of estrogen receptor (*ER*) genes was measured. The rainbow trout displays four nuclear ER isoforms, ERα1, ERα2, ERβ1 and ERβ2 [41] encoded by *esr1a*, *esr1b*, *esr2a* and *esr2b* genes, respectively. Attention has also been paid to genes involved in early sexual differentiation in rainbow trout [26,42,43]: *sdY* (sexually dimorphic on the Y chromosome), *dmrt1* (Doublesex and mab-3 related transcription factor 1) and *sox9a2* (sex determining region Y-box9alpha2). We also focus on key genes of the steroidogenesis, with *cyp11b2.1* (cytochrome P450, family 11, subfamily b, polypeptide 2.1), the gene encoding the enzyme 11β-hydroxylase involved in the 11-oxygenated androgen production; and the gonads and brain isoforms of the aromatase (*cyp19a1a*, *b*), the enzyme which synthesizes estradiol. Finally, the well-established estrogenic exposure marker vitellogenin gene (*vtg*) was assayed.

Materials and Methods

2.1. Ethics Statement

In this experiment, all animals were handled in strict accordance with the European Union's guiding principles in the Care and Use of Animals (Directive 2010/63/EU) and the protocol was approved by the Animal Ethics Committee of the University of Namur (Permit Number: 10/149).

2.2. Fish and sampling

Three thousand eggs of an all-male rainbow trout (*Oncorhynchus mykiss*) population were imported at the eyed stage [23 days post fertilization (dpf)] from the experimental fish farm of the National Institute of Agronomic Research (INRA) (PEIMA, Sizun, France). The embryos were incubated until hatching in the bottom of nests at a constant temperature of 10°C (pH 8, O_2: 9–10 mg/L, photoperiod LD 12:12). These nests consisted of rectangular empty cubes (15×12×15 cm) surrounded by a fine thread hung on the flanges of 110-L closed circulating tanks. Fifty four percents of the fish reached hatching, which is close to values observed in experimental conditions [44]. After yolk sac resorption (60 dpf), the fry were randomly distributed into the tanks (108 individuals per tank) at an average water temperature of 13°C. At the onset of first feeding [Day 0 = D0 at 60 dpf], the fish were submitted to one of 5 nominal concentrations of 17α-ethynylestradiol (purity ≥ 98%, Sigma-Aldrich, Germany) solubilized in ethanol: 0 (solvent control), 0.01 µg/L, 0.1 µg/L, 1 µg/L and 10 µg/L, with 3 tanks per condition. Maximum ethanol levels were 0.002%. To reduce EE2 loss (due to adsorption, metabolic and microbial breakdown), 80% of the water was removed and replaced every 3 days (including new addition of ethynylestradiol). Actual EE2 concentrations were measured in each tank at 6 time points using the Quantitative Ethynylestradiol Enzyme Immunoassay (EIA) Kit (Marloie, Belgium) according to the manufacturer's instructions. The fish were exposed chronically to EE2 for 76 days. At the end of the exposure, the fish weight and length were recorded, and the gonads and brain were collected from all fish. They were immediately frozen in liquid nitrogen and stored at −80°C until RNA extraction. The gonads from five to ten fish, as well as the brains from three fish per tank were pooled to reach enough material for further analyses. Moreover, the gonad pairs of six fish per tank were fixed in Bouin's solution for histological analysis. It is worth noting that most of gonad samples were kept for microarray analysis, whose results will be presented in further reports.

2.3. Histological analysis

The gonads were removed from Bouin's solution, dehydrated through a graded series of methanol, cleared with toluene, and embedded in paraffin. Sections (5 μm) were mounted on glass slides and stained with hematoxylin – eosin – safran trichrome (HES). To make sure ovotestis can be detected, transversal sections were taken over the entire gonad using at least 6 section planes. Each slide (minimum 6 per fish) was seen in its entirety to detect any signs of ovotestis. In total, 81 fish were analysed. Some slides were also stained with Masson's trichrome (hemalun, phloxin and light green) to visualize the presence of cortical alveoli. The staging of the germ cells was done using the criteria of Billard 1992 [45] for trout male and Grier 2007 [46] for trout female. Female control samples were obtained from another experiment (Segner et al., unpublished data). The fish were reared in clean water at a temperature of 7.5°C up to 180 dpf. Sampled fish were anaesthetized with MS 222 (buffered 3-aminobenzoic acid ethyl ester methanesulphonate, Argent Chemical Laboratories, Redmont, CA, USA). For histological analyses, fish were fixed in 4% buffered formalin. Samples were embedded in paraffin, 5 μm thick sections were prepared and stained by hematoxylin-eosin (HE).

Cautions were taken to avoid cross-contamination of samples during tissue collection or processing (e.g. cleaning of equipment between samples, use of different fixative containers, new solvents for dehydration,...) and the male fish were never in contact with female samples.

2.3.1. Terminology. There exists a variable use of terms to describe intersexuality observed in fish in the literature. Here, we adapted the terminology proposed by Hecker et al. (2006) [47], with the term "ovotestis" describing the presence of oocytes into either normal or degenerated testicular tissue. The term "intersex" is the broader term refering to the occurrence of either female gametes in a male gonad or male gametes in a female gonad, while ovotestis explicitely designates presence of oocytes into male gonads.

2.4. Sex steroids

The gonad concentrations of testosterone (T), Estradiol-17β (E2), and 11-keto-testosterone (11KT) were assayed by radio immunoassay (RIA), according to the protocol of Fostier and Jalabert (1986) [48]. Two pools containing 10 pairs of gonads per tank were assayed (n = 6 per treatment) separately. The steroids were extracted from gonads, following a procedure adapted from D'Cotta et al. (2001) [49]. Briefly, 1 g of frozen gonads were homogenized in 1 ml ethanol:H₂O solution (50:50). Homogenates were then centrifuged 15 min at 4000 g at 10°C, and supernatants were collected. The pellets were re-extracted in ethanol:H₂O solution (80:20). The supernatants were then extracted 3 times with dichloromethane, and the organic phase was conserved. These phases were evaporated to dryness and pellets were suspended in 200 μl ethanol. The extraction efficiency was measured with several testis samples spiked with 600 ng of E2. The mean efficiency (± SD) measured was 86.6±7%. All samples as well as the standards of E2, T and 11-KT (Sigma-Aldricht, USA) were assayed in duplicate. The anti-E2 and anti-T were supplied by the Hormonology Laboratory of Marloie (Belgium) and anti-11-KT was a gift from Dr. A. Fostier (INRA, Rennes,

France); the radioactive hormones were purchased from Amersham Pharmacia (UK). No data are shown for the 1 μg/L group because of a technical accident which led to high mortalities and impede the sampling due to a lack of biological material.

2.5. Semi-quantitative Real-time PCR (SQ-RT-PCR)

2.5.1. Total RNA extraction and RT. Total RNA from the gonads (2 pools per tank, containing 5 pairs of gonads each; in total n = 6 per concentration tested) and brain (3 pools per tank containing 3 brains each; in total n = 9 per concentration tested) were extracted using TRIzol reagent (Invitrogen, Life Technologies Europe B.V., Ghent, Belgium) as described previously [50]. Following extraction, samples were treated with DNAse (DNA-free kit, Ambion, Austin,USA) to avoid DNA contamination, according to the manufacturer's instructions. The total RNA concentration was determined using an ISOGEN NanoDrop 2000c spectrophotometer (Wilmington, Delaware, USA) and RNA quality was further controlled on a Bioanalyzer 2100 (Agilent). Only samples with a RIN (RNA Integrity Number) >8 were kept for further analysis. To obtain cDNA, 4 μg of total mRNA was reverse-transcribed using the RevertAid™ H Minus First Strand cDNA Synthesis Kit (Fermentas, Germany) according to the manufacturer's instructions. Negative reverse transcriptions controls were performed by omitting reverse transcriptase from the RT step in order to validate CT values and check we avoid genomic DNA amplification.

2.5.2. Primer design. Key genes involved in sex differentiation in gonads (the four ER isoforms, dmrt1, sox9a2, sdY, cyp11b2.1, cyp19a1a, vtg) and the brain (the four ER isoforms and cyp19a1b) were selected in order to investigate their expression under the different experimental conditions. Specific primers were selected from the literature or were newly designed using Primer3 software [51] (Table 1). The newly-designed primers respected, whenever possible, the following restriction parameters: length, 21–23 base pairs (bp); no more than four successive identical nucleotides; guanin-cytosin content, 30–70%; a maximum of two guanines or cytosines among the five 3′-end bases; no primer dimers; and a short amplicon size (70–150 bp). Moreover, each pair was chosen with at least one primer flanking an intron-exon boundary to prevent genomic amplification. The localization of introns/exons boundaries was made from Leroux et al. (1993) [52] annotation of the ER gene. To further assessed their identity, each primer was then blasted using BLASTN alignement tool [53] against each ER isoforms as well as the entire rainbow trout nucleotides collection. All primers were purchased from Eurogentec (Seraing, Belgium).

2.5.3. Quantitative Real-Time PCR. Real-time PCR was performed in 20 μl (5 μl of cDNA, 2.5 μl of each primer at 500 nM, 10 μl MasterMix 2x) with SYBR Green (Applied Biosystems, Foster City, California, USA) as an intercalating agent. Each measurement was performed in duplicate. The PCR conditions were: 10 min at 95°C, 40 cycles: 15 sec at 95°C, 1 min at 60°C. The specificity and identity of the RT-PCR products were checked by performing a dissociation curve (gradient from 60°C to 95°C) for each gene.

Relative quantifications were established by the comparative CT method (also known as the $2^{-\Delta\Delta Ct}$ method) [54]. This method assumed that the PCR efficiencies of target and reference genes are approximately equal [55], which was validated in our experiment. PCR efficiencies were calculated from high linearity standard dilution curves (Pearson correlation coefficient $R^2 >$ 0.985) slopes according to the equation $E = 10^{(-1/\text{slope})}$ (Table 1). Relative gene expressions are represented as the fold change in gene expression normalized to an endogenous reference gene (housekeeping gene) and relative to the untreated control (0 μg/L

Table 1. Primer sequences used for Q-PCR analysis.

Gene		Sequence	Reference sequence	Efficiency	Reference
esr1a	F	ACA-GGA-ATC-GTA-GGA-AGA-GCT-G	AJ242741	2.05	Newly designed
	R	CGT-AGG-GTT-TCT-CTC-TGT-CAC-C	AJ242741		
esr1b	F	ACC-AGG-GTG-AAA-GCT-GTC-TGC-TA	DQ177438	1.98	Newly designed
	R	CTA-AGA-GGG-ATA-GAG-GAA-AGG-AGA	DQ177438		
esr2a	F	AGA-CGG-TCA-TCT-CGC-TGG-AAG	DQ177439	1.95	Newly designed
	R	ACA-CTT-TGT-CAT-GCC-CAC-TTC-GTA	DQ177439		
esr2b	F	AGA-GGA-AGT-GAA-CTC-CTC-CTC-AGG	DQ248229	2.02	Newly designed
	R	GAT-AGT-AGC-ACT-GGT-TAG-TTG-CTG-GAC	DQ248229		
dmrt1	F	GGA-CAC-CTC-CTA-CTA-CAA-CTT-CA	AF209095	2.03	[77]
	R	GTT-CGG-CAT-CTG-GTA-TTG-TTG-GT	AF209095		
sox9a2	F	ATG-CAG-GTG-CCC-AAG-GCT-CA	AB006448	2.09	[43]
	R	CTC-TGG-CTG-GGG-CTC-ATA-TA	AB006448		
sdY	F	GTG-GTT-TTA-AGC-TCT-AGG-GAG-GA	AB626896	1.97	[26]
	R	GAG-TGA-TGA-GTC-TTG-TCC-AAA-C	AB626896		
cyp11b2.1	F	CTG-GGA-CAT-GTG-TCC-AGG-CG	AF179894	1.98	[96]
	R	CTG-GAT-CCT-GAA-ACA-CAT-CT	AF179894		
Vtg	F	ACC-CTG-AAC-CGG-TCT-GAA-G	BX084166	2.06	Le Gac et al. (personal communication)
	R	CAG-TAT-CTG-CTC-CAC-CAC-A	BX084166		
hprt1	F	AAG-CAG-CCC-CTG-TGT-TGT-GA	TC55247	2.0	Leder *et al.* (personal communication)
	R	CGG-TTT-AGG-GCC-TTG-ATG-TA	TC55247		
cyp19a1a	F	CTC-TCC-TCT-CAT-ACC-TCA-GGT-T	BX083177	1.98	[43]
	R	AGA-GGA-ACT-GCT-GAG-TAT-GAA-T	BX083177		
cyp19a1b	F	CTG-GCA-AAC-GGT-TCT-GAT-C	AJ311937	2.1	[101]
	R	TGA-TGG-ACA-GAG-TGT-CTG-G	AJ311937		

EE2), following these equations: $\Delta\Delta Ct = (C_{T,Target} - C_{T,Housekeeping})_{Test} - (C_{T,Target} - C_{T,Housekeeping})_{Control}$ and Fold change $= 2^{-\Delta\Delta Ct}$. *hprt1* (the hypoxanthine phosphoribosyltransferase 1, a nucleoside metabolism enzyme) was chosen as the housekeeping gene in our experiment. It was selected among several candidate housekeeping genes mentioned in the literature as relevant for gene expression measurement after EE2 exposure, after effective validation of its expression in our samples [42,56].

2.6. Statistical analysis

The results were analyzed by one-way analysis of variance (ANOVA) with linear regression and lack of fit to linearity with EE2 concentration as the fixed criterion and independent variable [57] using the Excel software. Homoscedasticity was checked by the Bartlett test. When heteroscedasticity was significant ($p < 0.05$), data were log-transformed. When the lack of fit to linear regression was significant ($p < 0.05$), the ANOVA1 was followed by Sheffe's post-hoc comparisons [57] ($p < 0.05$ threshold), to point out significant and physiologically relevant variations between conditions. All data are presented as the mean of the replicates and the error bars represent ± 2 standard errors of the mean (SEM), i.e. an approximation of the 95% confidence interval of the mean of each group. A Fisher's exact test was performed to analyse if any differences in frequencies in gonad morphologies observed in our study occured between the triplicate tanks.

Results

The mean EE2 concentrations ± SD were 0.08±0.06 µg/L; 1.62±1.74 µg/L and 9.88±5.06 µg/L. The 0.01 µg/L EE2 concentration was under the detection limit (set at 0.02 µg/L).

The exposure of trout fry to ethynylestradiol impaired their growth rate at the high concentrations used, with a significant drop in the size and weight of the fish exposed to 1 and 10 µg/L compared to the controls ($p < 0.05$). Fish size, weight and mortality observed following the 76 days of exposure are given in Table 2. No effect of fish densities were observed on fish size and weight.

3.1. Histology

Alterations of gonad morphology were observed for all EE2 concentrations tested. In general, histological analysis of gonad sections identified common features due to EE2 treatment, but also revealed features specific to each EE2 concentration. Three categories of fish gonad morphology could be distinguished, namely testicular, ovotestis, and ovarian-like. A description of each of these categories is given below. The proportion of fish belonging to each of these categories per experimental condition is presented thereafter.

3.1.1. Testicular morphology. Macroscopically, the gonads of all-male rainbow trout at 136 dpf appeared as paired and tubular elongated structures lying along the dorsal side of the abdominal body cavity. Histologically, the gonads displayed the histological features of a differentiated yet immature testis, as

Table 2. Fish size, weight and mortality.

EE2 concentration (μg/L)	Number of fish	Mean size (cm) ± SD	Mean weight (g) ± SD	Mean mortality (%) ± SD
0	289	6.3±1.49	8.1±0.65	10.8±6
0.01	283	6.3±1.32	8.01±0.57	12.7±6
0.1	270	6.05±0.3	7.8±0.66	16.7±5
1	101	2.93±0.7*	5.75±0.5*	68.8±9
10	208	2.58±0.4*	5.34±0.34*	35.8±3

*significantly different from control ($p < 0.05$).
Fish size, weight and mortality observed following 76 days of exposure (from 60 to 136 dpf) to several doses of ethynylestradiol (EE2). Mean values ± SD for 3 replicates tanks are provided.

described in the literature [25,45,58]. It was characterized by early seminiferous tubules or cords containing undifferentiated gonocytes (future spermatogonia) surrounded by several supporting cells (future Sertoli cells) in the interstitium between the tubules (Figure 1 A–D). This morphological feature was observed in all control fish (Figure 1 A–B), as well as in several fish exposed to EE2 at the lower concentration used (0.01 μg/L) (see section 3.1.5), covering the entirety (Figure 1 C–D) or areas of the gonad (see ovotestis morphology section 3.1.2). A normal ovarian morphology at the same developmental stage is given in Figure 1E–F.

In treated fish, the testis could show different degrees of degeneration (i.e. loss of tubular arrangement, loss of germ cell number and differentiation, presence of lacuna) which could occupy small areas of the testis or could extend over the whole tissue (see Figure 2 and Figures 3 E–F). Based on these considerations, several individuals display gonads containing germ cells without any recognizable differentiation into male or female phenotype. Since our study was conducted on a all-male rainbow trout population, they were specified as having "altered testicular" morphology (Figure 2).

3.1.2. Ovotestis morphology. Ovotestis gonads were observed only in the EE2 treatments. An ovotestis phenotype is characterized by the presence of oocytes within the testis. Depending on the number and maturation stage of the oocytes in the testis, three categories of ovotestis were distinguished:

– *Ovotestis 1:* the normal and degenerated testis structures reminiscent of ovigerous lamellae, i.e. blisters containing groups of meiotically dividing oogonia. No large previtellogenic follicles are seen, but the germ cells display meiotic stages. The majority of the gonads display a normal testicular morphology, with most sperm cells (gonocytes) being at the spermatogonial stage in cord structure (Figure 3 A–B). With respect to the female germ cells, ovotestis 1 shows similarities to the morphology of an early ovarian developmental stage [46,59,60].

– *Ovotestis 2:* presence of isolated oocytes within an otherwise normal and/or degenerated testis structure (Figure 3 C–D). The oocytes display the characteristics of primary oocytes (protoplasmic oocyte or previtellogenic follicles). From the diplotene phase onward, the oocytes are surrounded by early follicular cells [46]. The oocytes are dispersed among the gonocytes, albeit most of them appeared mainly in the peripheral part of the testis. The arrangement of oocytes in ovigerous lamellae, as is characteristic for ovaries, was not seen.

– *Ovotestis 3:* gonads are separated into an ovarian part, where oocytes but no spermatognia can be found, and a testicular

part. The latter shows a degenerative morphology (Figure 3 E–F). The oocytes are usually in the primary growth stage (Fig. 3 E–F). However, in some individuals, mature oocytes containing much vitellogenin are also observed (see section 3.1.4.2). For all samples belonging to this category, serial sections were made and confirmed that the asymmetry was not restricted to one single area within the gonad, but was consistently present throughout the entire gonad.

The classification above is based exclusively on the criteria "number of oocytes" and "maturation stage/organization of oocytes". The testicular alterations are neither described in details in this article nor are they considered in the ovotestis classification, since this was not within the scope of this study.

3.1.3. Ovarian-like morphology. Several fish subjected to the EE2 treatments displayed ovary-like gonads, i.e. they were apparently gender-converted (at least in terms of structure) (Figures 4). The gonads displayed the characteristics of an immature ovary, with an ovarian germinal epithelium forming ovigerous lamellae inside the ovarian cavity [25,59,61,62]. The mesenchymal stroma underlying the multi-layered epithelium was composed of fibroblasts, interstitial cells, leukocytes, blood capillaries and collagen fibers [25],[63],[46]. At least in some individuals, it appeared that the area of the stromal tissue (ST in figure 4) between the follicles was increased compared with that expected in a "normal" ovarian morphology of genetic females at this developmental stage.

3.1.4. Treatment-associated morphological particularities of the gonads. *Oviduct-like epithelioid structure:* Some fish exposed chronically to EE2 displayed an epithelial structure extending from the gonad. The epitheloid was composed of ciliated columnar cells lying upon thin loose vascular connective tissue (Figure 5A). The structure contained many circumvolutions, was always connected to the gonad, and displayed blind ends. It could reach huge proportions, sometimes exceeding the size of the gonads (Figure 5B). Oocytes were never observed in this struture. This oviduct-like structure was exclusively observed at the higher EE2 concentrations (0.1, 1 and 10 μg/L). The percentages of fish displaying this feature per concentration were: 70% (14/20) at 0.1 μg/L; 78% (7/9) at 1 μg/L and 83% (15/18) at 10 μg/L. The size of the structure appeared to increase with the increase in EE2 concentration, attaining huge proportions at 10 μg/L, compared to the gonad size (Figure 5B).

Vitellogenic oocytes: Surprisingly, we found vitellogenic follicles in several juvenile fish exposed to EE2 (Figure 6 A–D). Their size was between 0.8 and 1.5 mm. They were filled with cortical alveoli and vitellogenic deposits uniformly distributed in the ooplasm (Figure 6 A,C), or aggregated on its periphery (Figure 6 B,D). Their walls displayed features of vitellogenic follicles, with the zona

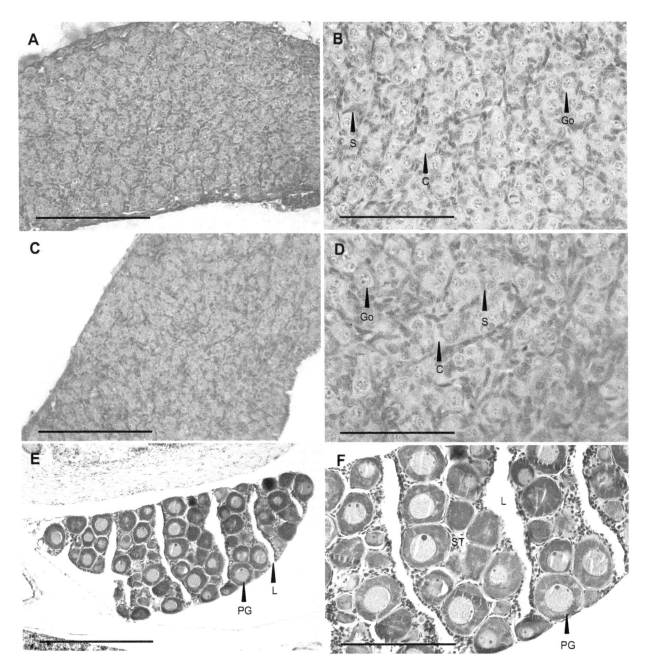

Figure 1. Testicular morphology. A–B: Rainbow trout testis at 136 days post fertilization (dpf), control fish (1360 degree-days). This picture shows the differentiation status of a control testis, with a dominance of gonocytes (Go, future spermatogonia) surrounded by supporting cell (S, future sertoli cell) organized in cords (C, future tubules). HES. Scale bar A = 100 μm, B = 50 μm. C–D: Rainbow trout testis at 136 dpf, fish exposed chronically to 0.01 μg/L of EE2 for 76 days. HES. Go: gonocyte; S: supporting cell. Scale bar C = 100 μm, D = 50 μm. E–F: **Normal ovarian morphology**. Rainbow trout early stage ovarian morphology, control fish (1350 degree-days). PG: primary growth oocyte; L: lamellae; ST: stroma. Scale bar E = 100 μm, F = 50 μm.

radiata surrounding the oocyte, covered by the granulosa layer and the thecal cells. However, the morphology of the folliclar layer deviated in two aspects from the normal morphology (Figure 6 A–B). In figure 6A, the granulosa is composed of a regular continuous cuboidal cell layer, as described in mature rainbow trout by Van den Hurk (1979) [62]; in figure 6B, however, the oocyte displays an acellular layer composed of microvilli-like structures representing the zona pellucida (chorion). These protrusions may originate from the follicular cells. The results of Masson's trichrome staining supported the presence of aberrant follicular morphology (Figure 6

C–D), and the phloxine staining suggested the presence of proteins in the cortical alveoli. The vitellogenic follicles were floating in the body cavity, separated from the gonad. Intermediary stages between the previtellogenic oocytes (stage 2) and the vitellogenic follicles (stage 5) were not found. In total, five of all fish examined displayed mid-vitellogenic follicles, among all the concentrations tested. At 0.01 μg/L (two follicles observed), one of the follicles appeared immediately adjacent to testicular tissue. These results were absolutely unexpected, as the fish were juvenile and no other

The levels of E2 and T (mean values of 6 replicates for all the concentrations tested $\pm$ 2 SEM) did not vary significantly between the concentrations of EE2 tested (4.76 ± 0.51 and 13.4 ± 1.7 ng/g, respectively).

3.3. Gene Expression

3.3.1. Gonads. We measured the expression of the four isoforms of estrogen receptor (*ER*) genes (*esr1a, esr1b, esr2a, esr2b*), as well as four "male genes" (*sdy, dmrt1, cyp11b* and *sox9a2*) and two "female genes" (*cyp19a1a* and *vtg*). The expression of all genes measured varied significantly according to the treatment. Moreover, within the range of concentrations tested, most of the genes tested showed a positive or negative linear response to EE2 (Figure 9 A–J).

Three of the four ER isoforms decreased linearly with the increase in EE2 levels ($p < 0.05$). They showed under expression of a similar amplitude, with a fold change between the control and 10 µg/L of EE2 of 5.9, 4.2, and 2.9-fold decrease in expression of *esr1a, esr2a* and *esr2b*, respectively (see expressions in log2 fold change in Figure 9 A, C, D). Only the *esr1b* isoform showed a different pattern of expression. It increased at the intermediate EE2 concentration with a 2.45 fold induction at 0.1 µg/L and then decreased significantly ($p < 0.01$) at 10 µg/L, with its expression reduced by 50% compared with the control.

We observed a linear decrease in the expression of the "male genes" *dmrt1* ($R^2 = 0.85$; $p < 0.001$) and *sox9a2* ($R^2 = 0.93$; $p < 0.001$) (Figure 9 E–F). Sharp underexpression was seen for *dmrt1* which was 30 times less expressed at 10 µg/L than in the control group. The expression of *sox9a2* varied to a lesser extent, its expression being reduced to a fold change of 3.4 at 10 µg/L compared with the control. *sdY* and *cyp11b2.1* expressions were strongly repressed by the EE2 treatment, with up to 38 and 343-fold decrease at 1 µg/L respectively (Figure 9 G–H). Both genes displayed lesser underexpression at the 10 µg/L concentration.

Expression of the "female genes" varied, as shown in Figures 9 G–H. The *vtg* gene was overexpressed compared with the control group, with a significant peak ($p < 0.001$) at 0.1 µg/L (fold change of 91.2). This overexpression was still significantly different from the control group but declined to a fold change of 39.9 and 30.9 at 1 and 10 µg/L, respectively. At 0.01 µg/L of EE2, the *vtg* overexpression reached a fold change of 4.41, which was not significantly different from the control group.

Gonad aromatase gene expression showed slight but non-significant overexpression at 0.01 and 0.1 µg/L, with fold changes of 1.8 and 1.3 compared with the control group, respectively. Its expression decreased significantly at the higher concentrations, being 4 and 9 times underexpressed at 1 and 10 µg/L, respectively.

3.3.2. Brain. We measured the expression of the four isoforms of *ER* genes, as well as the brain aromatase form, *cyp19a1b*. Only the *esr1a* isoform displayed a significant linear increase ($R^2 = 0.97$; $p < 0.001$) among the concentrations of EE2 tested, with an amplitude of overexpression of 1.8 between the control and the 10 µg/L groups (Figure 10 A). The variations observed in *esr2a* expression were less clear. There was no significant difference between the control group and all the concentrations tested. However, a small decrease in its expression at 0.1 and 1 µg/L of EE2 and an increase at 10 µg/L created a significant difference between these groups, the expression at 10 µg/L being higher by a factor of 0.8 ($p < 0.01$) compared with 0.1 and 1 µg/L (Figure 10 B). The changes in *esr1b* and *esr2b* expression did not differ significantly between the EE2 treatments (data not shown).

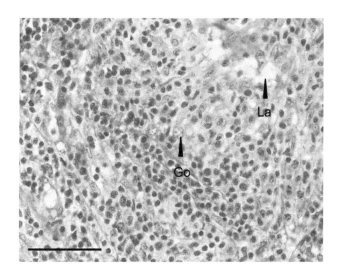

Figure 2. Altered testicular morphology. All-male rainbow trout testis at 136 dpf exposed chronically to 1 µg/L EE2. HES. Scale bar = 50 µm. This picture illustrates the degeneration of the testis observed in several fish exposed to EE2, such as a loss of tubular arrangement, loss of germ cell number (Go) and differentiation, presence of lacuna (La). There is no recognizable differentiation of the gonad into male or female phenotype.

signs of maturation were found in the gonads displaying these follicles.

3.1.5. Relative frequencies of the morphological gonad forms. Figure 7 presents the gonad morphology found at each percentage in the individual experimental treatments. No differences of frequencies in gonad morphologies were seen between replicate tanks (Fisher's exact test: $p = 0.85$ in the 0.01 µg/L group; $p = 0.94$ in the 0.1 µg/L group).

All fish from the control tanks belonged to the testicular category, with male fish displaying immature testes. The presence of female features within the testis appeared in the groups exposed to EE2 at 0.01 µg/L and higher. In the groups exposed to the lowest EE2 concentration (0.01 µg/L), nearly half of the group (47.4%) displayed intersex gonads. The three ovotestis categories were equally represented, with 15.8% of the fish belonging to each category. Several individuals displaying an altered testicular morphology were also observed, as well as one case of an ovarian-like morphology.

At higher concentrations, an increasing percentage of fish showed ovary-like gonads. At 0.1 µg/L, one third of the fish were reversed (30%), while ovotestis individuals still represented about 40%, with ovotestis 1 being the main stage represented (35%). Only one individual displayed ovotestis 3 (asymetric). No fish with histological testes were seen at this concentration. All the fish exposed to 1 and 10 µg/L of EE2 appeared reversed, with only one individual displaying an altered testicular morphology at 1 µg/L.

3.2. Sex Steroids

Testosterone (T), estradiol (E2) and 11-ketotestosterone (11KT) gonads levels were assayed to determine whether chronic EE2 exposure was able to disrupt the steroidogenesis in trout fry. At the concentrations tested, the levels of 11-ketotestosterone decreased linearly with the increase of the log of EE2 concentration ($R^2 = 0.94$; $p < 0.05$) (Figure 8). These levels ranged between 24.2 ± 5.2 ng/g in the control group and 13.9 ± 4.7 ng/g in gonads exposed to the highest EE2 concentration.

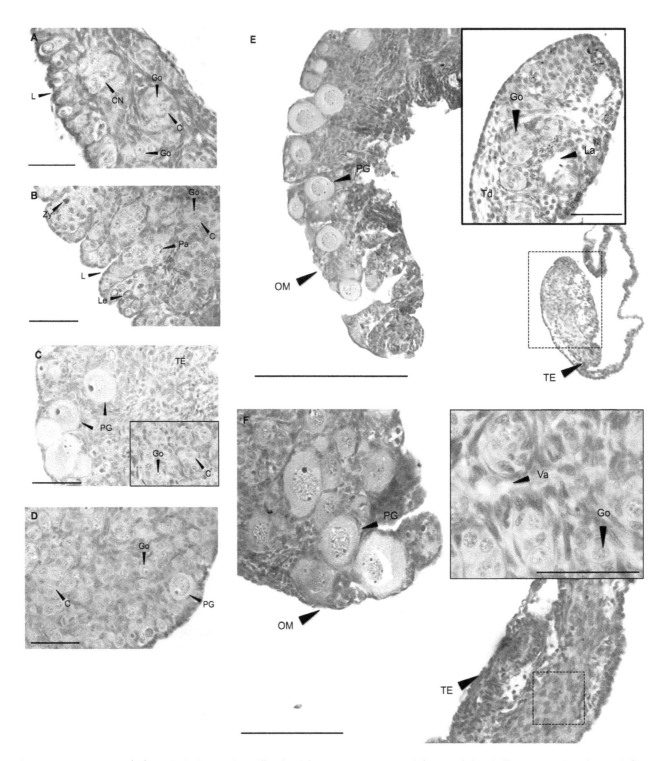

Figure 3. Intersex morphology. A–B: Ovotestis 1. All-male rainbow trout testis at 136 dpf exposed chronically to 0.01 µg/L and 0.1 µg/L for 76 days. HES. L: lamellae; C: cords of gonocytes (Go); CN: cell nest in different meiosis stages: Le: leptotene; Zy: zygotene; Pa: pachytene. Scale bar = 50 µm. **C–D: Ovotestis 2**. All-male rainbow trout testis at 136 days post fertilization (dpf) exposed chronically to 0.01 µg/L EE2 for 76 days. HES. Scale bar C = 50 µm, insert in C = 25 µm, D = 50 µm. Figure C shows oocytes appearance in an altered testis structure. The insert in Figure C shows the normal testis structure, with gonocytes (Go) organized in cords (C) observed in other areas of this gonad. Figure D shows oocyte appearance in a normal testis structure. PG: primary growth oocytes; TE: altered testicular tissue; FC: follicular cells. **E–F: Ovotestis 3**. All-male rainbow trout testis at 136 dpf exposed chronically to 0.01 µg/L for 76 days. HES. Inserts show the altered testis structure at higher magnification. Dotted line represents the enlarged tissue section. OM: ovarian-like morphology; PG: primary growth oocytes; TE: altered testicular tissue; Sg: spermatogonia, La: lacune; Td: tubule disorganization; Va: vacuolation. Scale bar E = 300 µm, F = 100 µm, inserts = 50 µm.

The expression of *cyp19a1b* increased slightly with the rise in EE2 concentration, and reached a significant peak at 10 µg/L with a fold change of 3.7 observed against the control group (p< 0.001) (Figure 10 C).

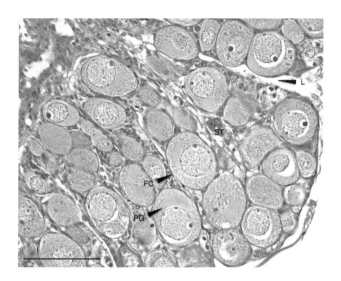

Figure 4. Ovarian-like morphology (sex-reversed fish). All-male rainbow trout testis at 136 dpf exposed chronically to 1 μg/L EE2 for 76 days. HES. FC: follicular cell; PG: primary growth oocyte; L: lamellae; ST: stroma. Scale bar = 100 μm.

Discussion

This study showed disruption of gonad morphology in rainbow trout male juveniles chronically exposed to EE2 (0.01 to 10 μg/L), together with changes in gonad sex steroid levels and gonad and brain gene expression. A concentration dependency of the effects was evident: at lower concentrations of EE2, ovotestis gonads predominated, while the gonads were sex converted at the higher concentrations. Intersexuality has been described using several terminology in the literature. Here, we refered to the term "ovotestis" to describe the occurrence of oocytes in a male gonad [47]. We use the more broad term "intersex" (occurrence of either female gametes in a male gonad or male gametes in a female gonad) to refer at the litterature. In gonochoristic species, intersex fish are fish with gonads displaying male and female characteristics simultaneously. They are observed at low prevalences in wild populations (*e.g.* in gudgeon [20,64]), including in the obvious absence of estrogen exposure [2,65]. The release of endocrine disruptive compounds into the environment has been shown to be an environmental inducer of intersex gonads through field and laboratory-based experiments [7,21,31]. The morphological appearence of intersex gonads appears to vary among fish species, but also with the type of steroid-active chemical and its concentration, the route and duration of exposure, and the timing of exposure (*i.e.* developmental stage of the fish) [65].

In rainbow trout, intersexuality has been rarely observed in the wild (only two cases of apparently natural intersex have been reported to date [66]), and there are no reports of intersex at contaminated sites in several trout species [27,33]. Several studies investigating the effects of effluents or EDCs (including EE2) did not observe ovotestis [37,67], and the sex ratios of exposed groups were unchanged compared with control groups. The few cases of disruption in gonad morphology of salmonids environmentally exposed to estrogens that have been reported in the literature include a high degree of vacuolation of the testis in brown trout (*Salmo trutta*) from a stream impacted by sewage effluent [34], and an intersex individual in brown trout exposed to a high dose of E2 (0.5 μg/L) at the juvenile stage [35]. For rainbow trout, inhibition of testicular growth was found after developmental exposure to estrogenic alkylphenols and EE2, but in this stduy gonad histology

was not examined [27]. Consistently with the laboratory-based study of Krisfalusi in 2000 [67], our study showed that gonad morphology in rainbow trout is sensitive to feminization by environmental estrogens. However, it confirmed that high levels of xenoestrogen exposure are necessary in trout fry to change gonad morphology. It also appears that not only the concentration, but also the timing of exposure is of great importance in this process. Likewise, the results of Krisfalusi and Nagler (see below) [67]) suggest that trout may be more sensitive if exposed prior to the fry stage.

In the present work, the EE2 treatment resulted in different forms of ovotestis, ranging from single oocytes in otherwise normal tetsicular tissue (ovotestis 2), to testes with areas ressembling early ovaries (ovotestis 1) over mixed testicular-ovarian gonads (ovotestis 3) to fully reversed ovarian-type gonads. To our knowledge, the first feature observed (ovotestis 1) has never been reported in rainbow trout or other species submitted to xenobiotics. A high proportion of the ovotestis fish belonged to this category (51%). This feature is characterized by early ovigerous lamellae and actively dividing oogonia grouped into nests at the periphery of the testis. This peripheral structure looks like ovary tissue at the early differentiating stage (6–9 weeks of development) as described by Cousin (1988) [61], which could correspond to delayed female tissue, as the fish were 14 weeks old at the sampling time. The second ovotestis category observed (Ovotestis 2) here has been extensively reported in studies on the morphological impact of EDCs in several fish species (for an extensive review, see [65], Table 6.2). Ovotestis has been observed in rainbow trout following exposure to high doses of exogenous hormones [35]. Interestingly, Krisfalusi and Nagler (2000) [67] observed ovotestis in genetically male rainbow trout juveniles after only two 2-hour immersions 7 days apart in 250 μg E2/L. They observed up to 63% intersex fish and found that the labile period in rainbow trout is between 44 and 51 dpf. Our results support this greater sensitivity of rainbow trout to exogenous steroid treatment during the early sex-differentiating period. The last feature observed (ovotestis 3) is similar to asymmetric intersex described in the literature in several fish species exposed to EDC (*e.g.* Zebrafish [68], Whitefish [69]). Cases of asymmetric gonads were reported during attempts to achieve inversion in juvenile rainbow trout treated with high doses of estrogens (30–60–120 μg/g estrone) [70], and in a *mal*-mutated family of rainbow trout [66]. Such cases can also occur in feral fish populations, although the reasons for this are not known (*e.g.* [71]). The relative frequency of the different types of gonads varied with the EE2 concentrations, suggesting that this represents a series of events, from initial, slight disturbances in normal germ cell differentiation in response to the presence of elevated estrogen signals, to the full conversion of gonad physiology. This interpretation was corroborated by the gene expression results and sex steroid analyses (see below).

The EE2 treatment did not simply shift the gonads from a testicular to an ovarian morphology, but also induced certain alterations in testis structure (see results section 3.1.2). Pathological alterations have been reported in testes and ovaries after exposure to EDCs in several fish species (for an extensive review, see [65], Tables 6.3–6.12), and could be due to a toxic effect of the treatment. Considering that a 96-h LC50 of 1.6 mg/L was reported for EE2 in rainbow trout [72], the alterations observed at doses 160 thousand times lower (i.e. 0.01 μg/L) attest of the powerfulness of EE2 action *in vivo*. A deeper histopathological analysis of the testicular morphology is beyond the scope of this study, and will be addressed in future studies. Here, we will discussed the major pathological features observed, described in the results section 3.1.4.

A

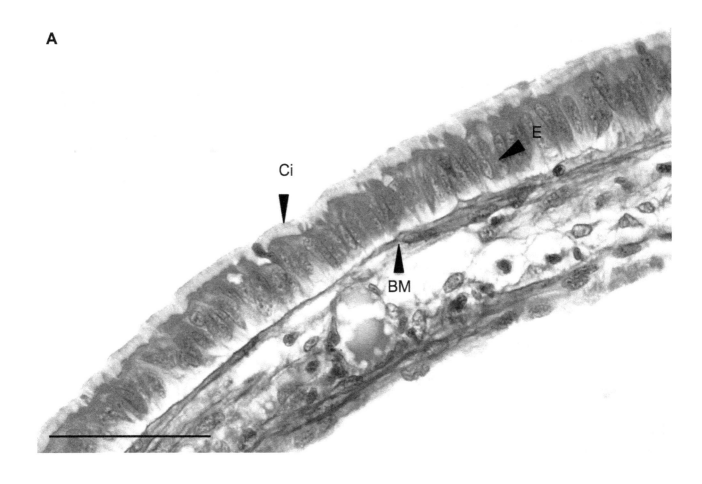

B

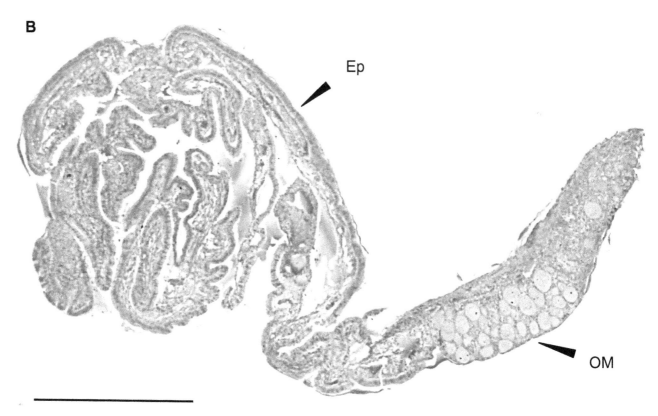

Figure 5. Proliferative oviduct-like epithelioid structure. All-male rainbow trout testis at 136 dpf exposed chronically to 1 and 10 µg/L EE2 for 76 days. HES. Ci: cils; E: columnar epithelial cells; BM: basement membrane; Ep: epithelioid; OM: ovarian-like morphology. Scale bars A = 50 µm; B = 500 µm.

The most intriguing pathological feature was the appearance of an epitheloid ciliated structure starting at the 0.1 µg/L concentration, which became more prominent with the increase in the EE2 doses. The "proliferative oviduct-like epithelioid" structure has never been reported in juvenile or mature rainbow trout females, either under natural or experimental (including toxic) conditions. The epitheloid designation is based on the fact that the tissue did not strictly correspond to the definition of an epithelium according to Grier *et al.* [63], as it did not border any body surface, lumen or duct and appeared to be vascularized. Moreover, Mac Millan [60] concluded that most epithelia of teleost oviducts are composed of ciliated cells interconnected by junctional complexes. However, salmonids display primitive ovarian features, the rudimentary ovarian cavity being continuous with the body cavity on its ventral side [46,73]. They have no oviducts and the ova are discharged directly into the coelom [46]. The "proliferative" appellation refers to the fact that the structure increased in size

with the increase in EE2 concentration. Several histopathologically oriented investigations of endocrine disruption have also inspected gonadal duct formation, and the results generally mention induction of this structure [7,12,74]. Our results are supported by the fact that estrogens play an essential role in stimulating growth of the female reproductive tract (including oviduct) [65]. Interestingly, Cousin [61] reported a high proliferation of the efferent duct ciliated epithelium in male rainbow trout juveniles exposed to 17α-methyltestosterone.

Another particular morphological feature in the gonads of EE2-exposed trout was the presence of mature follicles. Physiologically, rainbow trout develop mature follicles no earlier than after 2 years of age. Folliculogenesis (or oogenesis) is the complex process by which a diploid oogonia develops and matures into a haploid ova [60]. This process has been extensively described in fish, and the number of developmental stages described depends on the species studied and the authors [45,46,62,75,76]. In rainbow trout, follicle

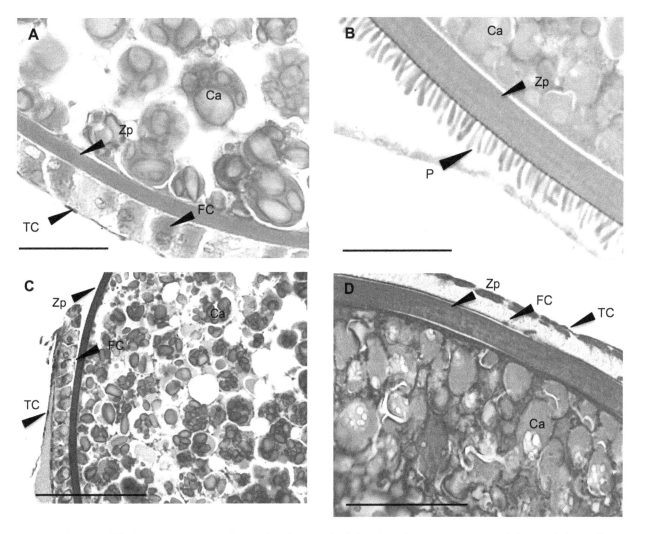

Figure 6. Vitellogenic follicles. A–B: Vitellogenic follicles found in gonads of all-male rainbow trout testis at 136 dpf exposed chronically to 0.01 and 10 µg/L EE2 for 76 days. HES. Zp: Zona pellucida; FC: follicular cells (granulosa); TC: thecal cells; CA: cortical alveoli; P: protrusions probably originated from the follicular cells. **C–D**: Masson's staining of a vitellogenic follicle. Scale bars 50 µm.

Gonads classification

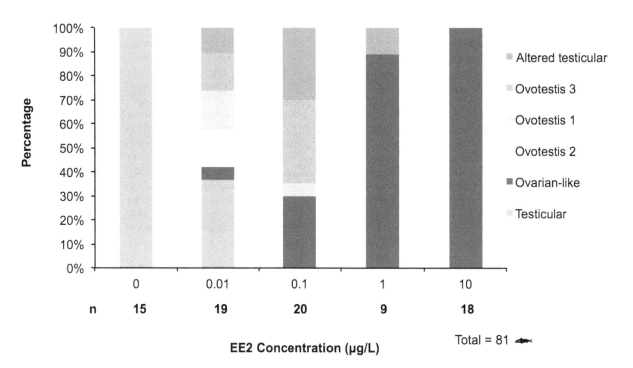

n   15   19   20   9   18

EE2 Concentration (µg/L)   Total = 81

Figure 7. Relative frequencies of the gonad mophological forms. This graph represents the percentage of fish belonging to each morphological category, per concentration of EE2 used in the chronic experiment. n refers to the number of fish analyzed per experimental concentration.

maturation has been divided into three main phases (previtello-genesis, exogenous vitellogenesis and maturation), with seven oocyte stages, depending on their histological features [62]. During this process, the oogonia undergo mitotic and meiotic divisions, become surrounded by follicular cells which specialize into granulosa and thecal cells, increase dramatically in size and accumulate yolk as the egg reserve during vitellogenesis. In mature fish, this process normally extends over nearly 8 months [25]. In

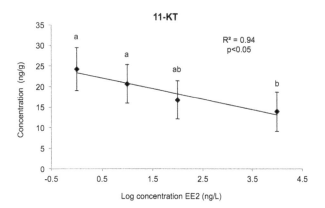

Figure 8. 11-ketosterone levels. This graph show the linear relationship between [11-ketosterone] and LOG[EE2] in rainbow trout fry gonads submitted chronically to increasing concentrations of EE2. For each group, data represents the mean ± 2 SEM from 6 replicates measured independantly. Each replicate consisted of a pool of 10 pairs of gonads.

juveniles, the ovary displays only stages 1 and 2 of previtellogenic oocytes [62], its size does not exceed 110 µm and it is surrounded by a thin follicule layer. The vitellogenic process only starts at the first maturation when fish are around 2 to 3 years old. In the present study, some isolated mid-vitellogenic oocytes were found in 4-month-old genetic males. This feature has never been reported thus far. Moreover, no intermediary follicular maturation stages were seen, and the mature follicles appeared isolated from the gonads. Thus, the action of EE2 seems to override oogenesis, maybe acting as a major actor in this maturation pathway. Whether these effects resulted from the action of EE2, or involved additional systemic processes implemented by the chronic treatment, cannot be determined from the data obtained in this study.

The expression of most of the genes tested was significantly regulated by the treatments and displayed a clear response pattern, with linear or optimal-response curved. In the testis, *cyp11b2.1* expression appeared strongly suppressed by the EE2 treatment. We also observed linear downregulation of *dmrt1* and *sox9a2* with increasing EE2 concentration, which is in line with the change of gonad morphology from testis to ovary. These findings also confirm the status of *cyp11b2.1*, *dmrt1* and *sox9a2* as early testicular markers [77,78]. However, unlike the conclusions of Vizziano in her study (2008 [78]), *dmrt1* response to EE2 was more marked than *sox9a2*. *sdY* has been recently described as a master sex-determining gene in rainbow trout [26]. To our knowledge, this is the first time the expression of this gene is investigated under chronic exposure to exogenous estrogenic hormone. The significant decrease observed in EE2-treated fish highlights a down-regulation of its expression by exogenous estrogenic treatment, in a

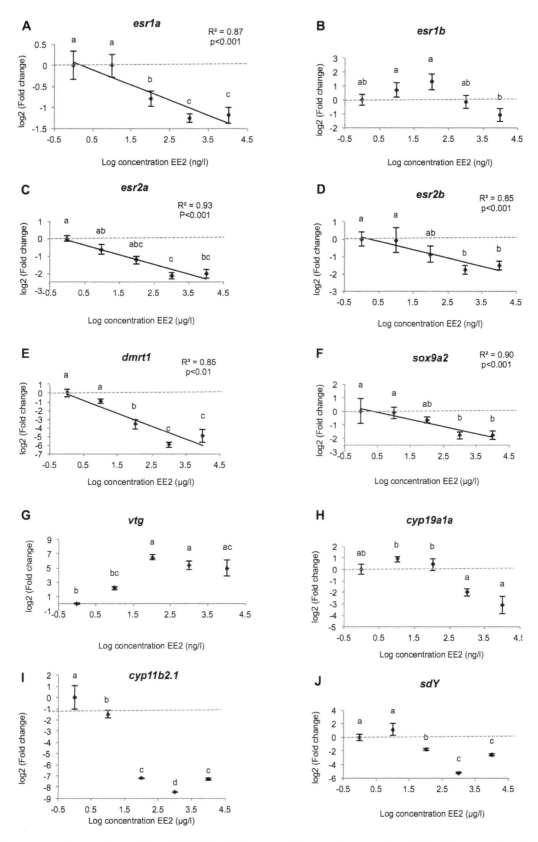

Figure 9. Genes expression profiles in the testis. Relationship between fold change (expressed as mRNA relative expression ratio with control group) of differentially expressed genes and LOG[EE2] in the testis of rainbow trout fry exposed chronically to increasing concentrations of EE2. For each group, data represents the mean ± 2 SEM from 6 replicates measured independantly. Each replicate consisted of a pool of 5 pairs of gonads. The letters a, b, c summarize the post hoc comparisons (p<0.05), the groups with the same letter being not significantly different. When the lack of fit to linear regression is not significant (p>0.05) the linear regression and associated R^2 are shown.

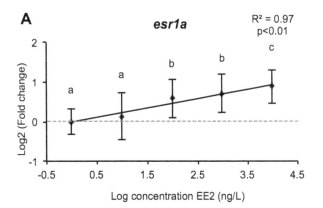

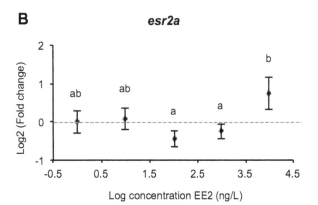

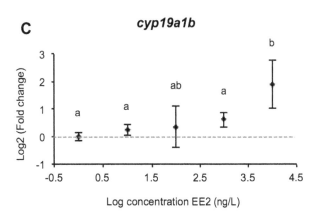

Figure 10. Genes expression profiles in the brain. Relationship between fold change (expressed as mRNA relative expression ratio with control group) of differentially expressed genes and LOG[EE2] in brains of juvenile male rainbow trout gonads exposed chronically to increasing concentrations of EE2. For each group, data represents the mean ± 2 SEM from 9 replicates measured independantly. Each replicate consisted of a pool of 3 brains. The letters a, b, c summarize the post hoc comparisons (p<0.05), the groups with the same letter being not significantly different. When the lack of fit to linear regression is not significant (p>0.05) the linear regression and associated R^2 are shown.

dose-dependant manner. This reinforces its role in testis differentiation. As this gene has been found to be mainly expressed during early testicular development, with a peak of expression between 60 and 90 dpf and a rapid falls thereafter, the relation between

estradiol and *sdY* expression should be investigated further, at early developmental stage.

The specific ovarian differentiation marker *cyp19a1a* and the *esr1b* isoform displayed slight overexpression at low EE2 doses and significant underexpression at higher ones. This pattern was also observed for gonadal *vtg*, but with a sharp 90-fold increase at 0.1 μg/L and a marked drop at higher doses. The increase observed at low concentrations for the estrogen-responsive genes in gonads and brain may represent a positive response to the estrogen signal, but the subsequent decline could attest to a toxic effect of the higher doses of EE2 used. This is consistent with the decline observed in the size and weight of the fish and the pathological morphological features observed in sex-reversed fish at these concentrations. This pattern of expression (at least for *cyp19a1a*) may also reflect a negative feedback in response to the presence of high doses of exogenous estrogens.

Nuclear estrogen receptors (*ERs*) act as transcription factors mediating the expression of estrogen-dependent genes. Being not strictly specific, they are the first targets to trigger xenoestrogen action. Four ER isoforms have been described in rainbow trout (ERα1–2; ERβ 1–2 encoded by *esr1a–b and esr2a–b* genes, respectively), with ubiquitous tissue distribution in juvenile fish, the highest ER mRNA levels being measured in the brain and testis [41]. Following xenoestrogen exposure, ER expression has been shown to increase in the gonads, brain and liver of several fish species [79–81], although in other fish species, no such induction has been described [82]. Our results are neither consistent with these results, nor with those obtained by Boyce Derricott (2009, 2010) who evaluated the expression of the entire ER family in rainbow trout juveniles exposed to a reversion dose of E2 and only found *esr1a* to respond, being upregulated in the liver and gonads [83,84]. However, while the upregulation of ER mRNA in the liver by estrogens is characteristic of oviparous vertebrate systems (*i.e.* the increase in ER is necessary for VTG uptake during vitellogenesis) [85], the regulation of ER transcription in the gonads and brain is still poorly understood. Indeed, several studies have reported that the expression of different ER isoforms can differ widely within the same species and that xenoestrogens can show varying time- and dose-dependent patterns of activation, which also differ from that of E2 [86,87]. Moreover, in rainbow trout, no sexual dimorphism is seen in the 4 ER expression patterns during gonad development [84]. Based on these considerations, we conclude that in rainbow trout fry, the responses of the different ER subtypes to estrogens vary between tissues, (*i.e.* the same receptor may respond differently if it is expressed in the gonads or brain) and most (but not all) ER isoforms respond to EE2 treatment in a dose-dependent manner. The ER response may vary as well with the developmental stage of the exposed fish, as different ER expression patterns have been observed in juvenile and mature fish exposed to exogenous estrogens [86].

The enzyme cytochrome p450 aromatase (*cyp19a1*), which converts androgens into estrogens, plays a key role in the control of ovarian and testicular differentiation. Its overexpression is needed to trigger and maintain ovarian development whereas its downregulation is necessary to induce testicular differentiation [88]. Several indications suggest that interference with the aromatase system could lead to malfunction of the reproductive system [89]. Two forms of this enzyme exist in fish, one mainly expressed in the gonads (*cyp19a1a*), and the other in the brain (*cyp19a1b*). Our results did not show any variation in gonad aromatase expression between control and EE2-treated fish. Only the brain aromatase expression showed an increase, at least at the highest concentrations. These observations would agree with the presence of ERE in the promoter region of *cyp19a1b*, while an ERE is lacking in the gonad form of aromatase [89]. Overall, exposure to xenoestrogens generally results in upregulation of *cyp19a1b* expression [90–93], whereas most studies conducted on fish feminization found no estrogen-inducible upregulation of *cyp19a1a* gene expression [78,90,94], with one exception in a study of EE2 induction of *cyp19a1a* mRNA in the testis of medaka [95]. Our results support the hypothesis that, in rainbow trout, feminization induced by exogenous estrogen treatment may not involve direct regulation by the aromatase pathway [88]. This is reinforced by the fact that neither E2 nor T levels significantly varied at all the concentrations tested in our study. On the other hand, the strong underexpression of *cyp11b2.1*, correlated with the decrease of 11-KT levels in EE2-exposed fish, draws attention to an important potential role of the 11β-hydroxylase in sex reversal in fish. This gene is expressed very early in differentiating gonads, with a sexual dimorphic pattern of expression visible from 45 dpf. Its expression has been also shown to be strongly and quickly suppressed by exogenous estrogenic treatment at very early developmental stage (from 53 dpf) [96]. Finally, although the initiation of ovarian development in female trout is mediated primarily by estradiol (E2), nuclear estrogen receptors (ERs), *cyp19a1a*, and, to an unknown extent, brain ERs and *cyp19a1b*, the lack of a correlation between the ER isoforms and *cyp19a1a–b* expression underscores the complexity of their relationship and the diversity of estrogen actions outside the sex differentiation process. Indeed, in the brain it has now been well established that estrogens play major regulatory roles in neurotrophic and neuroprotective pathways, besides their classical control of reproductive functions [97,98]. From this standpoint, the upregulation of *esr1a* observed in the brain, with no effects observed on *cyp19a1b* expression, could be related to these further functions of E2. Moreover, the gene expression results should be viewed with caution due to the involvement of those genes in transdifferentiation processes, as only one sampling point was analyzed, after a long exposure to the potent xenoestrogen EE2. Therefore, their expression pattern could reflect more complex responses involving other pathways and systems. Considering the sensitivity of rainbow trout to ovotestis induction at low doses of xenoestrogen, as highlighted in the present study, it would be interesting to study the mechanisms underlying this male-to-female transdifferentation processes in greater detail, focusing on low doses (below those used here) and very early developmental stages. Given that numerous genes should be implicated, and that the trout genome is being sequenced, it would be interesting to extend the study to the entire trout genome. It would also be of interest to study the reversibility of the morphological changes observed.

In conclusion, the fact that ovotestis has not been reported in field studies of rainbow trout does not mean that trout cannot develop ovotestis. The existing litterature highlights that the impact of estrogen exposure on gonad morphology, and therefore the propensity of a fish species to develop intersex/ovotestis, depend on a number of factors, such as the concentration of the xenoestrogen, the exposure duration and importantly the timing of exposure. The mode of sexual differentiation of a species, *e.g.* differentiated/undifferentiated gonochorist may be also of importance [99]. Interestingly, a study conducted by Lange et al. (2008) [100] on roach (*Rutilus rutilus*) showed that a single low EE2 concentration (4 ng/L) can lead to different effects depending on the exposure period/duration: a life-cycle exposure starting from fertilized eggs induced a 100% female population whereas exposure during the period of sexual differentiation induced intersex fish. Therefore both a certain estrogen level and exposure timing/duration are needed in roach to get an all-female population. From the results of our study it appears that trout is well able to develop ovotestis gonads, however, it needs exposure concentrations which are higher than what is usually found in the environment. Estrogens induce not only ovotestis but they can lead to complete testis-ovary conversion. The morphological changes are accompanied by corresponding changes in expression of testicular marker genes and sex steroid levels. No linear but rather complex relations were observed between ER/aromatase expression and estrogen exposure, indicating that these genes are not directly involved in the sex conversion process.

Acknowledgments

The authors are grateful to the Cell and Tissue Laboratory of the University of Namur (UNamur) for their support in the histological analysis. We also thank the URBE team, especially André Evrard, Jessica Douxfils and Sophie Massart, for their assistance during the in vivo experiments and sampling.

Author Contributions

Conceived and designed the experiments: SD ED PK. Performed the experiments: SD ML LD. Analyzed the data: SD BDM HS. Contributed reagents/materials/analysis tools: SD. Wrote the paper: SD.

References

1. Brennan J, Capel B (2004) One tissue, two fates: molecular genetic events that underlie testis versus ovary development. Nat Rev Genet 5: 509–521. Available: http://www.ncbi.nlm.nih.gov/pubmed/15211353. Accessed 12 April 2012.
2. Devlin RH, Nagahama Y (2002) Sex determination and sex differentiation in fish: an overview of genetic, physiological, and environmental influences. Aquaculture 208: 191–364.
3. Melamed P, Sherwood N, editors (2005) Hormones and their receptors in fish reproduction. Molecular aspects of fish and marine biology. Singapore.
4. Pandian TJ, Sheela SG (1995) Hormonal induction of sex reversal in fish. Aquaculture 138: 1–22. Available: http://linkinghub.elsevier.com/retrieve/pii/0044848695010750.
5. Piferrer F (2001) Endocrine sex control strategies for the feminization of teleost fish. Aquaculture 197: 229–281. Available: http://linkinghub.elsevier.com/retrieve/pii/S0044848601005890.
6. Mylonas CC, Fostier A, Zanuy S (2010) Broodstock management and hormonal manipulations of fish reproduction. Gen Comp Endocrinol 165: 516–534. doi:10.1016/j.ygcen.2009.03.007.
7. Van Aerle R, Nolan TM, Jobling S, Christiansen LB, Sumpter JP, et al. (2001) Sexual disruption in a second species of wild cyprinid fish (the gudgeon, *Gobio gobio*) in United Kingdom freshwaters. Environ Toxicol Chem 20: 2841–2847. Available: http://www.ncbi.nlm.nih.gov/pubmed/11764169.
8. Porter CM, Janz DM (2003) Treated municipal sewage discharge affects multiple levels of biological organization in fish. Ecotoxicol Environ Saf 54: 199–206. Available: http://www.ncbi.nlm.nih.gov/pubmed/12550098.
9. Aguayo S, Muñoz MJ, de la Torre A, Roset J, de la Peña E, et al. (2004) Identification of organic compounds and ecotoxicological assessment of sewage treatment plants (STP) effluents. Sci Total Environ 328: 69–81. Available: http://www.ncbi.nlm.nih.gov/pubmed/15207574. Accessed 2 April 2012.
10. Auriol M, Filali-Mcknassi Y, Tyagi RD, Adams CD, Surampalli RY (2006) Endocrine disrupting compounds removal from wastewater, a new challenge. Process Biochem 41: 525–539. Available: http://linkinghub.elsevier.com/retrieve/pii/S1359511305004162. Accessed 9 March 2012.
11. Routledge EJ, Sheahan D, Desbrouw C, Brighty GC, Waldock M, et al. (1998) Identification of estrogenic chemicals in STW effluent. 2. In vivo responses in trout and roach. Environ Sci Technol 32: 1559–1565.
12. Jobling S, Beresford N, Nolan M, Rodgers-Gray T, Brighty GC, et al. (2002) Altered sexual maturation and gamete production in wild roach (*Rutilus rutilus*) living in rivers that receive treated sewage effluents. Biol Reprod 66: 272–281.
13. Kidd KA, Blanchfield PJ, Mills KH, Palace VP, Evans RE, et al. (2007) Collapse of a fish population after exposure to a synthetic estrogen. Proc Natl Acad Sci U S A 104: 8897–8901. Available: http://www.pubmedcentral.nih.gov/articlerender.fcgi?artid=1874224&tool=pmcentrez&rendertype=abstract.

14. Legler J, Dennekamp M, Vethaak AD, Brouwer A, JH K, et al. (2002) Detection of estrogenic activity in sediment-associated compounds using in vitro reporter gene assays. Sci Total Environ 293: 69–83.

15. Brown KH, Schultz IR, Nagler JJ (2007) Reduced embryonic survival in rainbow trout resulting from paternal exposure to the environmental estrogen 17alpha-ethynylestradiol during late sexual maturation. Reproduction 134: 659–666. Available: http://www.pubmedcentral.nih.gov/articlerender. fcgi?artid = 2098700&tool = pmcentrez&rendertype = abstract. Accessed 2 May 2012.

16. Sumpter JP, Jobling S (1995) Vitellogenesis as a biomarker for estrogenic contamination of the aquatic environment. Environ Health Perspect 103 Suppl: 173–178. Available: http://www.pubmedcentral.nih.gov/articlerender. fcgi?artid = 1518861&tool = pmcentrez&rendertype = abstract.

17. Bon E, Barbe U, Nuñez Rodriguez J, Cuisset B, Pelissero C, et al. (1997) Plasma vitellogenin levels during the annual reproductive cycle of the female rainbow trout (Oncorhynchus mykiss): establishment and validation of an ELISA. Comp Biochem Physiol B Biochem Mol Biol 117: 75–84. Available: http://www.ncbi.nlm.nih.gov/pubmed/9180016.

18. Christiansen LB, Pedersen KL, Korsgaard B, Bjerregaard P (1998) Estrogenicity of xenobiotics in rainbow trout (Oncorhynchus mykiss) using in vivo synthesis of vitellogenin as a biomarker. Mar Environ Res 46: 137–140. Available: http://linkinghub.elsevier.com/retrieve/pii/S0141113697000470.

19. Burki R, Vermeirssen ELM, Körner O, Joris C, Burkhardt-Holm P, et al. (2006) Assessment of estrogenic exposure in Brown trout (Salmo trutta) in a swiss midland river: integrated analysis of passive samplers, wild and caged fish, and vitellogenin mRNA and protein. Environ Toxicol Chem 25: 2077. Available: http://doi.wiley.com/10.1897/05-545R.1. Accessed 2 April 2013.

20. Nadzialek S, Depiereux S, Mandiki SNM, Kestemont P (2011) In vivo biomarkers of estrogenicity: limitation of interpretation in wild environment. Arch Environ Contam Toxicol 60: 471–478. Available: http://www.ncbi.nlm. nih.gov/pubmed/20523976. Accessed 13 December 2012.

21. Jobling S, Nolan M, Tyler CR, Brighty G, Sumpter JP (1998) Widespread sexual disruption in wild fish. Environ Sci Technol 32: 2498–2506.

22. Lange A, Paull GC, Coe TS, Katsu Y, Urushitani H, et al. (2009) Sexual reprogramming and estrogenic sensitization in wild fish exposed to ethinylestradiol. Environ Sci Technol 43: 1219–1225. Available: http://www.ncbi.nlm. nih.gov/pubmed/19320183.

23. Hirakawa I, Miyagawa S, Katsu Y, Kagami Y, Tatarazako N, et al. (2012) Gene expression profiles in the testis associated with testis-ova in adult Japanese medaka (Oryzias latipes) exposed to 17α-ethinylestradiol. Chemosphere 87: 668–674. Available: http://www.ncbi.nlm.nih.gov/pubmed/22230730. Accessed 21 February 2013.

24. Takashima F (1980) Histological studies on the sex differentiation in Rainbow trout. Bull Japanese Soc Sci Fish 46: 1317–1322.

25. Upadhyay SN (1977) Morphologie des gonades immatures et étude expérimentale de l'induction de la gamétogenèse chez la truite arc-en-ciel juvénile (Salmo gairdnerii R.) Université Pierre et Marie Curie Paris VI.

26. Yano A, Guyomard R, Nicol B, Jouanno E, Quillet E, et al. (2012) An immune-related gene evolved into the master sex-determining gene in rainbow trout, Oncorhynchus mykiss. Curr Biol null. Available: http://dx.doi.org/10. 1016/j.cub.2012.05.045. Accessed 23 July 2012.

27. Jobling S, Sheahan D, Osborne J, Matthiessen P, Sumpter JP (1996) Inhibition of testicular growth in rainbow trout (Oncorhynchus mykiss) exposed to estrogenic chemicals. Environ Toxicol Chem 15: 194–202.

28. Verslycke T, Vandenbergh GF, Versonnen B, Arijs K, Janssen CR (2002) Induction of vitellogenesis in 17alpha-ethinylestradiol-exposed rainbow trout (Oncorhynchus mykiss): a method comparison. Comp Biochem Physiol C Toxicol Pharmacol 132: 483–492. Available: http://www.ncbi.nlm.nih.gov/pubmed/ 12223204.

29. Nagler JJ, Krisfalusi M, Cyr DG (2000) Quantification of rainbow trout (Oncorhynchus mykiss) estrogen receptor-alpha messenger RNA and its expression in the ovary during the reproductive cycle. J Mol Endocrinol 25: 243–251. Available: http://www.ncbi.nlm.nih.gov/pubmed/11013350.

30. Sheahan DA, Bucke D, Matthiessen P, Sumpter J, Kirby M, et al. (1994) The effects of low levels of 17alpha-ethynylestradiol upon plasma vitellogenin levels in male and female rainbow trout, Oncorhynchus mykiss (Walbaum) held at two acclimation temperatures. In: Muller R, Lloyd R, editors. Chronic Effects of Pollutants on Freshwater Fish. Oxford, U.K. 99–112.

31. Purdom CE, Hardiman PA, Bye VJ, Eno N., Tyler CR, et al. (1994) Estrogenic effects of effluents from sewage treatment works. J Chem Ecol 8: 275–285.

32. Schwaiger J, Mallow U, Ferling H, Knoerr S, Braunbeck T, et al. (2002) How estrogenic is nonylphenol? A transgenerational study using rainbow trout (Oncorhynchus mykiss) as a test organism. Aquat Toxicol 59: 177–189. Available: http://www.ncbi.nlm.nih.gov/pubmed/12127735.

33. Ackermann GE, Schwaiger J, Negele RD, Fent K (2002) Effects of long-term nonylphenol exposure on gonadal development and biomarkers of estrogenicity in juvenile rainbow trout (Oncorhynchus mykiss). Aquat Toxicol 60: 203–221. Available: http://www.ncbi.nlm.nih.gov/pubmed/12200086.

34. Bjerregaard LB, Madsen AH, Korsgaard B, Bjerregaard P (2006) Gonad histology and vitellogenin concentrations in brown trout (Salmo trutta) from Danish streams impacted by sewage effluent. Ecotoxicology 15: 315–327. Available: http://www.ncbi.nlm.nih.gov/pubmed/16739033. Accessed 2 May 2012.

35. Bjerregaard LB, Lindholst C, Korsgaard B, Bjerregaard P (2008) Sex hormone concentrations and gonad histology in brown trout (Salmo trutta) exposed to 17beta-estradiol and bisphenol A. Ecotoxicology 17: 252–263. Available: http://www.ncbi.nlm.nih.gov/pubmed/18320304. Accessed 20 March 2012.

36. Tarrant H, Mousakitis G, Wylde S, Tattersall N, Lyons A, et al. (2008) Raised plasma vitellogenin in male wild brown trout (Salmo trutta) near a wastewater treatment plant in Ireland. Environ Toxicol Chem 27: 1773–1779. Available: http://www.ncbi.nlm.nih.gov/pubmed/18315389.

37. Körner O, Vermeirssen ELM, Burkhardt-Holm P (2007) Reproductive health of brown trout inhabiting Swiss rivers with declining fish catch. Aquat Sci 69: 26–40. Available: http://link.springer.com/10.1007/s00027-006-0842-5. Accessed 25 March 2013.

38. Palace V, Evans R, Wautier K, Mills K, Blanchfield P, et al. (2009) Interspecies differences in biochemical, histopathological, and population responses in four fish species exposed to ethynylestradiol added to a whole lake. Can J Fish Aquat Sci 66: 1920–1935.

39. Johnstone R, Simpson T, Youngson A (1978) Sex reversal in salmonid culture. Aquaculture 13: 115–134.

40. Simpson T, Johnstone R, Youngston A (1976) Sex reversal in salmonids. Counc Explor Sea: 6.

41. Nagler JJ, Cavileer T, Sullivan J, Cyr DG, Rexroad C (2007) The complete nuclear estrogen receptor family in the rainbow trout: discovery of the novel ERalpha2 and both ERbeta isoforms. Gene 392: 164–173. Available: http:// w w w . p u b m e d c e n t r a l . n i h . g o v / a r t i c l e r e n d e r . fcgi?artid = 1868691&tool = pmcentrez&rendertype = abstract. Accessed 2 March 2012.

42. Cavileer T, Hunter S, Okutsu T, Yoshizaki G, Nagler JJ (2009) Identification of novel genes associated with molecular sex differentiation in the embryonic gonads of rainbow trout (Oncorhynchus mykiss). Sex Dev 3: 214–224. Available: http://www.ncbi.nlm.nih.gov/pubmed/19752601. Accessed 2 May 2012.

43. Baron D, Houlgatte R, Fostier A, Guiguen Y (2005) Large-scale temporal gene expression profiling during gonadal differentiation and early gametogenesis in rainbow trout. Biol Reprod 73: 959–966. Available: http://www.ncbi.nlm.nih. gov/pubmed/16014816. Accessed 10 March 2012.

44. Schubert S, Peter A, Burki R, Schönenberger R, Suter MJ-F, et al. (2008) Sensitivity of brown trout reproduction to long-term estrogenic exposure. Aquat Toxicol 90: 65–72. Available: http://www.ncbi.nlm.nih.gov/pubmed/ 18804294. Accessed 11 April 2012.

45. Billard R (1992) Reproduction in rainbow trout: Sex differentiation, dynamics of gametogenesis, biology and preservation of gametes. Aquaculture 100: 263–298.

46. Grier HJ, Uribe MC, Parenti LR (2007) Germinal epithelium, folliculogenesis, and postovulatory follicles in ovaries of Rainbow Trout, Oncorhynchus mykiss (Walbaum, 1792) (Teleostei, Protacanthopterygii, Salmoniformes). J Morphol 310: 293–310. doi:10.1002/jmor.

47. Hecker M, Murphy M, Coady K, Villeneuve DL, Jones PD, et al. (2006) Terminology of gonadal anomalies in fish and amphibians resulting from chemical exposures. Rev Env Contam Toxicol 187: 103–131.

48. Fostier A, Jalabert B (1986) Steroidogenesis in rainbow trout (Salmo gairdneri) at various preovulatory stages: changes in plasma hormone levels and in vivo and in vitro responses of the ovary to salmon gonadotropin. Fish Physiol Biochem 2: 87–99.

49. D'Cotta H, Fostier A, Guiguen Y, Govoroun M, Baroiller JF (2001) Aromatase plays a key role during normal and temperature-induced sex differentiation of tilapia Oreochromis niloticus. Mol Reprod Dev 59: 265–276. Available: http:// www.ncbi.nlm.nih.gov/pubmed/11424212.

50. Baron D, Fostier A, Breton B, Guiguen Y (2005) Androgen and estrogen treatments alter steady state messengers RNA (mRNA) levels of testicular steroidogenic enzymes in the rainbow trout, Oncorhynchus mykiss. Mol Reprod Dev 71: 471–479. Available: http://www.ncbi.nlm.nih.gov/pubmed/ 15858796. Accessed 17 October 2012.

51. Rozen S, Skaletsky HJ (2000) Primer3 on the WWW for general users and for biologist programmers. In: Krawetz S, Misener S, editors. Molecular Biology. Totowa: Humana Press. 365–386.

52. Le Roux M-G, Thézé N, Wolff J, Le Pennec J-P (1993) Organization of a rainbow trout estrogen receptor gene. Biochim Biophys Acta - Gene Struct Expr 1172: 226–230. Available: http://dx.doi.org/10.1016/0167-4781(93)90302-T. Accessed 23 August 2013.

53. Altschul SF, Gish W, Miller W, Myers EW, Lipman DJ (1990) Basic local alignment search tool. J Mol Biol 215: 403–410. Available: http://dx.doi.org/ 10.1016/S0022-2836(05)80360-2. Accessed 8 August 2013.

54. Schmittgen TD, Livak KJ (2008) Analyzing real-time PCR data by the comparative CT method. Nat Protoc 3: 1101–1108. Available: http://www. nature.com/doifinder/10.1038/nprot.2008.73. Accessed 26 October 2012.

55. Livak KJ, Schmittgen TD (2001) Analysis of relative gene expression data using real-time quantitative PCR and the 2(-Delta Delta C(T)) Method. Methods 25: 402–408. Available: http://www.ncbi.nlm.nih.gov/pubmed/11846609. Accessed 1 November 2012.

56. Filby AL, Tyler CR (2007) Appropriate "housekeeping" genes for use in expression profiling the effects of environmental estrogens in fish. BMC Mol Biol 8: 10. Available: http://www.pubmedcentral.nih.gov/articlerender. fcgi?artid = 1802086&tool = pmcentrez&rendertype = abstract. Accessed 5 March 2012.

57. Dagnelie P (1998) Statistique théorique et appliquée. De Boeck U.

58. Grier H (1993) Comparative organization of Sertoli cells including the Sertoli cell barrier. In: Russel L, Griswold M, editors. The Sertoli Cell. 704–739.

59. Lebrun C, Billard R, Jalabert B (1982) Changes in the number of germ cells in the gonads of the rainbow trout (*Salmo gairdneri*) during the first 10 post-hatching weeks. Reprod Nutr Dévelop 22: 405–412.

60. McMillan DB (2007) Fish Histology Female Reproductive Systems. Springer. Dordrecht.

61. Cousin M (1988) Effets de la méthyltestosterone sur la morphogenèse et la cyto-différenciation du tractus génital de la truite arc-en-ciel *Salmo gairdneri*: étude sur une population monosexe femelle Université de Rennes.

62. Van den Hurk R, Peute J (1979) Cyclic changes in the ovary of the Rainbow Trout, *Salmo gairdneri*, with special reference to sites of steroidogenesis. Cell Tissue Res 199: 289–306.

63. Grier HJ (2002) The germinal epithelium: its dual role in establishing male reproductive classes and understanding the basis for indeterminate egg production in female fishes. GCFI 53.

64. Kestemont P (1989) Etude du cycle reproducteur du goujon, *Gobio gobio* L. 2. Variations saisonnières dans l'histologie des testicules. J Appl Ichtyology 5: 111–121.

65. Dietrich DR, Krieger HO (2009) Histological analysis of endocrine disruptive effects in small laboratory fish. Dietrich DR, Krieger HO, editors New Jersey: Wiley.

66. Quillet E, Labbe L, Queau I (2004) Asymmetry in sexual development of gonads in intersex rainbow trout. J Fish Biol 64: 1147–1151. Available: http://doi.wiley.com/10.1111/j.1095-8649.2004.00373.x. Accessed 2 May 2012.

67. Krisfalusi M, Nagler JJ (2000) Induction of gonadal intersex in genotypic male rainbow trout (*Oncorhynchus mykiss*) embryos following immersion in estradiol-17beta. Mol Reprod Dev 56: 495–501. Available: http://www.ncbi.nlm.nih.gov/pubmed/10911399.

68. Örn S, Holbech H, Madsen TH, Norrgren L, Petersen GI (2003) Gonad development and vitellogenin production in zebrafish (*Danio rerio*) exposed to ethinylestradiol and methyltestosterone. Aquat Toxicol 65: 397–411. Available: http://linkinghub.elsevier.com/retrieve/pii/S0166445X03001772. Accessed 22 February 2013.

69. Bogdal C, Naef M, Schmid P, Kohler M, Zennegg M, et al. (2009) Unexplained gonad alterations in whitefish (*Coregonus* spp.) from Lake Thun, Switzerland: levels of persistent organic pollutants in different morphs. Chemosphere 74: 434–440. Available: http://www.ncbi.nlm.nih.gov/pubmed/18986675. Accessed 24 February 2013.

70. Billard R, Chevassus B, Jalabert B, Escaffre A-M, Carpentier M (1973) Tentative de contrôle expérimental du sexe chez la truite arc-en-ciel: production d'animaux stériles et d'hermaphrodites simultanés autofécondables. Colloque sur l'Aquaculture. Brest, France. 161–174.

71. Bernet D, Wahli T, Kueng C, Segner H (2004) Frequent and unexplained gonadal abnormalities in whitefish (central alpine *Coregonus* sp.) from an alpine oligotrophic lake in Switzerland. Dis Aquat Organ 61: 137–148. Available: http://www.ncbi.nlm.nih.gov/pubmed/15584420.

72. Caldwell DJ, Mastrocco F, Hutchinson TH, Länge R, Heijerick D, et al. (2008) Derivation of an aquatic predicted no-effect concentration for the synthetic hormone, 17 alpha-ethinylestradiol. Environ Sci Technol 42: 7046–7054. Available: http://www.ncbi.nlm.nih.gov/pubmed/18939525.

73. Jalabert B, Fostier A (2010) La truite arc-en-ciel De la biologie à l'élevage. QUAE.

74. Weber LP, Hill RL, Janz DM (2003) Developmental estrogenic exposure in zebrafish (*Danio rerio*): II. Histological evaluation of gametogenesis and organ toxicity. Aquat Toxicol 63: 431–446. Available: http://linkinghub.elsevier.com/retrieve/pii/S0166445X02002084. Accessed 19 February 2013.

75. Babin PJ, Cerdà J, Lubzens E (2007) The fish oocyte: from basic studies to biotechnological applications. Springer. Babin PJ, Cerdà J, Lubzens E, editors Dordrecht.

76. Lubzens E, Young G, Bobe J, Cerdà J (2010) Oogenesis in teleosts: how eggs are formed. Gen Comp Endocrinol 165: 367–389. Available: http://www.ncbi.nlm.nih.gov/pubmed/19505465. Accessed 10 March 2012.

77. Marchand O, Govoroun M, D'Cotta H, McMeel O, Lareyre JJ, et al. (2000) DMRT1 expression during gonadal differentiation and spermatogenesis in the rainbow trout, *Oncorhynchus mykiss*. Biochim Biophys Acta 1493: 180–187. Available: http://www.ncbi.nlm.nih.gov/pubmed/10978520.

78. Vizziano-Cantonnet D, Baron D, Mahè S, Cauty C, Fostier A, et al. (2008) Estrogen treatment up-regulates female genes but does not suppress all early testicular markers during rainbow trout male-to-female gonadal transdifferentiation. J Mol Endocrinol 41: 277–288. Available: http://www.ncbi.nlm.nih.gov/pubmed/18719050. Accessed 10 March 2012.

79. Katsu Y, Lange A, Urushitani H, Ichikawa R, Paull GC, et al. (2007) Functional associations between two estrogen receptors, environmental estrogens, and sexual disruption in the roach (*Rutilus rutilus*). Environ Sci Technol 41: 3368–3374. Available: http://www.ncbi.nlm.nih.gov/pubmed/17539551.

80. Filby AL, Tyler CR (2005) Molecular characterization of estrogen receptors 1, 2a, and 2b and their tissue and ontogenic expression profiles in fathead minnow (Pimephales promelas). Biol Reprod 73: 648–662. Available: http://www.ncbi.nlm.nih.gov/pubmed/15930325. Accessed 29 March 2012.

81. Marlatt VL, Martyniuk CJ, Zhang D, Xiong H, Watt J, et al. (2008) Auto-regulation of estrogen receptor subtypes and gene expression profiling of 17beta-estradiol action in the neuroendocrine axis of male goldfish. Mol Cell Endocrinol 283: 38–48. Available: http://www.ncbi.nlm.nih.gov/pubmed/18083300. Accessed 27 February 2013.

82. Chakraborty T, Shibata Y, Zhou L-Y, Katsu Y, Iguchi T, et al. (2011) Differential expression of three estrogen receptor subtype mRNAs in gonads and liver from embryos to adults of the medaka, *Oryzias latipes*. Mol Cell Endocrinol 333: 47–54. Available: http://www.ncbi.nlm.nih.gov/pubmed/21146584. Accessed 25 March 2013.

83. Boyce-Derricott J, Nagler JJ, Cloud JG (2009) Regulation of hepatic estrogen receptor isoform mRNA expression in rainbow trout (*Oncorhynchus mykiss*). Gen Comp Endocrinol 161: 73–78. Available: http://www.ncbi.nlm.nih.gov/pubmed/19084018. Accessed 22 February 2013.

84. Boyce-Derricott J, Nagler JJ, Cloud JG (2010) The ontogeny of nuclear estrogen receptor isoform expression and the effect of 17beta-estradiol in embryonic rainbow trout (*Oncorhynchus mykiss*). Mol Cell Endocrinol 315: 277–281. Available: http://www.pubmedcentral.nih.gov/articlerender.fcgi?artid=2814938&tool=pmcentrez&rendertype=abstract. Accessed 10 March 2012.

85. MacKay ME, Raelson J, Lazier CB (1996) Up-regulation of estrogen receptor mRNA and estrogen receptor activity by estradiol in liver of rainbow trout and other teleostean fish. Comp Biochem Physiol Part C Pharmacol Toxicol Endocrinol 115: 201–209. Available: http://dx.doi.org/10.1016/S0742-8413(96)00093-X. Accessed 8 August 2012.

86. Sabo-Attwood T, Kroll KJ, Denslow ND (2004) Differential expression of largemouth bass (*Micropterus salmoides*) estrogen receptor isotypes alpha, beta, and gamma by estradiol. Mol Cell Endocrinol 218: 107–118. Available: http://www.ncbi.nlm.nih.gov/pubmed/15130515. Accessed 25 February 2013.

87. Watson CS, Bulayeva NN, Wozniak AL, Alyea RA (2007) Xenoestrogens are potent activators of nongenomic estrogenic responses. Steroids 72: 124–134. Available: http://dx.doi.org/10.1016/j.steroids.2006.11.002. Accessed 16 March 2012.

88. Guiguen Y, Fostier A, Piferrer F, Chang C-F (2010) Ovarian aromatase and estrogens: a pivotal role for gonadal sex differentiation and sex change in fish. Gen Comp Endocrinol 165: 352–366. Available: http://www.ncbi.nlm.nih.gov/pubmed/19289125. Accessed 22 March 2012.

89. Cheshenko K, Pakdel F, Segner H, Kah O, Eggen RIL (2008) Interference of endocrine disrupting chemicals with aromatase CYP19 expression or activity, and consequences for reproduction of teleost fish. Gen Comp Endocrinol 155: 31–62. Available: http://www.ncbi.nlm.nih.gov/pubmed/17459383. Accessed 10 March 2012.

90. Kazeto Y, Place AR, Trant JM (2004) Effects of endocrine disrupting chemicals on the expression of CYP19 genes in zebrafish (*Danio rerio*) juveniles. Aquat Toxicol 69: 25–34. Available: http://www.ncbi.nlm.nih.gov/pubmed/15210295. Accessed 5 February 2013.

91. Kishida M, McLellan M, Miranda J, Callard G (2001) Estrogen and xenoestrogens upregulate the brain aromatase isoform (P450aromB) and perturb markers of early development in zebrafish (*Danio rerio*). Comp Biochem Physiol B Biochem Mol Biol 129: 261–268. Available: http://www.ncbi.nlm.nih.gov/pubmed/11399458.

92. Menuet A, Pellegrini E, Brion F, Gueguen M., Anglade I, et al. (2005) Expression and estrogen-dependent regulation of the zebrafish brain aromatase. J Comp Neurol 485: 304–320.

93. Lyssimachou A, Jenssen BM, Arukwe A (2006) Brain cytochrome P450 aromatase gene isoforms and activity levels in atlantic salmon after waterborne exposure to nominal environmental concentrations of the pharmaceutical ethynylestradiol and antifoulant tributyltin. Toxicol Sci 91: 82–92. Available: http://www.ncbi.nlm.nih.gov/pubmed/16484284. Accessed 28 February 2013.

94. Kortner TM, Mortensen AS, Hansen MD, Arukwe A (2009) Neural aromatase transcript and protein levels in Atlantic salmon (*Salmo salar*) are modulated by the ubiquitous water pollutant, 4-nonylphenol. Gen Comp Endocrinol 164: 91–99. Available: http://www.ncbi.nlm.nih.gov/pubmed/19467236. Accessed 22 February 2013.

95. Scholz S, Gutzeit H (2000) 17-alpha-ethinylestradiol affects reproduction, sexual differentiation and aromatase gene expression of the medaka (*Oryzias latipes*). Aquat Toxicol 50: 363–373. Available: http://www.ncbi.nlm.nih.gov/pubmed/10967398.

96. Vizziano D, Randuineau G, Baron D, Cauty C, Guiguen Y (2007) Characterization of early molecular sex differentiation in rainbow trout, *Oncorhynchus mykiss*. Dev Dyn 236: 2198–2206. Available: http://www.ncbi.nlm.nih.gov/pubmed/17584856. Accessed 2 May 2012.

97. Pellegrini E, Menuet A, Lethimonier C, Adrio F, Gueguen M-M, et al. (2005) Relationships between aromatase and estrogen receptors in the brain of teleost fish. Gen Comp Endocrinol 142: 60–66. Available: http://www.ncbi.nlm.nih.gov/pubmed/15862549. Accessed 15 April 2012.

98. Diotel N, Le Page Y, Mouriec K, Tong S-K, Pellegrini E, et al. (2010) Aromatase in the brain of teleost fish: expression, regulation and putative functions. Front Neuroendocrinol 31: 172–192. Available: http://www.ncbi.nlm.nih.gov/pubmed/20116395. Accessed 9 March 2012.

99. Segner H (2011) Reproductive and developmental toxicity in fishes. In: Gupta RC, editor. Reproductive and Developmental Toxicology. Elsevier Inc. 1145–1166. Available: http://dx.doi.org/10.1016/B978-0-12-382032-7.10086-4.

100. Lange A, Katsu Y, Ichikawa R, Paull GC, Chidgey LL, et al. (2008) Altered sexual development in roach (*Rutilus rutilus*) exposed to environmental concentrations of the pharmaceutical 17alpha-ethinylestradiol and associated

expression dynamics of aromatases and estrogen receptors. Toxicol Sci 106: 113–123. Available: http://www.ncbi.nlm.nih.gov/pubmed/18653663. Accessed 12 December 2012.

101. Dalla Valle L, Toffolo V, Vianello S, Ikuo H, Takashi A, et al. (2005) Genomic organization of the CYP19b genes in the rainbow trout (*Oncorhynchus mykiss* Walbaum). J Steroid Biochem Mol Biol 94: 49–55. Available: http://www.ncbi.nlm.nih.gov/pubmed/15862949. Accessed 10 March 2012.

2

Identification and Characterization of 63 MicroRNAs in the Asian Seabass *Lates calcarifer*

Jun Hong Xia, Xiao Ping He, Zhi Yi Bai, Gen Hua Yue*

Molecular Population Genetics Group, Temasek Life Sciences Laboratory, National University of Singapore, Singapore, Republic of Singapore

Abstract

Background: MicroRNAs (miRNAs) play an important role in the regulation of many fundamental biological processes. So far miRNAs have been only identified in a few fish species, although there are over 30,000 fish species living under different environmental conditions on the earth. Here, we described an approach to identify conserved miRNAs and characterized their expression patterns in different tissues for the first time in a food fish species Asian seabass (*Lates calcarifer*).

Methodology/Principal Findings: By combining a bioinformatics analysis with an approach of homolog-based PCR amplification and sequencing, 63 novel miRNAs belonging to 29 conserved miRNA families were identified. Of which, 59 miRNAs were conserved across 10–86 species (E value $\leq 10^{-4}$) and 4 miRNAs were conserved only in fish species. qRT-PCR analysis showed that miR-29, miR-103, miR-125 and several let-7 family members were strongly and ubiquitously expressed in all tissues tested. Interestingly, miR-1, miR-21, miR-183, miR-184 and miR-192 showed highly conserved tissue-specific expression patterns. Exposure of the Asian seabass to lipopolysaccharide (LPS) resulted in up-regulation of over 50% of the identified miRNAs in spleen suggesting the importance of the miRNAs in acute inflammatory immune responses.

Conclusions/Significance: The approach used in this study is highly effective for identification of conserved miRNAs. The identification of 63 miRNAs and determination of the spatial expression patterns of these miRNAs are valuable resources for further studies on post-transcriptional gene regulation in Asian seabass and other fish species. Further identification of the target genes of these miRNAs would shed new light on their regulatory roles of microRNAs in fish.

Editor: Baohong Zhang, East Carolina University, United States of America

Funding: This study is funded by Ministry of National Development for the project "To ensure self-sufficiency of safe seafood for Singapore through the development of aquaculture genomic tools for marker-assisted selective breeding of tropical marine foodfish". The funders had no role in study design, data collection and analysis, decision to publish, or preparation of the manuscript.

Competing Interests: The authors have declared that no competing interests exist.

* E-mail: genhua@tll.org.sg

Introduction

MicroRNAs (miRNAs) are short, endogenous noncoding RNAs (~22 nucleotides in length) found in animals, plants and virus [1,2,3]. miRNA expression can be regulated at transcriptional and post-transcriptional levels during RNA biogenesis [4]. Recent studies showed that miRNAs might have potentially enormous importance in the regulation of many fundamental biological processes, such as cardiogenesis and myogenesis [5,6,7], neurogenesis [8,9], hematopoiesis [10], stem cells, regeneration and homeostasis [11,12,13,14], proliferation, differentiation, cell fate determination, apoptosis, signal transduction, organ development [15,16,17,18,19,20,21], immunological and inflammatory disorders [22,23,24]. The down-regulation of the expression of specific mRNA targets by miRNAs accounted for the approximately 70% detectable changes at the mRNA levels of all regulated proteins, either by directing the cleavage of mRNAs or interfering with translation [2,25,26]. miRNAs make up 1–5% of all genes making them the most abundant class of regulators in genome [27,28].

Several approaches have been used to identify miRNAs. The first approach was through forward genetics such as the discovery of lin-4 in *Caenorhabditis elegans* [29,30]. The second was directional cloning and sequencing by constructing a cDNA library, which

has been used in *Arabidopsis thaliana* [31], rice *Oryza sativa* [32], rainbow trout *Oncorhynchus mykiss* [33] and zebrafish *Danio rerio* [34]. However, miRNAs expressing at a low level or only in a specific condition or specific cell types would be difficult to find with this approach [4]. Recently high-throughput sequencing strategies had greatly expanded the depth of small RNA cloning coverage [35,36]. However, the sequencing data from these strategies were not yet saturated, as reflected by many of the new identified miRNAs that were represented only once in the sequence database. Furthermore, since some libraries were derived from a limited amount of tissues, some miRNAs that were only expressed in specific adult tissues were not available [34,37]. Bioinformatics prediction was genome-wide and sequence-based computational predictions. This method was based largely on the phylogenetic conservation and the structural characteristics of miRNA precursors (pre-miRNA) [38] and/or known miRNA genes [39,40], enabling one to overcome the problem in directly cloning. On the other hand, the process required knowledge of the complete genome sequence, species-specific miRNAs could not be accurately identified without this information [37] and bioinformatically predicted miRNAs should be validated for their expression by northern blotting or sequencing. Over the past few years, thousands of miRNAs had been identified. With release

14, the miRBase sequence database had broken through the 10000 entries (http://www.mirbase.org/). Further studies showed that most of the miRNA sequences and the stems of miRNA hairpins were highly conserved across species although a number of lineage-specific miRNAs and species-specific miRNAs had been identified [39,41,42,43]. These studies suggested that the cloning of miRNAs for a particular species was plausible based on the conserved homologs among pre-miRNAs and/or mature miRNAs originating from related species.

There are over 30,000 fish species on the earth, representing 50% of all vertebrates [44]. Currently, the cloning and characterization of miRNAs from fishes have been carried out only for two model fish species (medaka *Oryzias latipes* [45] and zebrafish [34]) and one food fish species, rainbow trout [33]. The Asian seabass *Lates calcarifer* belonging to the family Latidae of the order Perciformes, is an important food resource in Southeast Asia. As a food fish species, the Asian seabass could be an excellent model organism for aquaculture-genomic studies due to its compact genome (700 Mb) and extremely high fecundity (1.7 million eggs/spawning) [46]. Therefore, the Asian seabass might offer a good system for understanding of fish biology, such as in infectious diseases, growth and organ development under aquatic environment. Recently massive mortalities of seabass caused by bacterial or viral infections had caused seriously economic losses [47,48,49]. Unfortunately, an effective approach to protect the fish from infections is still not developed to date. miRNAs might potentially have an enormous impact in the regulation of immunological and inflammatory disorders as well as growth development [16,22,23,24]. Identification of miRNAs and their target genes was an important step toward understanding the regulatory networks, gene silencing mechanisms and for practical use for gene manipulations in the Asian seabass. To date, no miRNA is available for the Asian seabass. Conserved miRNAs likely played an important role in regulating basic cellular and developmental pathways from lower to higher organisms [41]. Nevertheless, in comparison to model fish species with a fully sequenced genome, e.g., zebrafish, the in silico identification of conserved miRNAs families in the Asian seabass was not possible due to the absence of whole genomic information for the species.

We attempted to clone and identify conserved miRNAs from the Asian seabass as the first step toward understanding the regulatory roles of miRNAs in fish living under different environmental conditions. To identify conserved miRNAs, bioinformatics analysis of the conservation and structural similarity of pre-miRNA sequences across related model species was performed first. Then an approach of homolog-based PCR amplification and sequencing were carried out. In this study, 63 novel miRNAs in the Asian seabass were identified. The approach developed in this study was highly effective for identification of conserved miRNAs. Quantitative real-time RT-PCR (qRT-PCR) revealed differential expressions of these miRNAs in 8 organs (gill, brain, eye, muscle, liver, intestine, heart and kidney) and in spleen of the Asian seabass before and post a challenge with lipopolysaccharide (LPS). Our data supply the basis for the understanding of the functions of miRNAs in fish.

Results and Discussion

1. Identification of pre-miRNAs and mature miRNAs in the Asian seabass

To find conserved homlogs among fish species 623 pre-miRNA sequences from zebrafish (360), Fugu, *Fugu rubripes* (131) and *Tetraodon nigroviridis* (132) were retrieved from miRBase (the microRNA database; www.mirbase.org/). The conserved homo-

logs of the pre-miRNAs were investigated using commercial software Sequencher. One hundred and six pairs of primers (Table S1) were designed based on the conserved fragments among species. For cloning less conserved pre-miRNAs, additional 47 pairs of primers were designed only based on the zebrafish pre-miRNA sequences since no countparts were found in Fugu and Tetraodon releases (Table S1).

Based on a Switching Mechanism At 5′ end of RNA Transcript (SMART) -based method for cDNA library construction [50], one full-length cDNA library using pooled mRNAs from 9 tissues was constructed and used as template in following PCR reactions for amplification of putative pre-miRNAs. This library provided a good tool to identify miRNAs differentially expressed in these tissues. Eight hundred clones were sequenced and 786 reads were obtained after trimming of end and vector sequences and screening to eliminate sequences with low quality. These sequences were further clustered to identify unique sequences. Finally 322 unique sequences (Table S2) were obtained.

Sequence analyses using the BLASTn program revealed that 107 of the unique sequences showed high conservation (E value≤e^{-4}) with known pre-miRNAs (Table S2). By plotting blasted E value of the small RNAs against total number of unique sequences (Figure 1), we found that the most common conservation intervals were <e^{-2} (123 sequences), between e^{-2} and e^{-3} (63 sequences) and followed by <e^{-8} (48 sequences). Prediction of fold-back structures and energies performed with the DINAMelt server indicated that the stem-loop structure for many sequences were stable (with low folding energy) (Table S2) and lacked large internal loops or bulges. This is suggestive of the general characteristics of pre-miRNAs. One of the examples is shown in Figure 2. However, most of the unique sequences could not form a good stem-loop structure which might result from sequence artifacts (PCR errors) due to the formation of heteroduplex molecules, the error of Taq DNA polymerase [51] and/or nature of sequences since only partials of the pre-miRNAs were obtained in this study.

The unique sequences considered as putative seabass pre-miRNAs were then searched against the miRBase database to predict conserved mature miRNAs in the Asian seabass (designated as lcal-miRNAs). Based on the phylogenetic conservation with known pre-miRNAs and miRNAs and their stability of folding stem-loop structures, 108 sequences [64 unique sequences with high conservation (E value≤e^{-4}) and 44 with less conservation (E value>e^{-4})] were selected to predict conserved

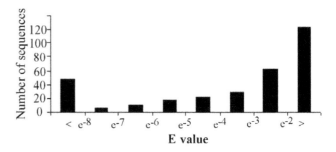

Figure 1. The distribution of the seabass miRNA precursor candidates in various E value intervals as identified by the BLASTn program showing conservation with known pre-miRNAs. Conservation of the 322 unique sequences that cloned from the seabass cDNA library with known pre-miRNAs was identified by the BLASTn program (E value). The most common conservation intervals were <e^{-2} (123 sequences), between e^{-2} and e^{-3} (63 sequences) and followed by <e^{-8} (48 sequences).

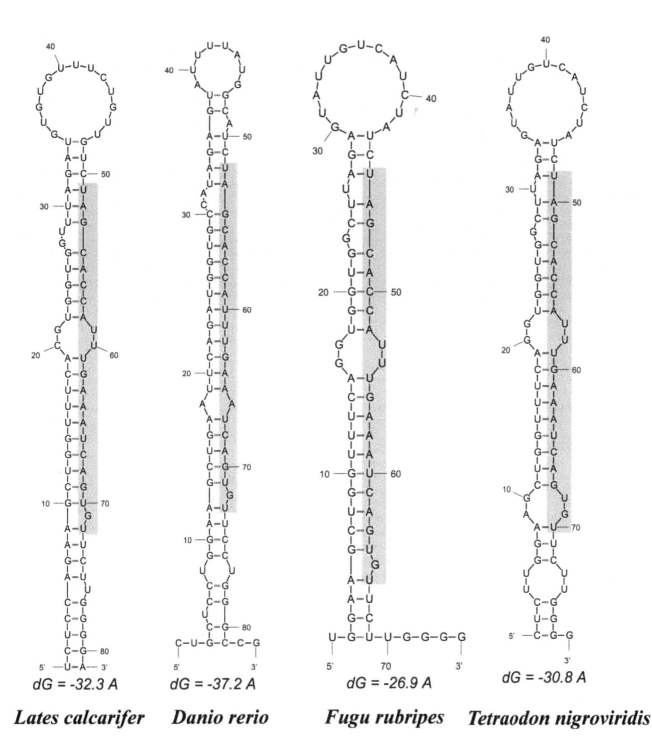

dG = -32.3 A   dG = -37.2 A   dG = -26.9 A   dG = -30.8 A

Lates calcarifer   *Danio rerio*   *Fugu rubripes*   *Tetraodon nigroviridis*

Figure 2. Conservation of the miR-29 family in Pisces. Predicted fold-back structures of mir-29 precursors from *Lates calcarifer, Danio rerio, Fugu rubripes, Tetraodon nigroviridis* and the folding energy (dG) for each fold-back structure are presented. The shadow in each precursor structure indicates the mature miR-29 sequence in the species.

mature miRNAs in the Asian seabass. These miRNAs might represent two classes of conserved miRNAs, i.e., miRNAs in broad conservation families and poor conservation families. Fold-back structures could be predicted for some of the putative pre-miRNAs in the Asian seabass.

To identify the nucleotides at the very 5′ and 3′ ends of the putative miRNAs by PCR, an adaptor-ligated miRNA cDNA library was constructed and used as template. The miRNA

sequence-specific primers and universal primers (Table S1)were used in following PCR reactions. Five hundreds of clones were sequenced. Finally, the nucleotides at the very 5′ or 3′ ends were validated for 151 putative conserved mature miRNAs. Of which, 112 miRNAs with the very 3′ end and 102 miRNAs with the very 5′ end were identified. Upon combined the data, 63 miRNAs with the nucleotides at both of the very ends were confirmed, representing novel mature miRNAs identified in the Asian seabass

(Table S3). The sizes of the 63 miRNAs were ranged from 19 to 25 bases. Of the 63 newly identified miRNAs, 48 began with a 5' uridine, which was a characteristic feature of miRNAs, 11 began with a 5' A, 2 began with a 5' C and G. Names of the seabass miRNAs were assigned based on the homologies between the miRNA and published miRNA sequences in the Sanger database; the isoforms from one family were labeled in alphabetical order (Table S3).

2. Conservation and evolution of the identified miRNAs

Many miRNAs were highly conserved among organisms [52]. For example, at least a third of *C. elegans* miRNAs had homologs in humans [53]. The conservation among species suggests that miRNAs represent a relatively old and important regulatory pathway [54]. miRNAs conservation could be used to identify the novel miRNAs for species without reference genome sequences. Based on the sequence conservation between the lcal-miRNAs and published miRNA sequences in Sanger database, the 63 novel lcal-miRNAs were classified into 29 conserved miRNA families with a range of 1 to 15 loci per family (Table S4). Since the mature miRNA sequences from one family were highly conserved, the PCR product amplified with one primer pair possibly contained several similar miRNAs from one family. We identified 15 members of let-7 family and 4 members of miR-124 family in the Asian seabass. To date, 18 let-7 isoforms were identified in the zebrafish, and 10 let-7 isoforms for the Tetraodon and the Fugu respectively according to the miRBase database (release 14). Additionally, six miR-124 isoforms in zebrafish and three isoforms in Tetraodon and Fugu respectively were registered in the miRBase. The identified numbers for these two miRNA families in this study were comparable to the numbers in model species. In animals and plants, miRNAs exist as multigene families [37], therefore, it was not surprising to get so many isoforms for these families in the study.

A sequence logo in bioinformatics was a graphical representation of the sequence conservation of nucleotides [55]. The sequence logos for 15 lcal-miRNA families with multiple miRNA sequences (≥2) were presented in Figure 3. Some of the sequences were found to have slight shifts in their 5' and 3' ends, such as family mir-21, mir-124, mir-126, mir-183, and mir-184, which is a common phenomenon in miRNA cloning and could be attributed to processing shifts or enzymatic modifications of miRNAs such as RNA editing, 3' nucleotide additions or sequencing artifacts [37]. Some miRNAs from one family differed only by a few base pairs not only on either side, but also in the middle region, e.g., let-7, mir-23, mir-29, mir-101 and mir-128. The two members in mir-199 family were highly divergent, since the two miRNAs might be produced from the 5' arm and the 3' arm of a pre-miRNA, respectively. Sequence logos showed most of the bases for miRNAs in one family were highly conserved.

Sequence similarity searches against the central miRNA registry also showed that most of the miRNAs were conserved across many species (Figure 4). Of which, 59 miRNAs were conserved across 10–86 species according to our conservation criteria (a blast E value≤10^{-4} for mature miRNA and precursor). For example, lcal-miR-20 was conserved in all lineages of Vertebrata, including Amphibia, Mammalia, Pisces and Aves. lcal-miR-29a,b and many members of the lcal-let-7 family were conserved in Bilateria, including Deuterostoma, Lophotrochozoa and Ecdysozoa. High conservation of the miRNA families suggested an evolutionary conserved function.

A number of lineage-specific miRNAs and species-specific miRNAs were discovered recently. For example, many miRNAs that were recently discovered in human and chimpanzee were not conserved beyond mammals, and ~10% were taxon-specific [39]. In our study we have also cloned some miRNAs that only conserved among closely related species. For example, 4 miRNAs (lcal-let-7l, lcal-miR-21a, lcal-miR-101b and lcal-miR-724a) were conserved only in fish species (E value≤10^{-4}), indicating these miRNAs were fish-specific and might play a key role in the evolutionary process of fish.

3. Expression patterns of the Asian seabass miRNAs

Information about the expression of a miRNA is useful for the understanding of its functions [56]. The expression of miRNAs was tightly regulated both in time and space [37]. Hence, to validate expression and assist with the determination of functions, the expression for the 63 miRNAs in 8 different organs (gill, brain, eye, muscle, liver, intestine, heart and kidney) and in spleen following 24 h post-challenge with LPS and control sample were examined by qRT-PCR (Figure 5). Each reaction was performed in triplicate. We selected lcal-miR103 as the reference gene in the analysis as suggested in Peltier et al. [57]. Six (lcal-miR-199a, lcal-miR-152, lcal-let-7a, b, c and d) of sequencing validated miRNAs did not provide positive results in all of the evaluated tissues. These might represent miRNAs with low abundance or expressed only in few cells or tissues or resulted from the low quality of primers used. Based on miRNA abundance, hierarchical clustering analysis showed that the expression profiles between brain and eye and between kidney and intestine were more similar (Figure 5).

Many of the conserved miRNAs were expressed ubiquitously in rainbow trout [33]. In the Asian seabass, miR-29, miR-103, miR-125 and several let-7 family members were strongly and ubiquitously expressed in all tissues tested (Figure 5). Our results suggest these miRNAs might play an important role in the regulation of constitutive processes in diverse tissues. However, miRNAs were also expressed in a tissue-specific manner which provided clues about their physiological functions [2]. For example, in zebrafish, many miRNAs were highly expressed at later stages of development [34,58]. In our study many miRNAs showed highly conserved tissue-specific expression patterns (Figure 5). For example, interesting expression was observed for miR-183-1 and miR-183-2, which were highly specifically expressed in eye and barely detectable in remaining tissues. Some studies demonstrated that miR-183 family members were expressed abundantly in specific sensory cell types in the eye, nose, and inner ear and contributed specifically to neurosensory development or function [59,60]. Previous studies had shown that miR-1 was one of the highly conserved miRNAs and was abundantly and specifically expressed in the heart and other muscular tissues in fish [5,27,61,62], *C. elegans* [63] and mouse [64]. Our study also showed that miR-1 was strongly expressed in muscle and heart, and weakly expressed in eye and intestine of the Asian seabass. Like other studies, our study suggests that miR-1 play an important role in regulation of muscle gene expression in the Asian seabass. In addition, miR-192 was highly expressed in intestine but moderate in liver and kidney. miR-184-1 and miR-184-2 were strongly expressed in eye and moderately expressed in muscle. Our expression data provide a basis for further understanding of regulatory roles of miRNAs in fish living under different aquatic conditions.

4. Potential functions of the identified miRNAs in an acute inflammatory response induced by LPS and vibrio bacteria

Our previous study had shown that exposure of Asian seabass at the age of 35 days post hatchery (dph) to LPS led to a dramatic

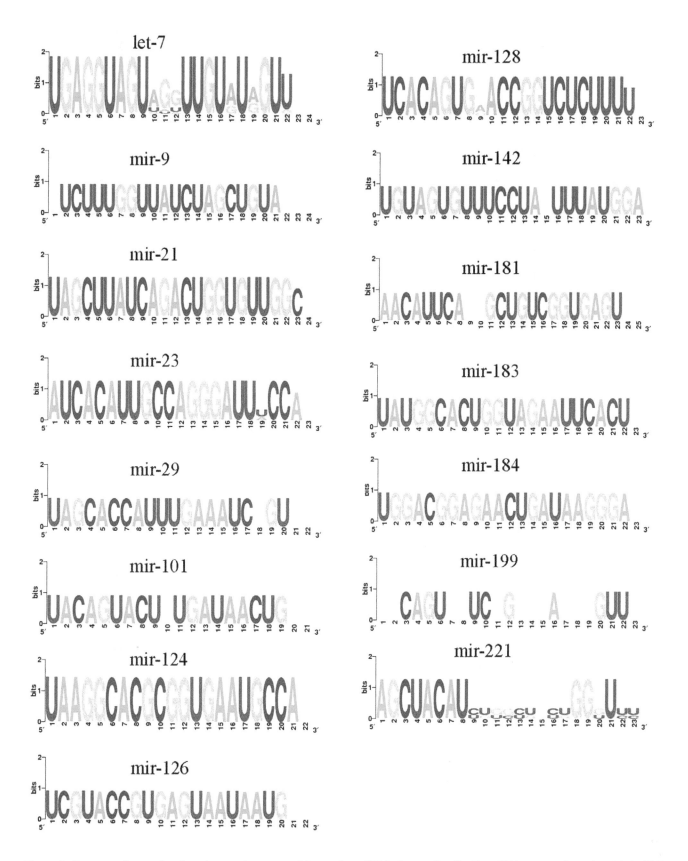

Figure 3. Sequence logos showing the most conserved bases for miRNAs in one family. The miRNA sequence logos for 15 lcal-miRNA families with multiple sequences (≥2) are presented. Each logo consists of stacks of symbols, one stack for each position in the sequence. The overall height of the stack indicates the sequence conservation at that position, while the height of symbols within the stack indicates the relative frequency of each nucleotide at that position. Some miRNAs from one family differ only by a few base pairs on either side and/or in the middle (e.g. let-7, mir-9, mir-21, mir-124 and mir-126); the mir-199 family is highly divergent, since the two miRNAs in that might be from the 5′ arm and from the 3′ arm of a pre-miRNA, respectively.

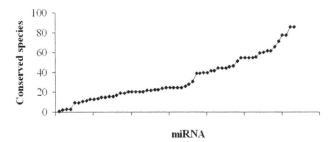

Figure 4. Conservation of the seabass mature miRNAs across species. The 63 miRNAs validated by PCR and sequencing for the Asian seabass were searched against the miRBase database to identify the conservation across species. Of which, 59 miRNAs were conserved across 10–86 species and 4 miRNAs were found to be conserved less than 4 species according to our conservation criteria (E value$\leq 10^{-4}$). Each small black rhombus represents an identified miRNA.

increase of the expression of 25 immune-related genes in spleen by inducing an acute inflammatory response at 24 h post challenge [65]. This study showed that 14 of the 63 miRNAs were found to express at a very low levels ($<1/100$ fold of the reference gene; Figure 5) in spleen. Examination of the mean expression changes among the rest 49 miRNAs showed that exposure to LPS resulted in a general elevation in mature miRNA levels. A similar result was found in Moschos et al. [66]. Following LPS challenge, they observed rapid and transient increase in both the mean (4.3-fold) and individual levels of miRNA expression (46 of 104 miRNAs) in mouse lung. In this study, the expression for 34 of the 63 miRNAs (54%) was increased and only 12 of the 63 miRNA genes (19%) were down-regulated in spleen of the Asian seabass. Highly differentiated expression (>1.5fold) was summarized in Table 1. The miR-21 was highly up-regulated and miR-101 was highly down-regulated after exposure to LPS.

Vibrio infection of fish can cause significant mortality in fish mariculture, e.g., seabass [67]. To explore the function of identified miRNAs in an acute inflammatory response caused by vibrio bacteria, the temporal expression patterns of lcal-miR-21a in three immune-related organs, the spleen, kidney and liver of *Vibrio harveyi*-challenged seabass were examined by qRT-PCR. Interleukin-1 beta (IL-1β) is an important mediator of the inflammatory response [68,69]. It was evident that the high level expression of IL-1β was caused by the presence of the pathogen in kidney (179 fold increased) at 1–3 hours post injection (hpi) and in spleen (234 fold increased) and liver (143 fold increased) at 6 hpi (Figure S1). This data indicated that an acute inflammatory response was induced in the challenged fish within 1 hpi. Further study by qRT-PCR analysis indicated that the miR-21 gene was also remarkably elevated in kidney (1.71 fold) at 3 hpi, spleen (2.21 fold) at 12 hpi and liver (4.65 fold) at 24 hpi (Figure S2).

The miR-21, being one of the most abundant miRNAs, was functioned as an anti-apoptotic factor and oncogene related to cell growth [33,70,71,72,73,74]. In our study the miR-21 was highly expressed in several tissues and highly up-regulated in spleen at 24 hour post challenge by LPS and in three immune-related organs of *Vibrio harveyi*-challenged seabass. These results were consistent with the findings of previous studies [72,73,74,75] demonstrating the importance of the miRNAs in acute inflammatory immune responses with protection against pathogen.

Conclusions

By combining a bioinformatics analysis with an approach of homolog-based PCR amplification and sequencing, 107 unique

sequences showing high conservation with known pre-miRNAs were obtained; and 63 novel miRNAs belonging to 29 conserved miRNA families were identified for the first time in the Asian seabass. The methods used in this study were effective in identifying a large number of highly conserved miRNAs as well as less conserved miRNAs, and could be applied to identify conserved miRNAs expressed at a low level that were difficult to clone by traditional methods in other fish species.

The pre-miRNAs cloned in this study provide the basis for future cloning of primary miRNAs and conducting functional analysis. The determination of the spatial expression patterns of these miRNAs is a valuable resource for further study on post-transcriptional gene regulation in Asian seabass and other fish species. Further identification of the target genes of these miRNAs could shed new light on their regulatory roles of miRNAs in fish.

Materials and Methods

Ethics statement

All handling of fishes was conducted in accordance with the guidelines on the care and use of animals for scientific purposes set up by the Institutional Animal Care and Use Committee (IACUC) of the Temasek Life Sciences Laboratory, Singapore.

Fish, LPS and *Vibrio harveyi* challenge and sampling

One hundred of individuals of small Asian seabass (15 dph) were transported from a commercial fish farm to the animal house at the Temasek Life Sciences Laboratory. The fishes were maintained in a large tank containing 500 L seawater at 25°C for three weeks of acclimation. Fishes were fed twice daily with pelleted feed (Zhongshan Tongyi, Taiwan). One day prior to challenge, 16 of healthy fishes of average weight 5 g were transferred to two smaller tanks holding 10 L of seawater. For 8 fishes in tank 1 each fish were injected intraperitoneally with 0.1 ml of 2 mg/ml of *Escherichia coli* LPS (Sigma-Aldrich, MO, USA) by dilution with phosphate buffered saline at room temperature. In tank 2 (control), a total of 8 fishes was received an intraperitoneal injection of 0.1 ml of phosphate buffered saline for each fish. Just before injection and sampling, the fishes were anaesthetized using AQUI-S® with a concentration of 15 mg/L (AQUI-S New Zealand Ltd, Lower Hutt, New Zealand). Eight fishes from each tank were sacrificed at 24 h post challenge. Spleen was taken for each fish from each tank and kept in Trizol reagent (Invitrogen, CA, USA) at −80°C until use. In addition, tissues including gill, brain, eye, muscle, liver, intestine, heart, spleen and kidney of 5 untreated seabass fishes were also taken and kept at −80°C.

To explore the temporal expression patterns of lcal-miR-21a in three immune-related organs, thirty Asian seabass (at the age of three months) were transferred to two tanks holding 200 L of seawater. For 15 fishes in test tank each fish were injected intraperitoneally with 0.1 ml of phosphate buffered saline dissolved culture pellet of *Vibrio harveyi* (~e10 copy/ml) at room temperature. In control tank, a total of 15 fishes was received an intraperitoneal injection of 0.1 ml of phosphate buffered saline for each fish. Just before injection and sampling, the fishes were anaesthetized using AQUI-S® with a concentration of 15 mg/L (AQUI-S New Zealand Ltd, Lower Hutt, New Zealand). Three fishes from each tank were sacrificed at 1, 3, 6, 12, 24 hpi. Spleen, kidney and liver were taken for each fish from each tank and kept in Trizol reagent (Invitrogen, CA, USA) at −80°C until use.

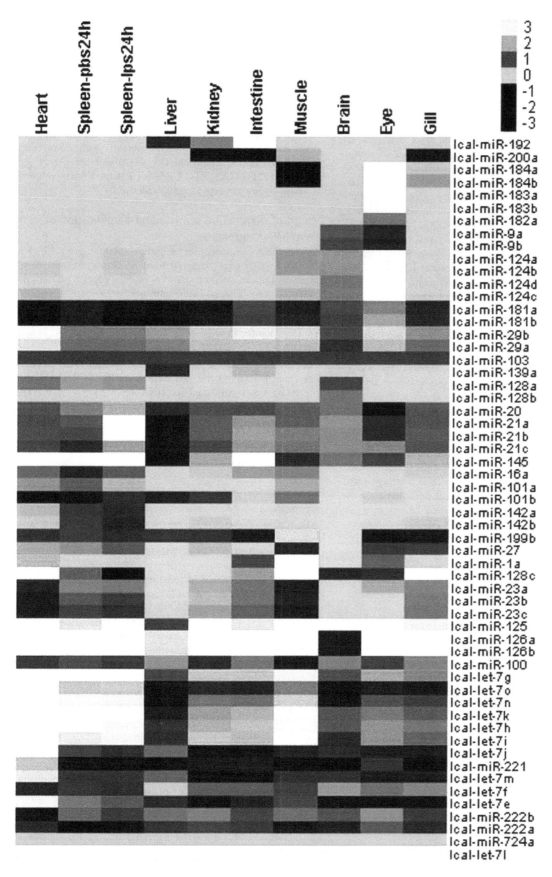

Figure 5. Expressions of the 57 conserved mature miRNAs in eight organs of the Asian seabass. The expression of 63 miRNA was determined in 8 organs (gill, brain, eye, muscle, liver, intestine, heart and kidney) and in spleen sampled at 24 hour post challenge with LPS and

control samples by real-time PCR. Six (Ical-miR-199a, Ical-miR-152, Ical-let-7a,b,c,d) of the sequencing validated miRNAs without positive results in all of the evaluated tissues were deleted from the data. Data were presented on a logarithmic scale. The relative expression of each gene shown in the figure was the average of triplicate real-time PCR reactions, normalized to Ical-miR-103 gene expression. Yellow shading indicated increased levels of expression, and gray shading represented decreased levels of expression relative to the center. Blue color denoted undetectable expression in tissues.

Construction of a full-length cDNA library and a miRNA library

Total RNA was isolated separately from gill, brain, eye, muscle, liver, intestine, heart, spleen and kidney of the Asian seabass using TRIzol (Invitrogen, CA, USA) according to the manufacturer's instructions. Purification of mRNA from total RNA was carried out using Oligotex mRNA Midi Kit (Qiagen, CA, USA). The resulting mRNA from 9 tissues of 5 fishes was mixed with equal quantity separately. One µg of the mixed mRNAs were used for cDNA library construction using a SMART-based method. The reverse transcription was carried out with PowerScript reverse transcriptase following the manufacturer's protocol (Clontech, CA, USA). A oligo(dT) primer [CDS III/3′ PCR Primer 5′-AAG CAG TGG TAT CAA CGC AGA GTA C(T)$_{30}$-3′] was used to prime the first-strand synthesis reaction, and the SMART IV Oligo (5′-AAG CAG TGG TAT CAA CGC AGA GTA CGC rGrGrG-3′) served as a short, extended template at the 5′ end of the mRNA. The resulting single strand cDNAs serving as a template was amplified by SMART-Oligo-IIA-primer (5′-AAG CAG TGG TAT CAA CGC AGA GT-3′) and the PCR products were kept at −80°C till use.

A miRNA library was constructed as described [76] with slight modification. In brief, enrichment of small RNAs from total RNA was performed with mirVana^{TM} miRNA isolation kit (Ambion, CA, USA) following the instruction manual and then was separated on a denaturing 15% polyacrylamide gel. The nucleotides from positions 18–25 bp were size fractionated. RNA was eluted overnight with 0.4 M NaCl at 4°C and recovered by ethanol precipitation with glycogen. The purified small RNAs were then ligated to a 3′ adaptor [5′-(Pu)uu AAC CGC GAA TTC CAG (idT)-3′; where lowercase letters indicate RNA, uppercase letters indicate DNA, Pu

denotes 5′-phosphorylated uridine, and idT represents 3′-inverted deoxythymidine.] and a 5′ adaptor (5′-GAC CAC GCG TAT CGG GCA CCA CGT ATG CTA TCG ATC GTG AGA TGG G-3′). Reverse transcription was performed with PowerScript reverse transcriptase (Clontech, CA, USA) and RT primer (5′-GAC TAG CTG GAA TTC GCG GTT AAA-3′). The resulting 1st strand cDNA was kept at −80°C till use.

Design of degenerate primers and amplification of pre-miRNA sequences

Pre-miRNA sequences of zebrafish (360), fugu (131) and Tetraodon (132) were retrieved from miRBase (the microRNA database release 14; http://www.mirbase.org/). All of the sequences were aligned with software Sequencher 4.9 (Gene Codes, MI, USA). The conserved fragments were used to design degenerate primers with the EditSeq program in DNAstar 7 (DNASTAR, WI, USA). The melting temperatures for these primers were designed at around 60°C.

For amplification of pre-miRNA sequences 153 pairs of degenerate primers (Table S1) were used. Single strand cDNA diluted 20 times was used as DNA template for PCR. The resulting PCR products were directly inserted into a pGEM-T vector (Promega, WI, USA) and transformed into E. coli strain XL-1 (Stratagene, CA, USA). For each product 4–6 clones were sequenced using BigDye chemicals and ABI 3730xl Genetic Analyzer (Applied Biosystems, CA, USA).

Prediction and determination of putative seabass pre-miRNAs

Trimming of vector sequences and low-quality regions from source sequences was performed using commercial software Sequencher 4.9 (Gene Codes, MI, USA). All trimmed sequences were used to form contigs. Singletons and consensus sequences of each contig were referred as unique sequences and were used to search the miRBase (the microRNA database release 14; http://www.mirbase.org/) [77,78,79] with parameters (BLASTn and stem-loop sequences) to find conservation of putative pre-miRNA sequences. Prediction of fold-back structures and energies were performed with the DINAMelt server (Prediction of Melting Profiles for Nucleic Acids; http://frontend.bioinfo.rpi.edu/applications/hybrid/quikfold.php) and with RNA 3.0 as energy rule. Based on these analyses, sequences with low folding energy (stable stem-loop structure) and conservation with known pre-miRNAs were considered as putative seabass pre-miRNAs and were then searched against the miRBase database (release 14) with parameters (BLASTn and mature miRNAs sequences) to predict conserved mature miRNAs in Asian seabass (designated as Ical-miRNAs).

Mapping boundary ends of putative miRNAs by PCR and sequencing

Sequence-specific primers with partial coverage of the putative miRNA sequences (Table S1) and general primers that matched to the adaptor sequences were used in the following PCR reactions. For mapping the 5′ end to the very nucleotide by PCR, gene-specific primers specific to the 3′ ends of putative miRNAs and a mRAP-5′ PCR primer (5′-GCG TAT CGG GCA CCA CGT ATG C-3′) were used; for mapping the 3′ end to the very

Table 1. Differential expression (>1.5 fold) of Ical-miRNAs in spleen of the Asian seabass at 24 h post challenge with LPS as revealed by real-time PCR.

miRNA expression	Folds of differential expression			
	1.5–2.0 fold	2.0–2.5 fold	2.5–3.0 fold	>3.0 fold
Up regulated	Ical-miR-192		Ical-miR-21c	Ical-miR-21a
	Ical-miR-27			Ical-miR-21b
	Ical-miR-124a			
	Ical-miR-124b			
	Ical-miR-124c			
	Ical-miR-124d			
	Ical-miR-222a			
Down regulated	Ical-miR-199b	Ical-miR-101a		
	Ical-miR-16a			
	Ical-miR-9a			
	Ical-miR-9b			
	Ical-miR-128c			

nucleotide by PCR, gene specific primers specific to the 5′ ends of putative miRNAs and a mRAP-3′ PCR primer (5′-GAC TAG CTT GGT GCC GAA TTC GCG GTT AAA-3′) were used. Briefly, 25 μl of reaction including 0.5 U HotStar Taq DNA Polymerase (Qiagen, CA, USA), 1×PCR Buffer, 100 μm dNTPs, 200 nmol of each primer and 1 μl diluted 1st strand miRNA cDNA solution was initially denatured at 95°C for 15 min, then amplified for 35 cycles (95°C, 30 s, 60°C, 30 s and 72°C, 20 s). The PCR products were separated on a 10% polyacrylamide gel. The fragments were size fractionated, purified and inserted into a pGEM-T vector (Promega, WI, USA) for sequencing. Automated base calling of the raw sequences for mapping of boundary ends of putative conserved miRNAs and removal of vector and adaptor sequences were performed with commercial software Sequencher 4.9. All trimmed sequences were used to search against the miRBase database (release 14) with BLASTn to find the conservation of putative miRNA sequences. Sequences between 19 and 25 bp in length with a blast E value$<e^{-4}$ were considered as miRNAs in Asian seabass. The miRNA sequence logos for 15 miRNA families with multiple miRNA sequences (≥2) were produced with program WebLogo Version 2.8 [55,80].

Analysis of expression of mature miRNAs through quantitative real-time RT-PCR (qRT-PCR) in Asian seabass

Gene-specific RT primers (Table S1) for 63 identified lcal-miRNAs were designed according to their mature miRNA sequences. Primers (lca-IL1b) for *IL-1β* gene were designed based on the seabass *IL-1β* EST sequence (Genbank acc. No. EX468370). The gene-specific RT primers were equally mixed used as RT primers for first strand cDNA synthesis. In brief, total RNA was isolated from gill, brain, eye, muscle, liver, intestine, heart and kidney of 3 seabass fishes and spleen from 3 LPS-challenged fishes and 3 PBS-treated fishes (control group) using TRIzol (Invitrogen, CA, USA). The concentration and purity of total RNA were examined using a NanoDrop ND-1000 Spectrophotometer (NanoDrop Technologies, NC, USA). Around 2 μg total RNA for each sample was treated with RNase-free DNase I (Promega, WI, USA) following the manufacturer's protocol. Reverse transcription was performed at 25°C for 10 minutes, then 42°C for 60 minutes with a final incubation at 75°C for 15 minutes using gene-specific RT primer cocktail listed in the supplementary material (Table S1), a 5′ adaptor (5′-GAC CAC GCG TAT CGG GCA CCA CGT ATG CTA TCG ATC GTG AGA TGG G-3′) and PowerScript reverse transcriptase (Clontech, DB, USA). The remaining reagents (buffer, dNTPs, dithiothreitol, RNase inhibitor, Thermoscript] were added as specified in the Thermoscript protocol. The reverse transcription of the tRNA from the spleen, kidney and liver of three *Vibrio harveyi*-challenged fishes and three control fishes sampled at each time point (1, 3, 6, 12, 24 hpi) were performed as above, but using a mixture of gene-specific RT primers and hexamer primers as RT primers for first strand cDNA synthesis.

Real-time quantitative PCR was performed with the iQ SYBR Green Supermix (Bio-Rad, CA, USA) as described by the manufacturer in an iQ^{TM}5 Real-Time PCR Detection Systems (Bio-Rad, CA, USA). PCR amplicons for gene-specific qRT-PCR primer pairs (Table S1) were validated by the presence of one peak as shown by the melting curve. The curve was generated by the

thermal denaturing protocol that followed each real-time PCR run. Briefly, 25 μl of reaction including 12.5 μl SYBR Green Supermix, 200 nmol each primer and 1 μl 10-times-diluted 1st strand cDNA solution was initially denatured at 95°C for 3 min, then amplified for 40 cycles (95°C, 5 s, 60°C, 10 s and 72°C, 20 s). PCR was performed in triplicates. Values shown in Figure 5 were the average of triplicate real-time PCR reactions, normalized to lcal-miR-103 gene expression. Gene expression datasets were analyzed by Cluster 3.0 which was originally developed by Michael Eisen, with parameters as hierarchical clustering, uncentered correlation, complete linkage (http://bonsai.ims.u-tokyo.ac.jp/~mdehoon/software/cluster/software.htm#ctv), and visualized in software Java TreeView [81]. Values shown in Figure S1 and S2 were the average of triplicate real-time PCR reactions, normalized to the test gene expression in the control at the respective time points.

Supporting Information

Figure S1 IL-1ß expression pattern in three immune-related organs of Vibrio harveyi-challenged seabass at different time points.

Figure S2 Lcal-miR-21a expression pattern in three immune-related organs of Vibrio harveyi-challenged seabass at different time points.

Table S1 Primer sequences used for cloning and analyzing the expression of the Asian seabass miRNAs.

Table S2 The 322 unique sequences identified by the Sequencher assembly program and their blast results against the miRBase database.

Table S3 Novel miRNAs in the Asian seabass that are homologous to known miRNAs from other species.

Table S4 Classification of the 63 newly cloned miRNAs in the Asian seabass into miRNA families.

Acknowledgments

We thank staff members of Marine Aquaculture Center, AVA for technical supports and Dr Mamta Chuahan for editing English.

Author Contributions

Conceived and designed the experiments: GHY JHX. Performed the experiments: JHX ZYB XPH. Analyzed the data: JHX. Contributed reagents/materials/analysis tools: GHY JHX. Wrote the paper: JHX GHY.

References

1. Lee RC, Feinbaum RL, Ambros V (1993) The *C. elegans* heterochronic gene lin-4 encodes small RNAs with antisense complementarity to lin-14. Cell 75: 843–854.
2. Sunkar R, Girke T, Jain PK, Zhu JK (2005) Cloning and Characterization of MicroRNAs from Rice. The Plant Cell 17: 1397–1411.
3. Jia W, Li Z, Lun Z (2008) Discoveries and functions of virus-encoded MicroRNAs. Chinese Science Bulletin 53: 169–177.
4. Kim VN, Nam JW (2006) Genomics of microRNA. TRENDS in Genetics 22: 165–173.

5. Zhao Y, Samal E, Srivastava D (2005) Serum response factor regulates a muscle-specific microRNA that targets Hand2 during cardiogenesis. Nature 436: 214–220.

6. Wang D, Chang P, Wang Z, Sutherland L, Richardson J, et al. (2001) Activation of cardiac gene expression by myocardin, a transcriptional cofactor for serum response factor. Cell 105: 851–862.

7. Iyer D, Chang D, Marx J, Wei L, Olson EN, et al. (2006) Serum response factor MADS box serine-162 phosphorylation switches proliferation and myogenic gene programs. Proceedings of the National Academy of Sciences 103: 4516–4521.

8. Giraldez AJ, Cinalli RM, Glasner ME, Enright AJ, Thomson JM, et al. (2005) MicroRNAs regulate brain morphogenesis in zebrafish. Science 308: 833–838.

9. Schratt GM, Tuebing F, Nigh EA, Kane CG, Sabatini ME, et al. (2006) A brain-specific microRNA regulates dendritic spine development. Nature 439: 283–289.

10. Chen CZ, Li L, Lodish HF, Bartel DP (2004) MicroRNAs modulate hematopoietic lineage differentiation. Science 303: 83–86.

11. Knight SW, Bass BL (2001) A role for the RNase III enzyme DCR-1 in RNA interference and germ line development in Caenorhabditis elegans. Science 293: 2269–2271.

12. Grishok A, Pasquinelli AE, Conte D, Li N, Parrish S, et al. (2001) Genes and mechanisms related to RNA interference regulate expression of the small temporal RNAs that control C. elegans developmental timing. Cell 106: 23–34.

13. Hatfield SD, Shcherbata HR, Fischer KA, Nakahara K, Carthew RW, et al. (2005) Stem cell division is regulated by the microRNA pathway. Nature 435: 974–978.

14. Reddien PW, Oviedo NJ, Jennings JR, Jenkin JC, Alvarado AS (2005) SMEDWI-2 is a PIWI-like protein that regulates planarian stem cells. Science 310: 1327–1330.

15. Reinhart BJ, Slack FJ, Basson M, Pasquinelli AE, Bettinger JC, et al. (2000) The 21- nucleotide let-7 RNA regulates developmental timing in Caenorhabditis elegans. Nature 403: 901–906.

16. Kapsimali M, Kloosterman WP, de Bruijn E, Rosa F, Plasterk RH, et al. (2007) MicroRNAs show a wide diversity of expression profiles in the developing and mature central nervous system. Genome Biology 8: R173.

17. Kloosterman WP, Plasterk RH (2006) The diverse functions of microRNAs in animal development and disease. Developmental Cell 11: 441–450.

18. Huppi K, Volfovsky N, Mackiewicz M, Runfola T, Jones TL, et al. (2007) MicroRNAs and genomic instability. Seminars in Cancer Biology 17: 65–73.

19. Ke XS, Liu CM, Liu DP, Liang CC (2003) MicroRNAs: key participants in gene regulatory networks. Current Opinion in Chemical Biology 7: 516–523.

20. Alvarez-Garcia I, Miska EA (2005) MicroRNA functions in animal development and human disease. Development 132: 4653–4662.

21. Lu J, Getz G, Miska EA, Alvarez- Saavedra E, Lamb J, et al. (2005) MicroRNA expression profiles classify human cancers. Nature 435: 834–838.

22. Sonkoly E, Stahle M, Pivarcsi A (2008) MicroRNAs and immunity: novel players in the regulation of normal immune function and inflammation. Seminars in Cancer Biology 18: 131–140.

23. Sonkoly E, Stahle M, Pivarcsi A (2008) MicroRNAs: novel regulators in skin inflammation. Clinical and Experimental Dermatology 33: 312–315.

24. Lindsay MA (2008) microRNAs and the immune response. Trends in Immunology 29: 343–351.

25. Selbach M, Schwanhausser B, Thierfelder N, Fang Z, Khanin R, et al. (2008) Widespread changes in protein synthesis induced by microRNAs. Nature 455: 58–63.

26. Baek D, Villén J, Shin V, Camargo FD, Gygi SP, et al. (2008) The impact of microRNAs on protein output. Nature 455: 64–71.

27. Zhao Y, Srivastava D (2007) A developmental view of microRNA function. TRENDS in Biochemical Sciences 32: 189–197.

28. Sonkoly E, Pivarcsi A (2009) Advances in microRNAs: implications for immunity and inflammatory diseases. Journal of Cellular and Molecular Medicine 13: 24–38.

29. Johnston RJ, Hobert O (2003) A microRNA controlling left-right neuronal asymmetry in Caenorhabditis elegans. Nature 426: 845–849.

30. Brennecke J, Hipfner D, Stark A, Russell R, Cohen S (2003) Bantam encodes a developmentally regulated microRNA that controls cell proliferation and regulates the proapoptotic gene hid in Drosophila. Cell 113: 25–36.

31. Sunkar R, Zhu JK (2004) Novel and stress-regulated microRNAs and other small RNAs from Arabidopsis. Plant Cell 16: 2001–2019.

32. Wang JF, Zhou H, Chen YQ, Luo QJ, Qu LH (2004) Identification of 20 microRNAs from Oryza sativa. Nucleic Acids Research 32: 1688–1695.

33. Ramachandra RK, Salem M, Gahr S, Rexroad III CE, Yao J (2008) Cloning and characterization of microRNAs from rainbow trout (Oncorhynchus mykiss): Their expression during early embryonic development. BMC Developmental Biology 8: 41.

34. Kloosterman WP, Steiner FA, Berezikov E, de Bruijn E, van de Belt J, et al. (2006) Cloning and expression of new microRNAs from Zebrafish. Nucleic Acids Research 34: 2558–2569.

35. Wei Y, Chen S, Yang P, Ma Z, Kang L (2009) Characterization and comparative profiling of the small RNA transcriptomes in two phases of locust. Genome Biology 10: R6.

36. Lu C, Tej SS, Luo S, Haudenschild CD, Meyers BC, et al. (2005) Elucidation of the small RNA component of the transcriptome. Science 309: 1567–1569.

37. Reddy AM, Zheng Y, Jagadeeswaran G, Macmil SL, Graham WB, et al. (2009) Cloning, characterization and expression analysis of porcine microRNAs. BMC Genomics 10: 65.

38. Ohler U, Yekta S, Lim LP, Bartel DP, Burge CB (2004) Patterns of flanking sequence conservation and a characteristic upstream motif for microRNA gene identification. RNA 10: 1309–1322.

39. Berezikov E, van Tetering G, Verheul M, van de Belt J, van Laake L, et al. (2006) Many novel mammalian microRNA candidates identified by extensive cloning and RAKE analysis. Genome Research 16: 1289–1298.

40. Bentwich I, Avniel A, Karov Y, Aharonov R, Gilad S, et al. (2005) Identification of hundreds of conserved and nonconserved human microRNAs. Nature Genetics 37: 766–770.

41. Glazov EA, Cottee PA, Barris WC, Moore RJ, Dalrymple BP, et al. (2008) A microRNA catalog of the developing chicken embryo identified by a deep sequencing approach. Genome Research 18: 957–964.

42. Coutinho LL, Matukumalli LK, Sonstegard TS, Van Tassell CP, Gasbarre LC, et al. (2007) Discovery and profiling of bovine microRNAs from immune-related and embryonic tissues. Physiological Genomics 29: 35–43.

43. Burnside J, Ouyang M, Anderson A, Bernberg E, Lu C, et al. (2008) Deep sequencing of chicken microRNAs. BMC Genomics 9: 185.

44. IUCN (2010) IUCN Red List of Threatened Species. Version 2010.2. <http://www.iucnredlist.org >. Downloaded on 29 June 2010.

45. Tani S, Kusakabe R, Naruse K, Sakamoto H, Inoue K (2010) Genomic organization and embryonic expression of miR-430 in medaka (Oryzias latipes): insights into the post-transcriptional gene regulation in early development. Genetics 449: 41–49.

46. Wang CM, Zhu ZY, Lo LC, Feng F, Lin G, et al. (2007) A microsatellite linkage map of Barramundi, Lates calcarifer. Genetics 175: 907–915.

47. Carson J, Schmidtke LM, Munday BL (2006) Cytophaga johnsonae: a putative skin pathogen of juvenile farmed barramundi, Lates calcarifer Bloch. Journal of Fish Diseases 16: 209–218.

48. Carson J, Schmidtke LM, Munday BL (1993) Cytophaga johnsoniae:a putative skin pathogen of juvenile farmed barramundi, Lates calcarifer Bloch. Journal of Fish Diseases 16: 209–218.

49. Bromage EG, Thomas A, Owens L (1999) Streptococcus iniae, a bacterial infection in barramundi Lates calcarifer. Diseases of Aquatic Organisms 36: 177–181.

50. Clontech (1998) SMART cDNA Library Construction Kit. Clontechniques XIII: 12–13.

51. Acinas SG, Sarma-Rupavtarm R, Klepac-Ceraj V, Polz MF (2005) PCR-induced sequence artifacts and bias: insights from comparison of two 16S rRNA clone libraries constructed from the same sample. Applied and Environmental Microbiology 71: 8966–8969.

52. Altuvia Y, Landgraf P, Lithwick G, Elefant N, Pfeffer S, et al. (2005) Clustering and conservation patterns of human microRNAs. Nucleic Acids Research 33: 2697–2706.

53. Lim LP, Lau NC, Weinstein EG, Abdelhakim A, Yekta S, et al. (2003) The microRNAs of Caenorhabditis elegans. Genes & Development 17: 991–1008.

54. Grosshans H, Slack FJ (2002) Micro-RNAs: small is plentiful. The Journal of Cell Biology 156: 17–21.

55. Schneider TD, Stephens RM (1990) Sequence Logos: A new way to display consensus sequences. Nucleic Acids Research 18: 6097–6100.

56. Lee I, Ajay SS, Chen H, Maruyama A, Wang N, et al. (2008) Discriminating single-base difference miRNA expressions using microarray Probe Design Guru (ProDeG). Nucleic Acids Research 36: e27.

57. Peltier HJ, Latham GJ (2008) Normalization of microRNA expression levels in quantitative RT-PCR assays: identification of suitable reference RNA targets in normal and cancerous human solid tissues. RNA 14: 844–852.

58. Wienholds E, Kloosterman WP, Miska E, Alvarez-Saavedra E, Berezikov E, et al. (2005) MicroRNA expression in zebrafish embryonic development. Science 309: 310–311.

59. Pierce ML, Weston MD, Fritzsch B, Gabel HW, Ruvkun G, et al. (2008) MicroRNA-183 family conservation and ciliated neurosensory organ expression. Evolution & Development 10: 106–113.

60. Li H, Kloosterman W, Fekete DM (2010) MicroRNA-183 Family Members Regulate Sensorineural Fates in the Inner Ear. The Journal of Neuroscience 30: 3254–3263.

61. Ason B, Darnell DK, Wittbrodt B, Berezikov E, Kloosterman WP, et al. (2006) Differences in vertebrate microRNA expression. Proceedings of the National Academy of Sciences 103: 14385–14389.

62. Lagos-Quintana M, Rauhut R, Lendeckel W, Tuschl T (2001) Identification of novel genes coding for small expressed RNAs. Science 294: 853–858.

63. Simon DJ, Madison JM, Conery AL, Thompson-Peer KL, Soskis M, et al. (2008) The microRNA miR-1 regulates a MEF-2-dependent retrograde signal at neuromuscular junctions. Cell 133: 903–915.

64. Lagos-Quintana M, Rauhut R, Yalcin A, Meyer J, Lendeckel W, et al. (2002) Identification of tissue-specific microRNAs from mouse. Current Biology 12: 735–739.

65. Xia JH, Yue GH (2010) Identification and analysis of immune-related transcriptome in Asian seabass Lates calcarifer. BMC Genomics 11: 356.

66. Moschos SA, Williams AE, Perry MM, Birrell MA, Belvisi MG, et al. (2007) Expression profiling in vivo demonstrates rapid changes in lung microRNA levels following lipopolysaccharide-induced inflammation but not in the anti-inflammatory action of glucocorticoids. BMC Genomics 8: 240.

67. Toranzo AE, Magariños B, Romalde JS (2005) A review of the main bacterial fish diseases in mariculture systems. Aquaculture 246: 37–61.
68. Bensi G, Raugei G, Palla E, Carinci V, Buonamassa DT, et al. (1987) Human interleukin-1 beta gene. Genes & Development 52: 95–101.
69. Mauviel A, Temime N, Charron D, Loyau G, Pujol J-P (1988) Interleukin-1 α and β induce interleukin-1 β gene expression in human dermal fibroblasts. Biochemical and Biophysical Research Communications 156: 1209–1214.
70. Cheng AM, Byrom MW, Shelton J, Ford LP (2005) Antisense inhibition of human miRNAs and indications for an involvement of miRNA in cell growth and apoptosis. Nucleic Acids Research 33: 1290–1297.
71. Chen PY, Manninga H, Slanchev K, Chien M, Russo JJ, et al. (2005) The developmental miRNA profiles of zebrafish as determined by small RNA cloning. Genes & Development 19: 1288–1293.
72. Chan JA, Krichevsky AM, Kosik KS (2005) MicroRNA-21 is an antiapoptotic factor in human glioblastoma cells. Cancer Research 65: 6029–6033.
73. Si ML, Zhu S, Wu H, Lu Z, Wu F, et al. (2007) miR-21-mediated tumor growth. Oncogene 26: 2799–2803.
74. Zhu S, Si ML, Wu H, Mo YY (2007) MicroRNA-21 targets the tumor suppressor gene tropomyosin 1 (TPM1). The Journal of Biological Chemistry 282: 14328–14336.
75. Lu TX, Munitz A, Rothenberg ME (2009) MicroRNA-21 is up-regulated in allergic airway inflammation and regulates IL-12p35 expression. The Journal of Immunology 182: 4994–5002.
76. Takada S, Mano H (2007) Profiling of microRNA expression by mRAP. Nature Protocols 2: 3136–3145.
77. Griffiths-Jones S, Saini HK, van Dongen S, Enright AJ (2008) miRBase: tools for microRNA genomics. pp D154–D158.
78. Griffiths-Jones S, Grocock RJ, van Dongen S, Bateman A, Enright AJ (2006) miRBase: microRNA sequences, targets and gene nomenclature. pp D140–D144.
79. Griffiths-Jones S (2004) The microRNA Registry. pp D109–D111.
80. Crooks GE, Hon G, Chandonia JM, Brenner SE (2004) WebLogo: A sequence logo generator. Genome Research 14: 1188–1190.
81. Saldanha AJ (2004) Java Treeview—extensible visualization of microarray data. Bioinformatics 20: 3246–3248.

3

Introduced Pathogens and Native Freshwater Biodiversity: A Case Study of *Sphaerothecum destruens*

Demetra Andreou[1,2]*, Kristen D. Arkush[3], Jean-François Guégan[4,5], Rodolphe E. Gozlan[1]

1 Centre for Conservation Ecology and Environmental Change, School of Applied Sciences, Bournemouth University, Fern Barrow, Poole, Dorset, United Kingdom, 2 Cardiff School of Biosciences, Biomedical Building, Museum Avenue, Cardiff, United Kingdom, 3 Argonne Way, Forestville, California, United States of America, 4 Maladies Infectieuses et Vecteurs : Écologie, Génétique, Évolution et Contrôle, Institut de Recherche pour le Développement, Centre National de la Recherche Scientifique, Universities of Montpellier 1 and 2, Montpellier, France, 5 French School of Public Health, Interdisciplinary Centre on Climate Change, Biodiversity and Infectious Diseases, Montpellier, France

Abstract

A recent threat to European fish diversity was attributed to the association between an intracellular parasite, *Sphaerothecum destruens*, and a healthy freshwater fish carrier, the invasive *Pseudorasbora parva* originating from China. The pathogen was found to be responsible for the decline and local extinction of the European endangered cyprinid *Leucaspius delineatus* and high mortalities in stocks of Chinook and Atlantic salmon in the USA. Here, we show that the emerging *S. destruens* is also a threat to a wider range of freshwater fish than originally suspected such as bream, common carp, and roach. This is a true generalist as an analysis of susceptible hosts shows that *S. destruens* is not limited to a phylogenetically narrow host spectrum. This disease agent is a threat to fish biodiversity as it can amplify within multiple hosts and cause high mortalities.

Editor: Howard Browman, Institute of Marine Research, Norway

Funding: This work was funded by the Department for Environment, Food and Rural Affairs (DEFRA), contract FC1176. The funders had no role in study design, data collection and analysis, decision to publish, or preparation of the manuscript.

Competing Interests: The authors have declared that no competing interests exist.

* E-mail: dandreou@bournemouth.ac.uk

Introduction

Introduction of non-native species is known to pose high risks to native biodiversity in particular through introducing exotic virulent pathogens to naïve wild populations [1–4]. In freshwater ecosystems, non-native species introductions have been shown to be closely associated with human activity and the aquaculture industry [5]. Aquaculture facilities are often connected to rivers, thereby potentially increasing the risk of disease transmission from farmed fish to sympatric wildlife.

Parasite life history traits such as host specificity can heavily influence the probability of parasite transfer with invasive species [4] as well as the probability of host switch to a new naïve host. For example, generalist parasites as opposed to highly host-specific parasites are highly likely to switch hosts as they are equipped to parasitize a wide range of hosts. A wide host range ensures that the parasite can persist within a community. [6–7].

The decline and local extinctions of the previously widespread sunbleak *Leucaspius delineatus* in mainland Europe could represent a compelling example of the impact of both non-native species introductions and their microbial agents [6]. *Leucaspius delineatus* is the only representative of this genus and is now on the red list of species for a range of European countries and extinct in Slovenia [8]. Gozlan et al. [6] have shown that the population decline of this native mainland European cyprinid could be linked to the introduction of the topmouth gudgeon *Pseudorasbora parva*, a non-native cyprinid originating from Asia that was accidentally introduced into Romanian aquaculture facilities [9]. In both semi-natural (pond) and laboratory experiments, Gozlan et al. [6] demonstrated that *L. delineatus* cohabited with *P. parva* failed to

reproduce and that their population experienced a dramatic decline. This work has also shown *P. parva* to harbour *Sphaerothecum destruens* [6] a protistan pathogen responsible for disease outbreaks in salmonids in North America [10–11].

Sphaerothecum destruens is a member of a new monophyletic clade at the boundary of animal-fungal divergence [12] which includes other significant pathogens of amphibians, e.g., *Amphibiocystidium ranae* [13], and of birds and mammals including humans, e.g., *Rhinosporidium seeberi* [14]. Previous work has established that *S. destruens* is not host specific and that a range of salmonid species are susceptible to the pathogen [6,15–16]. *S. destruens* causes chronic but steady mortality in both subadult and adult Atlantic *Salmo salar*, Chinook salmon *Oncorhynchus tshawytscha* and *L. delineatus* [6,10–11,15–17].

Detecting disease related mortality in the wild is biased towards pathogens causing simultaneous, short-lived high mortalities, such as the mortality patterns caused by viral infections. Chronic, steady mortality is often undetected and underreported although it can lead to equally high mortalities and devastating effects on populations. Despite the slow-growing nature of *S. destruens* in the fish after infection, parasitism ultimately results in host cell death and often causes widespread destruction of various tissues [15–17].

Sphaerothecum destruens has an extracellular, motile zoospore stage [18–19] which is triggered when spores are in contact with fresh water and may facilitate spread to new hosts which have been shown to be more susceptible during their reproductive period [20]. However, due to the nature of the disease (i.e. slow growing), there have been limited attempts to assess the parasite's prevalence in wild populations other than through cohabitation of wild individuals with susceptible species. Nonetheless, the presence of *S. destruens* was demonstrated in up to 32% of hatchery-produced adult late Fall run Chinook salmon returning to the Upper Sacramento River of California, USA [15] and 5% in a wild *L. delineatus* population in the UK [17].

The main concern that has arisen from the Gozlan et al. paper [6] is the risk *S. destruens* poses to European freshwater biodiversity. Its association with invasive fish species such as *P. parva*, a healthy carrier [6,21], presents a risk of disease transfer from wild invasive populations to sympatric populations of susceptible native fish and as such could have major implications for fish conservation and aquaculture in Europe. Our objective was to determine the susceptibility of native cyprinid species (carp *Cyprinus carpio*, bream *Abramis brama* and roach *Rutilus rutilus*) to allopatric *S. destruens* and evaluate the risk posed to European fish biodiversity. In order to better elucidate the risks associated with *S. destruens*, a meta analysis of genetic distance between susceptible fish host species and susceptibility to the parasite was performed and used to assess the generalist nature of the pathogen.

Results

Experimental exposure to *S. destruens* led to significantly higher mortalities in *A. brama*, *C. carpio* and *R. rutilus* groups as compared to controls (Log rank test; *A. brama*: Chi-square = 10.6, d.f. = 1, $P<0.05$; *C. carpio*: Chi-square = 5.18; d.f. = 1; $P<0.05$; *R. rutilus*: Chi-square = 26.96; d.f. = 1; $P<0.05$). *A. brama* experienced high mortalities over a period of 23 days following exposure to *S. destruens* (mean mortality 53%; Figures 1, 2). The parasite was detected (by nested polymerase chain reaction [PCR]) in the kidney, liver and intestine of *A. brama* mortalities in the treatment groups with an overall prevalence of 75% (Table 1). All *A. brama* mortalities in the control group were also tested for the presence of *S. destruens* (nested PCR; kidney, liver, intestine) and were found negative for the parasite.

Experimentally-exposed *C. carpio* experienced an 8% mortality rate between 49 and 92 days post exposure (d.p.e.) (Figure 1). *Sphaerothecum destruens* DNA was detected in the kidney and intestine of *C. carpio* mortalities and sampled fish of the treatment group. Parasite DNA was detected in the intestine of two out of ten *C. carpio* sampled at 28 d.p.e. resulting in 20% prevalence in these individuals and in one out of five mortalities (Table 1). Mortality in

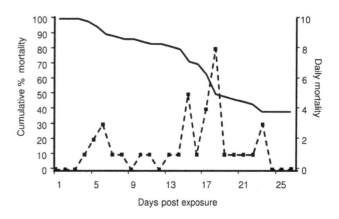

Figure 2. Mortality pattern in *Abramis brama* **as a result of infection with** *Sphaerothecum destruens.* The cumulative percentage mortality in the treatment groups (n = 60 individuals in total) and daily mortalities are presented for 26 days post exposure with *S. destruens.*

R. rutilus challenged with *S. destruens* was 37% (Figure 1) and the majority of mortalities occurred between 20 and 50 d.p.e. *S. destruens* DNA was detected in the kidney, liver and intestine of one of twenty-two *R. rutilus* mortalities at 23 d.p.e., resulting in a parasite prevalence of 5% in that species. Parasite DNA was not detected in the gills and gonads of the 13 *R. rutilus* mortalities analyzed (Table 1).

Sphaerothecum destruens DNA was not detected in the kidney, liver and intestine (by nested PCR) at six months post exposure or at the end of the experiment in both the treatment and control groups of all three cyprinids. Mean length and weight for the three species at the onset of the experiment were: 7.1 cm and 8.3 g for *A. brama*; 8.2 cm and 8.4 g for *R. rutilus*; and 7.4 cm and 7.0 g for *C. carpio*. There was no significant difference in body condition (Mann Whitney U test; *A. brama* $P = 0.257$, *C. carpio* $P = 0.457$, *R. rutilus* $P = 0.511$) between treatment and control groups across all species.

Overall, there was no significant correlation between the genetic distance and susceptibility matrices (Mantel statistic r = −0.0837, $P = 0.67$). Although not significant, a negative relationship between genetic distance and susceptibility appears to be present for the cyprinid family (Figure 3).

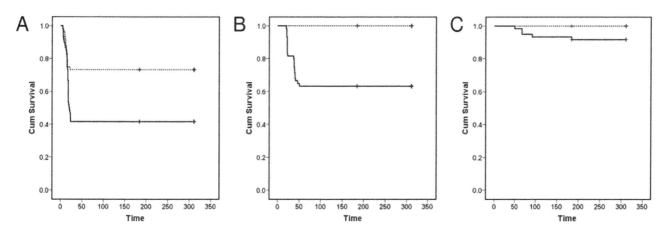

Figure 1. Kaplan-Meier survival curves for *Abramis brama, Rutilus rutilus* **and** *Cyprinus carpio* **following infection with** *Sphaerothecum destruens.* Cumulative proportion of (A) Bream *Abramis brama*, (B) Roach *Rutilus rutilus* and (C) Carp *Cyprinus carpio* surviving following exposure to *S. destruens*. Treatment fish (solid line) were exposed to an average concentration of 8.6×10⁴ *S. destruens* spores ml⁻¹ whilst control fish (dotted line) were sham exposed. Time: days post exposure.

Table 1. *Sphaerothecum destruens* prevalence mortalities of *Abramis brama, Cyprinus carpio* and *Rutilus rutilus* exposed to the parasite via bath immersion.

Species	Organs					Overall prevalence
	K	L	I	Gi	Go	
A. brama (n = 32)	75 (24/32)	63 (20/32)	34 (11/32)	n/t	n/t	75 (24/32)
C. carpio (n = 5)	20 (1/5)	0 (0/5)	20 (1/5)	n/t	n/t	20 (1/5)
R. rutilus (n = 22)	5 (1/22)	5 (1/22)	5 (1/22)	0 (0/13)	0 (0/13)	5 (1/22)

Overall prevalence (%) and organ specific prevalence is provided per species. The proportion of fish testing positive for *S. destruens* is also provided. Organs tested included the kidney (K), liver (L), intestine (I), gill (Gi) and gonad (Go). n: number of mortalities. n/t: not tested for *S. destruens*.

Discussion

Our results characterise *S. destruens* as a generalist pathogen, with a range of potential host species as demonstrated by experimental exposures (Figures 1, 2; Table 1). In this study, *S. destruens* was detected in *A. brama, C. carpio* and *R. rutilus* following experimental infection with the parasite. *A. brama* experienced mortalities exceeding 50% when exposed to the parasite with 75% of these being positive for *S. destruens* in at least two of the three organs tested (Figure 1, Table 1). These results show that *A. brama* is highly susceptible to *S. destruens*. Mortality rate in the treatment group of *C. carpio* was considerably lower (8%), with lower infection prevalence (20%), suggesting that *C. carpio* is less susceptible to the parasite. However, following 28 d.p.e, there were only 50 *C. carpio* in the treatment and control groups. This could have potentially biased the estimated *S. destruens* prevalence.

In contrast, *R. rutilus* experienced high mortalities when exposed to *S. destruens* but with equivocal conclusions regarding its susceptibility as *S. destruens* was only detected at a prevalence of 5%. The observed discrepancy between mortality and disease prevalence could be due to parasite levels in the organs tested being lower than the nested PCR detection limit and/or differences in parasite tropism in this species; with *S. destruens* being more prevalent in organs other than those tested.

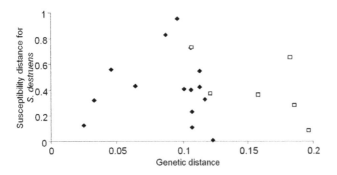

Figure 3. Host phylogeny and susceptibility to *Sphaerothecum destruens*. Genetic distance between all known susceptible species to *S. destruens* was plotted against the susceptibility distance to *Sphaerothecum destruens* for all the species combinations. The two families, Cyprinidae (□) and Salmonidae (◆) show different relationship patterns between genetic and susceptibility distances. Genetic distances were based in the pairwise analysis of ten Cytochrome b sequences. Analyses were conducted using the Tajima-Nei [30] method in MEGA4 [31]. All positions containing gaps and missing data were eliminated from the dataset. There were a total of 249 positions in the final dataset.

Although there were significantly higher mortalities in the groups exposed to *S. destruens* compared to control groups, fish died in the sham-exposed *A. brama* group. Most of these mortalities occurred during the first 15 d.p.e. (n = 11; 0/12 tested positive for *S. destruens*). This could be due to stress following the sham exposure. Mortalities were also observed during this period for the treatment group (treatment mortalities in first 15 d.p.e.: n = 12; 8/12 [67%] tested positive for *S. destruens*), however, in the case of the treatment group mortalities continued to increase past day 15.

Sphaerothecum destruens was not detected in the sampled *A. brama, C. carpio,* and *R. rutilus* at six and 11 months p.e. While prior exposure to *S. destruens* has never been reported from the source populations, surviving fish from the exposed group could have been naturally resistant to the parasite, being either refractory to initial infection or able to clear early stages of parasitism. Alternatively, surviving individuals might have entered a carrier state or developed a latent infection [22]. The absence of *S. destruens* in the surviving, experimentally-exposed fish likely suggests that either sterile immunity or resistance has occurred. There is no experimental evidence to suggest that the nested PCR used here is capable of detecting the carrier state [23]. Although not the focus of the current study, absence of carrier state should be confirmed via cohabitation of surviving fish with naïve individuals.

The low, but steady, mortality caused by this parasite (illustrated in Figure 2) highlights the danger in under detecting *S. destruens*'s related mortalities in the wild. Moribund fish are usually susceptible to predation and are thus not detected in the wild. In addition, occasional low mortality levels are often attributed to natural mortality and are thus not reported to the relevant agencies. Thus, although populations progressively decline and eventually collapse as shown in previous studies [6], the pathogen causing this decline remains undetected. It is also important to note that the factors leading to disease and mortality in a laboratory setting could differ from the ones acting in the wild. In the wild, the probability that individuals will be exposed to the minimum infectious dose could vary greatly and thus the impact on wild populations could be less severe compared to that in a laboratory setting. This highlights the need for longitudinal studies using wild populations which have or have not been exposed to *S. destruens* to assess whether this pathogen exerts a population level effect.

Among generalist parasites, some will preferentially exploit parasites from the same phylogenetic lineage whereas others appear to use a random set of locally available hosts [24–25]. The apparent lack of phylogenetic influence on host susceptibility to *S. destruens* suggests that this parasite belongs to the latter type of generalist parasites (Figure 3). It is possible that by exploiting a broader phylogenetic range of hosts, the parasite will use a number of locally available hosts and in doing so will maximise its survival and range expansion opportunities [24].

The effect of generalism on pathogenicity is unpredictable and is often not considered [23]. Generalist parasites can infect and cause high mortalities in hosts in which they do not have to persist indefinitely as long as they can persist in a reservoir host and even in the environment. Thus, in the absence of a strong host-parasite co-evolution generalist parasites can cause disease outbreaks. Such outbreaks can vary in frequency and magnitude with detrimental effects on the susceptible host (e.g. *Escherichia coli* O157 in humans) [25].

Sphaerothecum destruens is a true generalist with a highly invasive cyprinid (*P. parva*) as a reservoir host [6]. The rapid spread of *P. parva* through Europe via the aquaculture trade increases the risk of *S. destruens* introduction to a multitude of naïve fish communities enhancing the possibility for range expansion by this infectious parasite. The documented susceptibility and high mortalities in both salmonid and cyprinid species place *S. destruens* as a high risk parasite for freshwater biodiversity. In addition, these findings provide further illustration of the impacts of allochthonous infectious diseases on native fauna highlighting the risks associated with animal (and plant) trade at larger scales.

Materials and Methods

Ethics statement

All animal procedures followed strict guidelines set forward by the Home Office, UK. The project was approved by the Bournemouth University ethics committee and was performed under the Home Office licence no. 80/1979.

Sources of parasite, spore purification and DNA extraction

Sphaerothecum destruens spores used in this challenge were originally isolated from wild *L. delineatus* [6] and were cultured *in vitro* in *Epithelioma papulosum cyprini* cells [26] as described in Andreou et al. [18].

The tissues collected from sampled experimental fish and mortalities included the kidney, liver, posterior intestine, gill and gonad (if present). DNA was extracted from each tissue separately (15 mg each) using the Qiagen DNeasy 96 Blood & Tissue kit (rodent tail protocol). All steps were performed according to the manufacturer's guidelines, with an overnight incubation at 55°C and elution volume of 150 µl. Extracted DNA was quantified in a spectrophotometer at 260 nm (NanoDrop ND-1000; Labtech) and stored at −70°C until further analysis.

Sphaerothecum destruens was detected using a nested PCR amplifying a segment of the 18S rRNA gene [23]. PCR products were migrated on a 1.5% agarose gel which was post-stained with ethidium bromide (0.5 µg/ml). An individual was scored positive if *S. destruens*-specific DNA was amplified from any of its organs tested.

Fish source

Abramis brama and *R. rutilus* were supplied by the Calverton fish farm (Environment Agency, Calverton, Nottingham, UK). *Cyprinus carpio* originated from Water Lane fish farm (Burton Bradstock, Bridport). All fish were approximately one year old (1+) at the time of exposure to *S. destruens* spores. There has been no report of *S. destruens* infection in any of the farms. A total of 120 fish (60 exposed to *S. destruens* and 60 as controls) were used per species during the challenge experiments. The weight and length of ten randomly sampled fish per species were recorded at the onset of the experiment. Fish were kept in quarantine for 30 days prior to challenge with *S. destruens*.

Fish were fed twice a day with 1% of their body weight with CypriCo Crumble Astax (protein 53%, fat 13%, crude fibre 0.6%, ash 10.7%, astaxanthin 80 mg/kg; supplemented with vitamins A, D3, E and C; Coppens, Netherlands). All tanks had 25% of their water exchanged weekly and were checked for mortalities three times per day. Dead fish were collected and dissected immediately. Tissue samples were preserved for molecular analysis to test for the presence of *S. destruens*.

Experimental Infection with *Sphaerothecum destruens*

Fish from *A. brama*, *C. carpio* and *R. rutilus* were divided into six replicate 70 L tanks each containing 20 fish per tank. Each tank had its own biological filter and was aerated using an air pump. Water temperature was kept constant at 20°C and the photoperiod was maintained at 16 h light and 8 h dark. The treatment group (n = 60) was divided into three holding tanks (n = 20 per tank). Similarly, the control group (n = 60) was divided into three holding tanks (n = 20 per tank).

The exposure protocol followed was adapted from the protocol used to expose *O. tshawytscha* to *S. destruens* [23]. Fish were exposed to *S. destruens* spores (average concentration 8.6×10^4 spores ml^{-1}) in eight litres of de-chlorinated 20°C water aerated using an air pump. Control fish were sham exposed. For each species, three separate exposures to *S. destruens* were performed at three-day intervals. Exposures were maintained for four hours. The end of the third exposure was considered as Time 0.

The sampling strategy for the treatment groups is detailed in Table 2. Sampling at six months and at the end of the experiment (11 months) was identical for the treatment and control groups. Fish were euthanized with an overdose of 2-phenoxyethanol and their weight and length was recorded. Fulton's condition index K_F

Table 2. Sampling strategy for the treatment groups *Abramis brama*, *Cyprinus carpio* and *Rutilus rutilus*.

Species	Mortalities			Sampled fish (6-months p.e.)			Surviving fish at 11 months p.e.		
	K, L, I	Gi,Go	N	K, L, I	Gi, Go	N	K, L, I	Gi, Go	n
A. brama	×		32	×	×	5	×		23
C. carpio*	×		5	×		5	×		40
R. rutilus	×	×**	22	×	×	5	×		33

List of organs and organ numbers which have been tested for the presence of *Sphaerothecum destruens* DNA. Organs tested included the kidney (K), liver (L), intestine (I), gill (Gi) and gonad (Go). n: number of fish sampled.
*: at 28 d.p.e. the liver, kidney and intestine of 10 *C. carpio* were tested for *S. destruens*.
**: gill and gonad tissues were analyzed in only 13 of the 22 *R. rutilus* mortalities.

Table 3. GenBank sequences *Sphaerothecum destruens* prevalence values used in genetic and susceptibility distances.

Species	Cytochrome b	Prevalence	Exposure method	Ref.
Chinook salmon (*Oncorhynchus tshawytscha*)	AF392054	100, 71	Injection Water immersion	10, 15, 23
Coho salmon (*Oncorhynchus kisutch*)	AF165079	98	Injection	15
Rainbow trout (*Oncorhynchus mykiss*)	L29771	42.5	Injection	15
Atlantic salmon (*Salmo salar*)	AF133701	75	Disease outbreak (in aquaculture)	11
Brown trout (*Salmo trutta*)	X77526	43.3	Injection	15
Brook trout (*Salvelinus fontinalis*)	AF154850	2.6	Injection	15
Carp (*Cyprinus carpio*)	X61010	20	Water immersion	*1
Bream (*Abramis brama*)	Y10441	75	Water immersion	*
Roach (*Rutilus rutilus*)	Y10440	5	Water immersion	*
Sunbleak (*Leucaspius delineatus*)	Y10447	67, 40, 38	Cohabitation with *Pseudorasbora parva*	6, 17, 20

Mean prevalence was calculated and used where multiple prevalence values were available for a species. The infection method used is also provided.
(*) Current study.

was calculated with the following formula [27]. Tissue samples were harvested and stored at −70°C for molecular analysis.

Statistical analysis

All statistical analyses were performed using SPSS 14.0 (SPSS Inc., Chicago, Illinois, USA) unless otherwise stated. Statistical significance was accepted when $P \leq 0.05$. Standard deviation of the mean was calculated. Disease prevalence was calculated as: (number of *S. destruens* positive fish/total number of fish tested)×100. Survival analysis (Kaplan–Meier survival curves and log rank tests) were calculated for the three cyprinid species investigated. The genetic and susceptibility distance matrices were correlated using the Mantel test available from the Vegan package [28] in R [29].

Determining genetic and susceptibility distances

In order to investigate the phylogenetic influence of the host on the susceptibility to the parasite, genetic and susceptibility distances between susceptible host species were calculated. Genetic distance between susceptible species to *S. destruens* were calculated using the Tajima Nei genetic distance [30] using the software MEGA version 4 [31]. The Cytochrome b genetic marker was used to calculate genetic distances and sequences were obtained from NCBI GenBank (Table 3).

Susceptibility distance was defined as the difference in susceptibility to *S. destruens* between known susceptible species to the parasite and was calculated by subtracting *S. destruens* prevalence values for all possible pairs of fish species. Mean prevalence values were used for species with more than one reported *S. destruens* prevalence value. Prevalence values for the Salmonidae were obtained from Hedrick *et al.* [11], Arkush *et al.* [15] and Mendonca and Arkush [23]. For example, in the case of the *O. tshawytscha* – *O. mykiss* pair, *O. tshawytscha* had a mean *S. destruens* prevalence of 85.5% and *O. mykiss* a 42.5% prevalence giving a susceptibility distance of 43% or 0.43 (Figure 3).

Acknowledgments

The authors would like to thank the Environment Agency Calverton Fish Farm for providing fish.

Author Contributions

Conceived and designed the experiments: DA REG KDA. Performed the experiments: DA REG. Analyzed the data: DA REG JFG. Contributed reagents/materials/analysis tools: KDA JFG. Wrote the paper: DA KDA REG JFG.

References

1. Daszak P, Cunningham AA, Hyatt AD (2000) Wildlife ecology - Emerging infectious diseases of wildlife - Threats to biodiversity and human health. Science 287: 443–449.
2. Gozlan RE, Peeler EJ, Longshaw M, St-Hilaire S, Feist SW (2006) Effect of microbial pathogens on the diversity of aquatic populations, notably in Europe. Microbes Infect 8: 1358–1364.
3. Price PW (1990) Host populations as resources defining parasite community organization. In: Esch GB A, Aho J, eds. Parasite communities: patterns and processes. London: Chapman and Hall Ltd. pp 21–40.
4. Torchin ME, Lafferty KD, Dobson AP, McKenzie VJ, Kuris AM (2003) Introduced species and their missing parasites. Nature 421: 628–630.
5. Gozlan RE (2008) Introduction of non-native freshwater fish: is it all bad? Fish Fish 9: 106–115.
6. Gozlan RE, St-Hilaire S, Feist SW, Martin P, Kent ML (2005) Biodiversity - Disease threat to European fish. Nature 435: 1046–1046.
7. Peeler EJ, Oidtmann BC, Midtlyng PJ, Miossec L, Gozlan RE (2011) Non-native aquatic animals introductions have driven disease emergence in Europe. Biol Invasions 13: 1291–1303.
8. Lelek A (1987) The freshwater fishes of Europe: Threatened fishes of Europe: AULA-Verlag Wiesbaden. 343 p.
9. Gozlan RE, Andreou D, Asaeda T, Beyer K, Bouhadad R, et al. (2010) Pan-continental invasion of *Pseudorasbora parva*: towards a better understanding of freshwater fish invasions. Fish Fish 11: 315–340.
10. Harrell LW, Elston RA, Scott TM, Wilkinson MT (1986) A significant new systemic-disease of net-pen reared Chinook salmon (*Oncorhynchus tshawytscha*) Brood Stock. Aquaculture 55: 249–262.
11. Hedrick RP, Friedman CS, Modin J (1989) Systemic infection in Atlantic salmon *Salmo salar* with a *Dermocystidium*-like species. Dis Aquat Organ 7: 171–177.
12. Mendoza L, Taylor JW, Ajello L (2002) The class Mesomycetozoea: A group of microorganisms at the animal-fungal boundary. Annu Rev Microbiol 56: 315–344.
13. Pereira CN, Di Rosa I, Fagotti A, Simoncelli F, Pascolin R, et al. (2005) The pathogen of frogs *Amphibiocystidium ranae* is a member of the order dermocystida in the class Mesomycetozoea. J Clin Microbiol 43: 192–198.
14. Silva V, Pereira CN, Ajello L, Mendoza L (2005) Molecular evidence for multiple host-specific strains in the genus Rhinosporidium. J Clin Microbiol 43: 1865–1868.
15. Arkush KD, Frasca S, Hedrick RP (1998) Pathology associated with the rosette agent, a systemic protist infecting salmonid fishes. J Aquat Anim Health 10: 1–11.

16. Paley RK, Andreou D, Bateman KS, Feist SW (2012) Isolation and culture of *Sphaerothecum destruens* from Sunbleak (*Leucaspius delineatus*) in the UK and pathogenicity experiments in Atlantic salmon (*Salmo salar*). Parasitology Available on CJO doi:10.1017/S0031182012000030.

17. Andreou D, Gozlan RE, Stone D, Martin P, Bateman K, et al. (2011) *Sphaerothecum destruens* pathology in cyprinids. Dis Aquat Organ 95: 145–151.

18. Andreou D, Gozlan RE, Paley R (2009) Temperature influence on production and longevity of *Sphaerothecum destruens* zoospores. J Parasitol 95: 1539–1541.

19. Arkush KD, Mendoza L, Adkison MA, Hedrick RP (2003) Observations on the life stages of *Sphaerothecum destruens* n. g., n. sp., a mesomycetozoean fish pathogen formally referred to as the rosette agent. J Eukaryot Microbiol 50: 430–438.

20. Andreou D, Hussey M, Griffiths SW, Gozlan RE (2011) Influence of host reproductive state on *Sphaerothecum destruens* prevalence and infection level. Parasitology 138: 26–34.

21. Gozlan RE, Whipps CM, Andreou D, Arkush KD (2009) Identification of a rosette-like agent as *Sphaerothecum destruens*, a multi-host fish pathogen. Int J Parasitol 39: 1055–1058.

22. Thrusfield M (2007) Veterinary epidemiology. Oxford: Blackwell Publishing.

23. Mendonca HL, Arkush KD (2004) Development of PCR-based methods for detection of *Sphaerothecum destruens* in fish tissues. Dis Aquat Organ 61: 187–197.

24. Krasnov BR, Khokhlova IS, Shenbrot GI, Poulin R (2008) How are the host spectra of hematophagous parasites shaped over evolutionary time? Random choice vs selection of a phylogenetic lineage. Parasitol Res 102: 1157–1164.

25. Woolhouse MEJ, Taylor LH, Haydon DT (2001) Population biology of multihost pathogens. Science 292: 1109–1112.

26. Fijan N, Sulimanovic D, Bearzotti M, Muzinic D, Zwillenberg LO, et al. (1983) Some properties of the *Epithelioma papulosum cyprini* (Epc) Cell-Line from Carp *Cyprinus carpio*. Ann Inst De Virologie 134: 207–220.

27. Ostlund-Nilsson S, Curtis L, Nilsson GE, Grutter AS (2005) Parasitic isopod *Anilocra apogonae*, a drag for the cardinal fish *Cheilodipterus quinquelineatus*. Mar Ecol-Prog Ser 287: 209–216.

28. Oksanen J, Kindt R, Legendre P, O'Hara B, Simpson GL, et al. (2009) Vegan: community ecology package. R package version 115-2. Available: http://CRANR-projectorg/package=vegan. Accessed 2009 Nov 10.

29. R DCT (2009) R: a language and environment for statistical computing. Vienna, Austria: R Foundation for Statistical Computing, Available: http://CRAN.R-project.org Accessed 2009 Nov 10.

30. Tajima F, Nei M (1984) Estimation of Evolutionary Distance between Nucleotide-Sequences. Mol Biol Evol 1: 269–285.

31. Tamura K, Dudley J, Nei M, Kumar S (2007) MEGA4: Molecular evolutionary genetics analysis (MEGA) software version 4.0. Mol Biol Evol 24: 1596–1599.

4

Sea Louse Infection of Juvenile Sockeye Salmon in Relation to Marine Salmon Farms on Canada's West Coast

Michael H. H. Price[1,2]*, Stan L. Proboszcz[3], Rick D. Routledge[4], Allen S. Gottesfeld[5], Craig Orr[3], John D. Reynolds[6]

1 Department of Biology, University of Victoria, Victoria, Canada, 2 Raincoast Conservation Foundation, Sidney, Canada, 3 Watershed Watch Salmon Society, Coquitlam, Canada, 4 Department of Statistics and Actuarial Science, Simon Fraser University, Burnaby, Canada, 5 Skeena Fisheries Commission, Hazelton, Canada, 6 Earth to Ocean Research Group, Department of Biology, Simon Fraser University, Burnaby, Canada

Abstract

Background: Pathogens are growing threats to wildlife. The rapid growth of marine salmon farms over the past two decades has increased host abundance for pathogenic sea lice in coastal waters, and wild juvenile salmon swimming past farms are frequently infected with lice. Here we report the first investigation of the potential role of salmon farms in transmitting sea lice to juvenile sockeye salmon (*Oncorhynchus nerka*).

Methodology/Principal Findings: We used genetic analyses to determine the origin of sockeye from Canada's two most important salmon rivers, the Fraser and Skeena; Fraser sockeye migrate through a region with salmon farms, and Skeena sockeye do not. We compared lice levels between Fraser and Skeena juvenile sockeye, and within the salmon farm region we compared lice levels on wild fish either before or after migration past farms. We matched the latter data on wild juveniles with sea lice data concurrently gathered on farms. Fraser River sockeye migrating through a region with salmon farms hosted an order of magnitude more sea lice than Skeena River populations, where there are no farms. Lice abundances on juvenile sockeye in the salmon farm region were substantially higher downstream of farms than upstream of farms for the two common species of lice: *Caligus clemensi* and *Lepeophtheirus salmonis*, and changes in their proportions between two years matched changes on the fish farms. Mixed-effects models show that position relative to salmon farms best explained *C. clemensi* abundance on sockeye, while migration year combined with position relative to salmon farms and temperature was one of two top models to explain *L. salmonis* abundance.

Conclusions/Significance: This is the first study to demonstrate a potential role of salmon farms in sea lice transmission to juvenile sockeye salmon during their critical early marine migration. Moreover, it demonstrates a major migration corridor past farms for sockeye that originated in the Fraser River, a complex of populations that are the subject of conservation concern.

Editor: Brock Fenton, University of Western Ontario, Canada

Funding: The work was funded by: The Coastal Alliance for Aquaculture Reform (www.farmedanddangerous.org), David and Lucile Packard Foundation (www.packard.org), Gordon and Betty Moore Foundation (www.moore.org), Patrick Hodgson Family Foundation, Ritchie Foundation, Sandler Family Foundation, SOS Marine Conservation Foundation (www.saveoursalmon.ca), Tom Buell Endowment Fund, a Natural Sciences and Engineering Research Council of Canada (NSERC) Discovery Grants to RD Routledge and JD Reynolds (www.nserc-crsng.gc.ca), and an NSERC Industrial Postgraduate Scholarship to MHH Price. The funders had no role in study design, data collection and analysis, decision to publish, or preparation of the manuscript.

Competing Interests: The authors have declared that no competing interests exist.

* E-mail: pricem@uvic.ca

Introduction

Pathogens are growing threats to wildlife [1,2]. The spread of infectious pathogens commonly occurs when humans bring wildlife into increased contact with infected domestic animals [3,4]. Ensuing epizootics have devastated wild populations, as illustrated by the transmission of rabies from domestic dogs to wild carnivores [5,6], *Pasteurella* from domestic to wild sheep [7], and *Crithidia bombi* from commercial to wild bumble bees [4].

Caligid sea lice (mainly *Lepeophtheirus salmonis* and *Caligus* spp.) are the most widespread marine parasites affecting domestic and wild fish, and have now emerged as important pathogens in many

coastal marine areas [8–10]. Sea lice feed on surface tissues of their hosts, which can lead to many problems especially for small juvenile fish [8,11]. Sea lice can compromise osmoregulation [12], induce behavioral changes that increase predation risk [13], reduce growth rates and, in sufficient numbers, result in host death [9,14,15]. Sea lice also have been shown to serve as vectors for the spread of fish diseases [16,17].

The transmission of pathogens to wildlife frequently occurs where host populations are concentrated into dense aggregations [6,18]. The recent global expansion of marine salmon farming is one such situation in which concentrated reservoir populations may dramatically alter the natural transmission dynamics of salmonid host-parasite systems [9,19–21]. In natural systems, migratory allopatry (the spatial separation of age classes) of wild salmon creates a barrier to parasite transmission [22]. Conversely, salmon farms hold domestic fish, mainly Atlantic salmon (*Salmo salar*), in high densities for months in the same location (i.e., 15–30 kg/m^3 for up to 24 months) [23]. These crowded conditions facilitate parasite and disease transmission within the farm, and enable exponential population growth of pathogens and release to the surrounding environment [24,25]. Juvenile wild salmon swimming past salmon farms are frequently infected with sea lice [21,26], and studies have implicated sea lice from farms in the decline of some wild salmonid populations in Europe and North America [9,27,28].

Recent research has raised concern that sea lice from salmon farms may infect juvenile sockeye salmon (*Oncorhynchus nerka*) in an area of Canada's west coast between Vancouver Island and the mainland known as the Discovery Islands [29]. This region is home to the northeast Pacific's largest salmon farm industry and hosts one of the largest migrations of salmon in the world (primarily to and from the Fraser River) [30]. Sockeye is the Pacific Ocean's most economically and culturally important salmon species, and several populations from the Fraser River are endangered [31]. Productivity of Fraser River sockeye has been declining since the early1990s, with 2009 being the lowest on record, prompting the Canadian government to launch a Judicial Inquiry to investigate the cause of the decline and identify imminent threats to their survival [32]. The early marine phase of sockeye remains one of the least understood [33], yet has received the most attention in the search for answers to declining sockeye productivity [34]. Thus, determining whether sockeye are at risk to sea lice transmission from salmon farms during their early marine migration is highly relevant to conservation and management efforts.

In this study we examined parasite infection of wild juvenile sockeye from two geographically separated regions of Pacific Canada: one with salmon farms, and one without. Within the farm region, we compared infection rates on fish from locations that vary in their exposure to farms. We used molecular genetics techniques to determine the origins of the fish, and we employed mixed-effects modelling to examine factors that best explain sea lice abundance.

Materials and Methods

Ethics statement

All juvenile salmon were humanely euthanized in accordance with Fisheries and Oceans Canada's national guidelines, under permit XR 21 2007–2008. Study approval by academic ethics committees was not necessary as no academic institution was involved during the data collection.

Study area and sampling

We collected juvenile sockeye from marine waters surrounding the Discovery Islands, an area containing 18 active salmon farms, from April 22 to June 15, 2007 (n = 381) and May 31 to July 3, 2008 (n = 510), and from the north coast of British Columbia, an area without salmon farms, from May 26 to July 5, 2007 (n = 369; Figure 1). Up to five replicate sets of samples were obtained from each site, each year, in the Discovery Islands (1–50 juvenile sockeye salmon per sample), and during 2007 on the north coast (1–129 juvenile sockeye salmon per sample). We used a beach seine (50 m long, 1.5 m deep, 6 mm mesh) among the Discovery Islands to capture sockeye, and a surface trawl-net (18 m long, 5 m opening, 4.6 m deep) on the north coast. The trawl-net was fitted

with a rigid holding box at the far end designed for live capture and to minimize the loss of scales and ectoparasites [35]. We recorded sea surface salinity and temperature during each sampling event in both regions using a YSI-30 SCT meter. Fish were immediately frozen and labeled for subsequent laboratory analyses in which individual fish were thawed and assayed for sea lice using a dissecting microscope. Species of motile (i.e., sub-adult and adult) stages of sea lice were directly identified by morphology [36,37]; younger copepodid and chalimus stage lice were removed from the fish, mounted on permanent slides and examined under a compound microscope for determination based on detailed morphology [36,37].

Genetic analyses

We proportionally sampled previously frozen tissues for genetic determination in the Discovery Islands from juveniles retained at each capture location, per sampling event, each year (i.e., 1/3 from 2007, n = 92; 1/5 from 2008, n = 114), and placed them individually in vials of 95% ethanol. We collected fresh tissue from all sockeye (n = 478) on the north coast, and placed them individually in vials of 99% ethanol. Tissue samples from both regions were analyzed at the Fisheries and Oceans Canada (DFO) molecular genetics laboratory in British Columbia. DNA was extracted from tissue [38], and samples were analyzed for polymerase chain reaction products at 14 microsatellite loci [39]. We considered amplification at a minimum of 7 loci as adequate for estimating stock origin as previous surveys of the microsatellite variation in Fraser River sockeye at 6 loci indicated differentiation among populations [38]. Individuals were assigned to source populations using mixed stock analysis techniques employing Bayesian mixture modeling [40] using the software program cBayes. Stock proportions were determined by comparing one mixture (north coast 2007) to a baseline comprising 227 sockeye populations, and two mixtures (Discovery Islands 2007 and 2008) to a baseline comprising 85 sockeye populations [39,41]. The reported stock composition estimates with corresponding standard deviations were derived from combined posterior distributions using the last 1 000 iterations from 10 Monte Carlo Markov runs of 20 000 iterations.

Statistical analyses

To test for spatial patterns in sea lice on sockeye, we organized capture locations within the Discovery Islands based on whether each site was: upstream (a position on the juvenile sockeye migration route where fish likely had not passed a salmon farm), or downstream (a position where fish must have passed at least one salmon farm), given the net movement of juvenile sockeye through the region [42]; downstream collection sites are encircled within Figure 1. The ocean environment surrounding the Discovery Islands is estuarine, with a net-northward flow predominating during the months of our study [43]. Fish captured downstream of a salmon farm could only have arrived at that location by swimming past a salmon farm, and our results on genetic origins of the fish substantiated this. However, sockeye caught at two sites considered upstream of a salmon farm may have swum past a farm before capture because of fish movements or strong tidal currents, and the close proximity to a farm. Although we consider these occurrences infrequent, they may have contributed to the observed variability in louse infection levels observed at these sites. We placed collection sites from the north coast in a third category: no farms.

Marine Harvest Canada (MHC) is the only salmon farm company to report sea louse average abundance; raw sea louse data were not reported publicly at the time of our study. We used

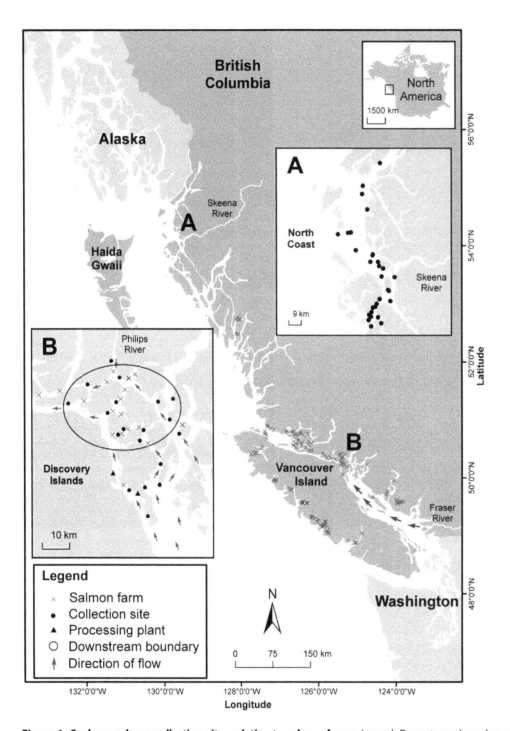

Figure 1. Sockeye salmon collection sites relative to salmon farms. Legend: Downstream boundary encircles all sockeye collection sites situated downstream of at least one salmon farm given the direction of prevailing oceanic flow and migration direction; all other collection sites are considered upstream.

average *Caligus clemensi* abundance and *L. salmonis* motile abundance provided online to estimate sea louse trends on six MHC farms in the Discovery Islands during 2007–2008; sea louse data were not provided for the other 12 farms operating in the region. For periods without reported information, we calculated average abundance using the previous and subsequent values.

We performed exploratory analyses to probe for obvious spatial clusters in louse abundances for *L. salmonis* and *C. clemensi*. We used the SAS Cluster procedure with Ward's method for calculating distances between clusters; one capture site upstream of farms emerged as a clear outlier. Because such outliers can exercise undue influence on inferences based on regression-style statistical models [44], yet can also provide important insight, we singled out this site for special consideration.

We used mixed-effects modelling for formal analyses of sea lice abundances, with a random effect associated with sockeye sampling events. We performed a separate analysis for total abundance of each louse species. We used a generalized linear mixed modelling approach using SAS GLIMMIX procedure, and we specified a Poisson error distribution for lice on individual fish within a capture event. We calculated denominator degrees of freedom with a Satterthwaite [45] approximation. We included salinity, temperature, year, and position relative to salmon farms as fixed factors, as these are thought to most influence lice levels on juvenile salmon [10,46]; position in the Discovery Islands area was set to 0 for upstream sites and 1 for downstream sites, and in the north coast to 2 for no farms. Specifically, we hypothesized that fish from locations downstream of farms would have higher louse abundance than upstream sites, that these would in turn be higher than on the north coast where there are no farms, and that high temperature and salinity would also be correlated with high lice loads (because sea louse growth in lab-based trials depends strongly on temperature and salinity [8]). This approach permitted us to test these factors simultaneously for potential influence on lice abundances. We also explored the potential contribution from an additional random factor associated with sampling sites (nested within exposure class); however, this random factor failed to contribute a significant component to the variance, and we omitted it from the final versions of the models. Finally, we ran analyses with and without the outlier site excluded. Because results were broadly similar, and due to the statistical problems of including the outlier site (mentioned above and in the Discussion section), we report findings with the outlier excluded.

We ran the complete suite of $2^4 - 1 = 15$ models of all subsets of the four factors on total abundance of each louse species. Because the methodology underlying GLIMMIX is based on approximations, which can generate misleading values of Akaike's Information Criterion and its variants, we used other methods to compare competing models. Specifically, we identified models for which (i) each included factor was significant, and placed further emphasis on the subset of these models for which (ii) any model containing these factors plus at least one more contained at least one factor that was not significant. That is, when we tried to add another factor, either the extra factor or a previous one already in the model became non-significant. These criteria sometimes produced more than one viable model; however, such ambiguities are to be anticipated given the correlations amongst all factors in these models (which ranged from 0.144 to 0.547). All analyses were generated using SAS/STAT software, V-9.1 (SAS Institute Inc., 2000–2004).

Results

Genetic analyses confirmed that the majority of juvenile sockeye on the north coast were from the Skeena, Nass, and adjacent watersheds (98.3% combined), and thus they were unlikely to have been influenced by salmon farms further south before capture (Table 1; Figure 1). Conversely, all sockeye migrating through the Discovery Islands region were either from the Fraser River (85%) or nearby Johnstone and Queen Charlotte Strait rearing lakes (15%), and may have been influenced by salmon farms depending on their location.

Sea louse abundances on the north coast for *C. clemensi* and *L. salmonis* combined were an order of magnitude lower than in the Discovery Islands (Table 2). Within the Discovery Islands, *C. clemensi* was the principal louse species infecting sockeye in both years, and most abundant on fish downstream of salmon farms (Figure 2). The maximum infection intensity of *C. clemensi* was highest downstream of farms in 2007 (28 lice per fish) compared to upstream sites (16 lice per fish), and equal throughout the region in 2008 (9 lice per fish).

Excluding sockeye caught at the outlier site among the Discovery Islands in 2008, which hosted the highest levels of

either louse species during that year, *L. salmonis* was most abundant on juveniles downstream of salmon farms, and more abundant in 2008 compared to 2007 (Figure 3). In correspondence with the hypothesized contributions of salmon farms to these wild fish, MHC farms hosted more *C. clemensi* during the out-migration period in 2007 than 2008, and more *L. salmonis* in 2008 than 2007 (Figure 4).

Mixed-effects modelling showed some variation in results depending on louse species. Position relative to farms was consistently significant in all models for total abundance of *C. clemensi* in which it was included. Furthermore, whenever this factor was included, none of the others was significant; thus, the top model was clearly the one containing only this factor ($p < 0.0001$). The ratio of *C. clemensi* total abundance between upstream and downstream categories was estimated by this model at 2.80 with 95% confidence intervals of 1.03 and 7.68. This ratio is significantly larger than 1 ($p = 0.044$), and *C. clemensi* abundances were significantly and substantially larger in the Discovery Islands than on the north coast ($p \leq 0.0022$; Figure 2).

For total abundance of *L. salmonis*, year was consistently significant in every model in which it appeared ($p < 0.017$), although position relative to farms and salinity were also significant on their own ($p < 0.001$; Table 3). Two models satisfied our selection criteria: (i) year + position relative to farms + temperature, and (ii) year + salinity + temperature; hence, the effects of position relative to farms and salinity appear confounded in these models. According to the former model, the total abundance of *L. salmonis* was significantly lower on the north coast than at each of the upstream and downstream sites in the Discovery Islands area ($p \leq 0.0035$), but there was no significant difference between upstream and downstream sites ($p = 0.26$). Transformed estimates derived from the least squares means for this model and their standard errors are plotted in Figure 3.

Discussion

We have demonstrated a potential role of open net-pen salmon farms in transmission of sea lice to wild juvenile sockeye salmon. Most juvenile sockeye assessed for sea lice originated either in the Fraser or Skeena watershed, thus providing a novel comparison of sea louse infection between Canada's largest sockeye rivers. Moreover, our genetics results demonstrate a major migration corridor past farms for fish that originated in the Fraser River, a complex of populations that have been the subject of concern due to declining productivity since the early 1990s, and a collapse in 2009 followed by a substantial rebound in 2010.

Juvenile sockeye salmon in both regions were primarily infected by *C. clemensi*, which is consistent with juvenile pink and chum salmon in areas without salmon farms in the north Pacific [22,49]. The predominance of *C. clemensi* routinely shifts to *L. salmonis* for pink and chum in regions with intensive salmon farming [21,29,47], and this was shown for those species in the Discovery Islands during the years of our study [26]. Most of the sockeye we examined among the Discovery Islands were caught in mixed schools with *L. salmonis*-infected juvenile pink and chum. Thus, the predominance of *C. clemensi* on sockeye upstream of farms suggests that sockeye either show higher resistance to *L. salmonis*, or heightened susceptibility to *C. clemensi*; alternatively, perhaps *C. clemensi* has a preference for sockeye, or *L. salmonis* prefers juvenile pink and chum salmon. This warrants future experimental work.

Juvenile sockeye migrating along the north coast hosted an order of magnitude fewer sea lice than those migrating through the Discovery Islands. Wild juvenile salmon in Europe and North America consistently host low levels of sea lice during their early

Table 1. Stock proportion estimates and standard deviations for genetically identified juvenile sockeye salmon.

Stock Origin	North Coast 2007 Estimate (SD)	Discovery Islands 2007 Estimate (SD)	Discovery Islands 2008 Estimate (SD)
Chilko Lake (Fraser River)	0.0 (0.0)	22.8 (4.7)	26.9 (3.9)
Quesnel Lake (Fraser River)	0.0 (0.1)	33.4 (5.2)	3.1 (1.9)
Shuswap Lake (Fraser River)	0.0 (0.1)	0.0 (0.2)	57.9 (4.1)
Other Fraser River	0.0 (0.2)	5.4 (2.8)	11.0 (2.7)
Washington & Oregon	0.0 (0.0)	0.0 (0.2)	0.0 (0.1)
West coast Vancouver Island	0.0 (0.1)	0.0 (0.2)	0.1 (0.4)
Johnstone & Queen Charlotte Straits	0.0 (0.1)	37.8 (4.9)	0.6 (0.6)
Queen Charlotte Strait to Skeena estuary	2.2 (0.9)	0.0 (0.5)	0.0 (0.4)
Skeena River estuary	3.1 (0.9)	0.0 (0.2)	0.0 (0.2)
Babine Lake (Skeena River)	85.0 (1.9)	0.0 (0.2)	0.0 (0.1)
Other Skeena River	7.7 (1.4)	0.0 (0.2)	0.0 (0.1)
Nass River	0.9 (1.2)	0.0 (0.2)	0.0 (0.2)
Queen Charlotte Islands	0.2 (0.4)	0.0 (0.5)	0.0 (0.3)
Southeast Alaska	0.7 (0.6)	0.6 (0.9)	0.3 (0.6)

marine migration in areas without salmon farms [22,48,49], though brief localized outbreaks have occurred [50,51]. Louse parasitism of juveniles is frequently higher for sustained periods in regions with salmon farming [27,47,52]. Factors beyond the absence of farm salmon on the north coast may have contributed to the significantly lower lice levels on sockeye compared to the Discovery Islands. In particular, differences in lice levels may be due to our use of different sampling gear or different environmental conditions, though we did incorporate the two key conditions known to affect sea louse infection levels into our analyses: salinity and temperature. Our analyses show that the lower infection rates for *C. clemensi* on the north coast cannot be explained by salinity and temperature alone. The primary strength of our study was the comparison of infection levels before and after fish had been exposed to salmon farms within the Discovery Islands.

Parasitism of sockeye by *C. clemensi* in the Discovery Islands was higher on juveniles downstream of salmon farms than on those upstream of farms. These findings are consistent with previous research on juvenile pink and chum salmon in this region, and elsewhere in the north Pacific [26,29]. Farm data provide further evidence that *C. clemensi* was abundant on farm salmon while juvenile sockeye migrated through the region, particularly during the higher infection year of 2007 [53,54] (see our Figure 4). Although the position of sockeye relative to salmon farms was the only significant factor to explain our data, we need to consider alternative explanations. First, the spatial distribution of upstream/downstream collection sites assumes a northbound migration. Juveniles caught downstream of farms were consistently larger than upstream sockeye, which may be evidence for extended residency time (i.e., increased exposure to sea lice, which may lead to epizootics [55]). Juveniles that spent longer in the marine environment would host greater proportions of motile stage lice, as lice would have had more time to develop. However, juveniles downstream of farms primarily hosted larval stage lice, which suggests they were infected recently by a local source. Moreover, juveniles from different populations within the Fraser River are not of equal size, and they vary in their migration timing

Table 2. Summary statistics and sea louse infection rates on juvenile sockeye.

Region	Position to Salmon Farms	Year	Total Fish	Fork Length	Mass	Salinity	Temperature	Caligus clemensi P [a]	A [b]	I [c]	Nm [d]	Lepeophtheirus salmonis P [a]	A [b]	I [c]	Nm [d]
North Coast	No farms	2007	369	8.17 cm	5.21 g	16.97‰	9.80°C	0.09	0.17	1.97	0.97	0.01	0.01	1.00	1.00
Discovery Islands	Upstream	2007	163	7.26 cm	3.91 g	25.42‰	10.79°C	0.29	1.10	3.83	0.92	0.05	0.05	1.00	0.78
	Downstream	2007	218	7.76 cm	5.08 g	27.38‰	10.94°C	0.84	4.83	5.72	0.95	0.09	0.09	1.05	1.00
	Upstream	2008	60	8.98 cm	8.15 g	25.98‰	14.72°C	0.40	0.95	2.31	0.72	0.05	0.05	1.00	0.33
	Downstream	2008	400	10.30 cm	12.04 g	28.47‰	9.64°C	0.62	1.61	2.60	0.55	0.21	0.30	1.42	0.31
	Outlier	2008	50	9.22 cm	8.50 g	30.00‰	9.00°C	0.92	3.60	4.42	0.70	0.42	0.64	1.52	0.94

[a]Louse prevalence.
[b]Louse abundance.
[c]Louse intensity.
[d]Proportion of combined non-motile life stages (copepodid and chalimus I to IV).
Legend: All morphometric and abiotic values represent the mean, except sea lice infection rates.

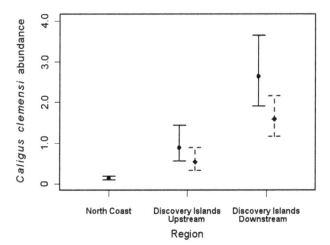

Figure 2. Annual estimates of *Caligus clemensi* abundance on sockeye salmon. Legend: North Coast region is without salmon farms, Discovery Islands upstream region encompasses sockeye collection sites upstream of all salmon farms given the direction of prevailing oceanic flow and migratory direction, and Discovery Islands downstream represents all collection sites downstream of farms for 2007 (solid line) and 2008 (dotted line). Estimates were obtained by back-transforming least-squares means; error bars, by back-transforming the least-squares means ±1 standard error.

through our study region (M. Price unpublished data); thus, size may not be a simple metric for residency time and deserves further examination. Second, because *C. clemensi* is a generalist parasite, non-salmonids such as Pacific herring (*Clupea pallasi*) may have been a local source for lice (as has been hypothesized elsewhere [51]). We also consider this unlikely to account for *C. clemensi* increases on sockeye downstream of farms, as pelagic fishes would need to assume a similar spatial distribution (i.e., more fishes

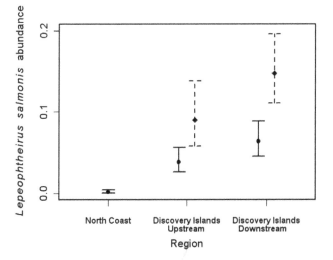

Figure 3. Annual estimates of *Lepeophtheirus salmonis* abundance on sockeye salmon. Legend: North Coast region is without salmon farms, Discovery Islands upstream region encompasses sockeye collection sites upstream of all salmon farms given the direction of prevailing oceanic flow and migratory direction, and Discovery Islands downstream represents all collection sites downstream of farms for 2007 (solid line) and 2008 (dotted line). Estimates were obtained by back-transforming least-squares means; error bars, by back-transforming the least-squares means ±1 standard error.

downstream of farms) over consecutive years, and there is no evidence for this.

Similar to *C. clemensi*, parasitism of sockeye by *L. salmonis* was higher in the Discovery Islands than the north coast, and lice levels further increased for juveniles downstream of salmon farms. Notably, the year of highest infection among the Discovery Islands was the opposite for each louse species infecting sockeye: *L. salmonis* was most abundant in 2008, *C. clemensi* was most abundant in 2007, and farm salmon in this region showed similar inter-annual trends for each species. Our mixed-effects modelling further showed that migration year best explained *L. salmonis* total abundance, indicating significant inter-annual variation in *L. salmonis* abundance on sockeye that is consistent with farm salmon. Farm salmon hosted lice well before sockeye began migrating through the region, and are the most likely source of infection.

Sockeye among the Discovery Islands were most infected with *L. salmonis* at the outlier site compared to all other sites. This site was approximately 8 km upstream from a farm salmon processing facility where large numbers of live sea lice, primarily nauplii, have recently been recorded from the effluent (A. Morton unpublished data). Tidal currents here (i.e., Discovery Passage) can transport particles this distance in a single tide-cycle [43], which suggests that the processing facility may have been a source for lice on sockeye. This also suggests that other 'upstream' locations may have been exposed to farm-origin lice (and may explain the significantly higher lice levels on sockeye at all upstream sites compared to the north coast), but to a lesser degree than downstream locations. Alternatively, this single location may have been home to a large congregation of resident fishes that were heavily infected with sea lice. Although we caught only sockeye during this single capture event, we have caught juvenile pink and chum salmon with relatively low lice levels at that location previously. Note that while we cannot justify including this outlier site in our formal statistical tests because it is inconsistent with the model assumptions, when we included the outlier in the analysis (the invalidity of the inferences notwithstanding), the primary conclusions remained essentially the same. Hence, this unique observation, though it does not critically impinge on the results of the study, is important in that it suggests the need for heightened attention towards the potential role of processing plants in sea lice dynamics.

Does *C. clemensi* pose a threat to sockeye salmon? Research to date has not examined the effects of this sea louse on wild juvenile Pacific salmonids, though significant fin damage by larval stage lice has been documented [50]. *Caligus clemensi* is smaller than *L. salmonis*, and is thought to cause less mechanical damage to juvenile pink and chum salmon [9,14,22]. Moreover, juvenile sockeye are larger and have developed scales at the time of ocean entry compared to juvenile pink and chum; thus, it is unlikely that the average number of *C. clemensi* observed on sockeye (2–3 lice/fish) would cause direct mortality for healthy fish. However, evidence is mounting that marine parasites, such as sea lice, can induce behavioral changes that may result in higher mortality rates for hosts [13,56]. The transition from freshwater to marine environments is one of the most physiologically demanding phases for salmon [57], and overall marine survival appears to depend on rapid early marine growth [58]. Even low levels of parasitic infection may be harmful during this critical period. Moreover, the presence and abundance of sea lice on juvenile sockeye may be a proxy for other farm-origin pathogens. Given the high intensities of *C. clemensi* observed on some juveniles in this study (i.e., up to 28 lice/fish), concern is justified, and research should be undertaken to understand the extent of threat posed.

There is considerable interest in understanding the factors that affect survival of juvenile sockeye in the marine environment, and

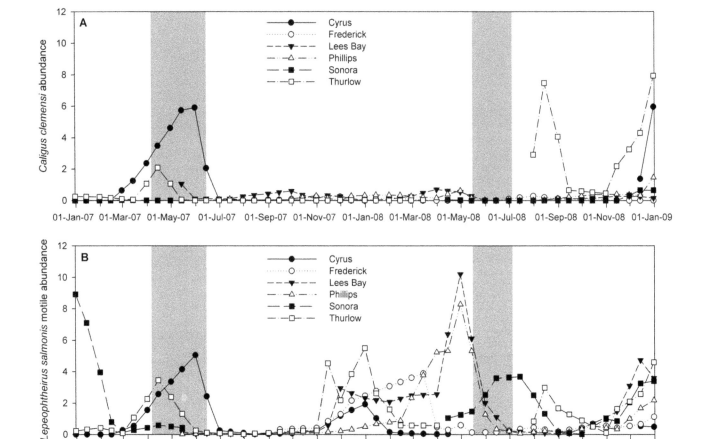

Figure 4. Sea louse abundance over time on Atlantic salmon on named salmon farms in the Discovery Islands. Legend: *Caligus clemensi* at top, and *Lepeophtheirus salmonis* at bottom. Period of sockeye collection during 2007 and 2008 in shaded grey.

specifically whether salmon farms are contributing to declines. Sockeye productivity in many Canadian river systems has declined over the last decade, including the Skeena River; thus multiple

Table 3. Mixed effects models, and associated variance component estimates and standard errors for *Lepeophtheirus salmonis* total abundance on sockeye.

Model	Variance Component (SE)
Intercept only	1.2445 (0.3829)
Year + position to farms + temperature	0.2352 (0.1402)
Year + salinity + temperature	0.2528 (0.1550)
Salinity + temperature	0.3002 (0.1714)
Year + position to farms	0.3022 (0.1558)
Position to farms + temperature	0.3153 (0.1764)
Year + salinity	0.3697 (0.1879)
Position to farms	0.4354 (0.1969)
Salinity	0.4560 (0.2124)
Year	0.6538 (0.2691)

Legend: all factors in these models are statistically significant (p<0.05).

contributing factors other than farm-origin parasites are likely responsible for reduced sockeye productivity. However, unlike most other systems, Fraser River sockeye experienced a record-low return in 2009, triggering a federal Judicial Inquiry [32]. Although the effect of sea louse parasitism on juvenile sockeye acting in isolation may arguably be small, it could be important when combined with multiple stressors [59]. Negative impacts of salmon farms on wild populations have been indicated in other parts of the world [9,10,60], and in juvenile pink, and coho salmon populations on the west coast of Canada [28,61]. A recent study found no correlation between numbers of lice on farms and adult pink salmon returns in the Broughton Archipelago, which is located between our southern and northern sites [21]. This study, based on a nine-year time series, lacked full statistical comparisons of productivity in regions without salmon farms. Another recent study that included such comparisons reported significant declines in productivity of pink salmon in relation to salmon farms [62].

Our evidence suggests that salmon farms are elevating parasite levels on Fraser River sockeye during their critical early marine migration; to establish the link more definitively between farms and wild fish would require collaborative work with the salmon farm industry as has begun in Europe and the Broughton Archipelago [21,63]. Ultimately, risks to wild salmon posed by salmon farms can be more easily mitigated than the far-reaching effects on ocean productivity of climate change and ocean acidification. Options already recommended include removal of farm salmon from migration routes of juvenile sockeye from the Fraser [64], and transitioning of salmon farms to closed-containment facilities [65]. At minimum, the Discovery Islands' migration corridor requires a co-ordinated aquaculture management plan to minimize the exposure of wild juvenile sockeye to sea lice.

Acknowledgments

This work is dedicated to S Bergh. We thank J Eriksson, S Latham, A Morton, A Woods, C Carr-Harris of the Skeena Fisheries Commission, the Pacific Salmon Commission, Fisheries and Oceans Pacific Biological Station, Watershed Watch Salmon Society, and Raincoast Conservation Foundation for technical and in-kind support, and M Krkosek for review.

Author Contributions

Conceived and designed the experiments: MP RR AG. Performed the experiments: MP AG. Analyzed the data: MP RR. Contributed reagents/materials/analysis tools: MP SP RR AG. Wrote the paper: MP SP RR AG CO JR.

References

1. Macdonald DW, Laurenson MK (2006) Infectious disease: inextricable linkages between human and ecosystem health. Biol Cons 131: 143–150.
2. Thirgood S (2009) New perspectives on managing wildlife diseases. J Appl Ecol 46: 454–456.
3. Dobson A, Foufopoulos J (2001) Emerging infectious pathogens of wildlife. Phil Trans R Soc Lon B 356: 1001–1012.
4. Otterstatter MC, Thomson JD (2008) Does pathogen spillover from commercially reared bumble bees threaten wild pollinators? PLoS ONE 3: e2771.
5. Power AG, Mitchell CE (2004) Pathogen spillover in disease epidemics. Am Nat 164: S79–S89.
6. Daszak P, Cunningham AA, Hyatt AD (2000) Emerging infectious diseases of wildlife - Threats to biodiversity and human health. Science 287: 443–449.
7. Jessup DA, Boyce WM, Clarke RK (1991) Diseases shared by wild, exotic and domestic sheep. In: Renecker LA, Hudson RJ, eds. Wildlife production: conservation and sustainable development University of Alaska. pp 438–445.
8. Costello MJ (2006) Ecology of sea lice parasitic on farmed and wild fish. Trends in Para 22: 475–483.
9. Costello MJ (2009) How sea lice from salmon farms may cause wild salmonid declines in Europe and North America and be a threat to fishes elsewhere. Proc R Soc Lon B 276: 3385–3394.
10. Krkosek M (2010) Sea lice and salmon in Pacific Canada: ecology and policy. Front. Ecol Environ 8: 201–209.
11. Pike AW, Wadsworth SL (2000) Sea lice on salmonids: their biology and control. Adv Para 44: 234–337.
12. Bjorn PA, Finstad B (1997) The physiological effects of salmon lice infection on sea trout post smolts. Nor J Freshw Res 73: 60–72.
13. Krkosek M, Connors B, Mages P, Peacock S, Ford H, et al. (2010) Fish farms, parasites, and predators: implications for salmon population dynamics. Ecol Appl. In press.
14. Morton A, Routledge RD (2005) Mortality rates for juvenile pink salmon Oncorhychus gorbuscha and chum O. keta salmon infested with sea lice Lepeophtheirus salmonis in the Broughton Archipelago. Alaska Fish Res Bull 11: 146–152.
15. Krkosek M, Lewis MA, Morton A, Frazer LN, Volpe JP (2006) Epizootics of wild fish induced by fish farm. Proc Natl Acad Sci U S A 103: 15506–15510.
16. Nese L, Enger R (1993) Isolation of Aeromonas salmonicida from salmon lice, Lepeophtheirus salmonis and marine plankton. Dis Aquat Org 16: 79–81.
17. Nylund A, Hovland T, Hodneland K, Nilsen F, Løvik P (1994) Mechanisms for transmission of infectious salmon anemia (ISA). Dis Aquat Org 19: 95–100.
18. McCallum H, Dobson AP (1995) Detecting disease and parasite threats to endangered species and ecosystems. Trends Ecol Evol 10: 190–194.
19. Orr C (2007) Estimated sea louse egg production from marine Harvest Canada farmed Atlantic salmon in the Broughton Archipelago, British Columbia, 2003-2004. N Am J Fish Manage 27: 187–197.
20. Fraser NL (2009) Sea-cage aquaculture, sea lice, and declines of wild fish. Cons Biol 23: 599–607.
21. Marty GD, Saksida SM, Quinn TJ (2010) Relationship of farm salmon, sea lice, and wild salmon populations. Proc Natl Acad Sci U S A doi: 10.1073/pnas.1009573108.
22. Krkosek M, Gottesfeld A, Proctor B, Rolston D, Carr-Harris C, et al. (2007) Effects of host migration, diversity and aquaculture on sea lice threats to Pacific salmon populations. Proc R Soc Lon B 274: 3141–3149.
23. Marine Harvest Corporate (2008) Sustainability report. Oslo, Norway.
24. Murray AG, Peeler EJ (2005) A framework for understanding the potential for emerging diseases in aquaculture. Prev Vet Med 67: 223–235.
25. Murray AG (2008) Using simple models to review the application and implications of different approaches used to simulate transmission of pathogens among aquatic animals. Prev Vet Med 88: 167–177.
26. Price MHH, Morton A, Reynolds JD (2010) Evidence of farm-induced parasite infestations on wild juvenile salmon in multiple regions of coastal British Columbia, Canada. Can J Fish Aquat Sci 67: 1925–1932.
27. Heuch PA, Bjorn PA, Finstad B, Holst JC, Asplin L, et al. (2005) A review of the Norwegian National Action Plan Against Salmon Lice on Salmonids: the effect on wild salmonids. Aquaculture 246: 79–92.

28. Krkosek M, Ford JS, Morton A, Lele S, Myers RA, et al. (2007) Declining wild salmon populations in relation to parasites from farm salmon. Science 318: 1772–1775.
29. Morton A, Routledge R, Krkosek M (2008) Sea lice infestation of wild juvenile salmon and herring associated with fish farms off the east-central coast of Vancouver Island, British Columbia. N Am J Fish Manage 28: 523–532.
30. Hartt AC, Dell MB (1986) Early oceanic migrations and growth of juvenile Pacific salmon and steelhead trout. Int N Pac Fish Com Bull 46: 1–105.
31. International Union for the Conservation of Nature (2008) Pacific sockeye salmon (Oncorhynchus nerka) added to IUCN red-list. Available: http://www.iucnredlist.org/apps/redlist/details/135301/0 via the internet. Accessed 30 December 2010.
32. Cohen Commission (2010) Commission of inquiry into the decline of Sockeye salmon in the Fraser River. Available: http://www.cohencommission.ca/en/TermsOfReference.php via the internet. Accessed 30 December 2010.
33. Welch DW, Melnychuk MC, Rechisky ER, Porter AD, Jacobs MC, et al. (2009) Freshwater and marine migration and survival of endangered Cultus Lake sockeye salmon (Oncorhynchus nerka) smolts using POST, a large-scale acoustic telemetry array. Can J Fish Aquat Sci 66: 736–750.
34. Peterman RM, Marmorek D, Beckman B, Bradford M, Mantua N, et al. (2010) Synthesis of evidence from a workshop on the decline of Fraser River sockeye. Available: http://www.psc.org/pubs/FraserSockeyeDeclineWorkshopIntro.pdf via the internet. Accessed 30 December 2010.
35. Holst JC, McDonald A (2000) FISH_LIFT: a device for sampling live fish with trawls. Fish Res 48: 87–91.
36. Kabata Z (1972) Development stages of Caligus clemensi (Copepoda: Caligidae). J Res Board Can 29: 1571–1593.
37. Johnson SC, Albright LJ (1991) The developmental stages of Lepeophtheirus salmonis (Krflyer, 1837) (Copepoda: Caligidae). Can J Zool 69: 929–950.
38. Withler RE, Le KD, Nelson RJ, Miller KM, Beacham TD (2000) Intact genetic structure and high levels of genetic diversity in bottlenecked sockeye salmon, Oncorhynchus nerka, populations of the Fraser River, British Columbia, Canada. Can J Fish Aquat Sci 57: 1985–1998.
39. Beacham TE, Lapointe M, Candy JR, McIntosh B, MacConnachie C, et al. (2004) Stock identification of Fraser River sockeye salmon using microsatellites and major histocompatibility complex variation. Trans Am Fish Soc 133: 1117–1137.
40. Pella J, Masuda M (2001) Bayesian methods for analysis of stock mixtures from genetic characters. Fish Bull 99: 151–167.
41. Beacham TE, Candy JR, McIntosh B, MacConnachie C, Tabata A, et al. (2005) DNA-level variation of sockeye salmon in Southeast Alaska and the Nass and Skeena Rivers, British Columbia, with applications to stock identification. N Am J Fish Manage 25: 763–776.
42. Groot C, Cooke K (1987) Are the migrations of juvenile and adult Fraser River sockeye salmon (Oncorhynchus nerka) in near-shore waters related? In: Smith HD, Margolis L, Wood CC, eds. Sockeye salmon (Oncorhynchus nerka) population biology and future management Can Spec Pub Fish Aquat Sci 96: 53–60.
43. Thomson RE (1981) Oceanography of the British Columbia coast. Can Spec Pub Fish Aquat Sci 56.
44. Kleinbaum DG, Kupper LL, Nizam A, Muller KE (2008) Applied regression analysis and other multivariable methods, fourth edition. Belmont, CA. 287 p.
45. Satterthwaite FE (1946) An approximate distribution of estimates of variance components. Biom Bull 2: 110–114.
46. Brooks KM (2005) The effects of water temperature, salinity, and currents on the survival and distribution of the infective copepodid stage of sea lice (Lepeophtheirus salmonis) originating from Atlantic salmon farms in the Broughton Archipelago of British Columbia, Canada. Rev Fish Sci 13: 177–204.
47. Krkosek M, Lewis MA, Volpe JP (2005) Transmission dynamics of parasitic sea lice from farm to wild salmon. Proc R Soc Lon B 272: 689–696.
48. Gottesfeld AS, Proctor B, Rolston LD, Carr-Harris C (2009) Sea lice, Lepeophtheirus salmonis, transfer between wild sympatric adult and juvenile salmon on the north coast of British Columbia, Canada. J Fish Dis 32: 45–57.
49. Morton A, Routledge R, Peet C, Ladwig A (2004) Sea lice (Lepeophtheirus salmonis) infection rates on juvenile pink (Oncorhynchus gorbuscha) and chum salmon

(*Oncorhynchus keta*) in the nearshore marine environment of British Columbia, Canada. Can J Fish Aquat Sci 61: 147–157.

50. Parker RR, Margolis L (1964) A new species of parasitic copepod, *Caligus clemensi* sp. Nov. (Caligoida: Caligidae), from pelagic fishes in the coastal waters of British Columbia. J Fish Res Bd Can 21: 873–889.

51. Beamish RJ, Wade J, Pennell W, Gordon E, Jones S, et al. (2009) A large, natural infection of sea lice on juvenile Pacific salmon in the Gulf Islands area of British Columbia, Canada. Aquaculture 297: 31–37.

52. Tully O, Gargan P, Poole WR, Whelan KF (1999) Spatial and temporal variation in the infestation of sea trout *Salmo trutta L.* by the caligid copepod *Lepeophtheirus salmonis* (Kroyer) in relation to sources of infection in Ireland. Parasitology 119: 41–51.

53. British Columbia Ministry of Agriculture and Lands (2007) Fish health program annual report. Available: http://www.agf.gov.bc.ca/ahc/fish_health/fish_health2007.pdf via the internet. Accessed 30 December 2010.

54. British Columbia Ministry of Agriculture and Lands (2008) Fish health program annual report. Available: http://www.agf.gov.bc.ca/ahc/fish_health/Fish_Health_Report_2008.pdf via the internet. Accessed 30 December 2010.

55. Krkosek M, Morton A, Volpe JP, Lewis MA (2009) Sea lice and salmon population dynamics: effects of exposure time for migratory fish. Proc R Soc Lon B 276: 2819–2828.

56. Webster SJ, Dill LM, Butterworth K (2007) The effect of sea lice infestation on the salinity preference and energetic expenditure of juvenile pink salmon (*Oncorhynchus gorbuscha*). Can J Fish Aquat Sci 64: 672–680.

57. Quinn TP (2005) The behavior and ecology of Pacific salmon and trout. Vancouver: UBC Press. 378 p.

58. Beamish RJ, Mahnken C, Neville CM (2004) Evidence that reduced early marine growth is associated with lower marine survival of coho salmon. Trans Am Fish Soc 133: 26–33.

59. Finstad B, Kroglund F, Strand R, Stefansson SO, Bjorn PA, et al. (2007) Salmon lice or suboptimal water quality - reasons for reduced postsmolt survival? Aquaculture 273: 374–383.

60. Ford JS, Myers RA (2008) A global assessment of salmon aquaculture impacts on wild salmonids. PLoS Biol 6: e33.

61. Connors BM, Krkosek M, Ford J, Dill LM (2010) Coho salmon productivity in relation to salmon lice from infected prey and salmon farms. J Appl Ecol 47: 1372–1377.

62. Krkosek M, Hilborn R (2011) Sea lice (*Lepeophtheirus salmonis*) infestations and the productivity of pink salmon (*Oncorhynchus gorbuscha*) in the Broughton Archipelago, British Columbia, Canada. Can J Fish Aquat Sci 68: 17–29.

63. Penston MJ, Millar CP, Davies IM (2008) Reduced *Lepeophtheirus salmonis* larval abundance in a sea loch on the west coast of Scotland between 2002 and 2006. Dis Aquat Org 81: 109–117.

64. Statement from Think Tank of Scientists (2009) Adapting to change: managing Fraser sockeye in the face of declining productivity and increasing uncertainty. Available: http://www.sfu.ca/cstudies/science/resources/1273690566.pdf via the internet. Accessed 07 January 2011.

65. Legislative Assembly of British Columbia (2007) Special committee on sustainable aquaculture: final report volume one. Available: http://www.leg.bc.ca/CMT/38thparl/session-3/aquaculture/reports/Rpt-AQUACULTURE-38-3-Volume1-2007-MAY-16.pdf via the internet. Accessed 30 December 2010.

5

Possible Involvement of Cone Opsins in Distinct Photoresponses of Intrinsically Photosensitive Dermal Chromatophores in Tilapia *Oreochromis niloticus*

Shyh-Chi Chen[1]*, **R. Meldrum Robertson**[1,2], **Craig W. Hawryshyn**[1,2]

1 Department of Biology, Queen's University, Kingston, Ontario, Canada, **2** Centre for Neuroscience Studies, Queen's University, Kingston, Ontario, Canada

Abstract

Dermal specialized pigment cells (chromatophores) are thought to be one type of extraretinal photoreceptors responsible for a wide variety of sensory tasks, including adjusting body coloration. Unlike the well-studied image-forming function in retinal photoreceptors, direct evidence characterizing the mechanism of chromatophore photoresponses is less understood, particularly at the molecular and cellular levels. In the present study, cone opsin expression was detected in tilapia caudal fin where photosensitive chromatophores exist. Single-cell RT-PCR revealed co-existence of different cone opsins within melanophores and erythrophores. By stimulating cells with six wavelengths ranging from 380 to 580 nm, we found melanophores and erythrophores showed distinct photoresponses. After exposed to light, regardless of wavelength presentation, melanophores dispersed and maintained cell shape in an expansion stage by shuttling pigment granules. Conversely, erythrophores aggregated or dispersed pigment granules when exposed to short- or middle/long-wavelength light, respectively. These results suggest that diverse molecular mechanisms and light-detecting strategies may be employed by different types of tilapia chromatophores, which are instrumental in pigment pattern formation.

Editor: Karl-Wilhelm Koch, University of Oldenburg, Germany

Funding: This study was supported by an NSERC (Natural Sciences and Engineering Research Council) Discovery Grant, the Canada Research Chairs program, Canada Foundation for Innovation (CWH). The funders had no role in study design, data collection and analysis, decision to publish, or preparation of the manuscript.

Competing Interests: The authors have declared that no competing interests exist.

* E-mail: 6sc35@queensu.ca

Introduction

The ability to detect and respond to the visual environment plays a critical role in the survival of animals. Changing the body color pattern is a strategy frequently used by lower vertebrates and invertebrates, allowing them to respond to biotic and abiotic stimuli. These color-changing mechanisms greatly impact physiological and behavioural aspects of animals, including UV protection [1,2], thermoregulation [3,4], concealed communication [5,6], camouflage [7,8], mate choice [9,10], aggressive signaling [11,12], and social status display [13,14]. Chromatophores, dermal specialized pigment cells, are thought to be the primary agents that shape body patterns in many animals. Chromatophores are typically categorized into six classes based on their internal structure and pigment colors: melanophores, erythrophores, xanthophores, cyanophores, leucophores, and iridophores [15–17]. They are capable of responding morphologically (in number or size) or physiologically (through translocation of inner pigment granules, i.e. aggregation or dispersion) in order to execute long-term or immediate color pattern changes, respectively [18,19]. Integumentary color change is achieved in three ways: 1) through pigment granule motility, 2) through reflective plates within chromatophores, and 3) through coordination of different chromatophore classes and/or other tissues [20,21]. The translocation of pigment granules within chromatophores was shown to be mediated by the action of microtubules and microfilaments, as well as various molecular motors [22,23].

In teleosts, this translocation is controlled by the sympathetic nervous system [24,25] and regulated by hormones such as prolactin, catecholamines, melatonin, noradrenaline (NA), melanin-concentrating hormone (MCH), and alpha-melanocyte-stimulating hormone (α-MSH) [9,14,26–30]. For example, erythrophores in Nile tilapia (*Oreochromis niloticus*) and swordtail (*Xiphophorus helleri*), pigment aggregation can be triggered by MCH, while dispersion can be mediated by MSH [28,29]. Pharmacological evidence suggests that the increase or decrease in intracellular cAMP level regulated by G_s or G_i protein plays an important role in signal transduction pathways of hormone-mediated pigment granule movements in chromatophores [29,31].

Recently, studies have shown that light can also trigger the color-changing process, and that these photoresponses depend on the intensity and wavelength of the light stimulus [32–35]. Chromatophore spectral sensitivity has been measured from several teleosts. For example, the most effective wavelength to induce melanophore dispersion of *Oryzias latipes* and *Zacco temmincki* are near 415 and 525 nm, respectively [36,37]. These results show that maximum chromatophore photosensitivity varies across species. Moreover, the light-induced translocation of intracellular pigment granules (i.e. retrograde vs. anterograde) is not identical in the same chromatophore class. For instance, xanthophores under illumination disperse in *Trematomus bernacchii*, but aggregate in *Oryzias latipes* [38,39].

Figure 1. An adult male Nile tilapia *Oreochromis niloticus* (A) and a close-up of the caudal fin (B) showing pigmentation traits.

Opsins have been suggested to be involved in chromatophore photoresponses, as putative opsins have been detected in skin tissues [32,33,40]. However, this hypothesis relies mainly on molecular evidence, e.g. opsin gene expression in integumentary tissues [32,33]. Moreover, chromatophore photoresponses were proposed to be generated through a pathway other than the G_t protein-activated cGMP signaling pathway in retinal photoreceptors. Thus, chromatophore dispersion and aggregation were

suggested to occur through the G_s or G_i protein-activated cAMP signal transduction pathway, similar to that found in hormonal regulation of chromatophores [31,32,41]. Nevertheless, the relationship between visual pigments and chromatophore photoresponses remains unclear.

An opsin protein in combination with a chromophore forms a visual pigment. Different classes of opsins vary in amino acid sequences leading to the change of spectral absorption properties

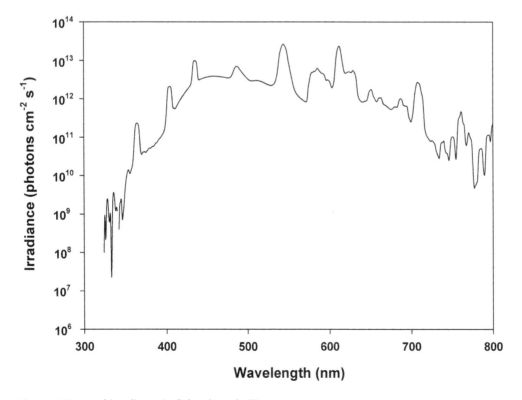

Figure 2. Spectral irradiance in fish culture facility.

of visual pigments. In vertebrates, five classes of retinal opsin genes have been identified: one rod opsin (*RH1*) and four classes of cone opsins including short wavelength sensitive 1 (*SWS1*), short wavelength sensitive 2 (*SWS2*), rod opsin-like (*RH2*), and long wavelength sensitive (*LWS*) opsins [42]. Due to gene duplication, Nile tilapia (*Oreochromis niloticus*) possesses 7 cone opsin genes with distinct peak absorbances (λ_{max}): *SWS1* (360 nm), *SWS2b* (423 nm), *SWS2a* (456 nm), *RH2b* (472 nm), *RH2aβ* (518 nm), *RH2aα* (528 nm), and *LWS* (561 nm) [43]. In the present study, we investigated the relationship between cone opsin expression and tilapia chromatophore photoresponses. Using RT-PCR, cone opsin expression was detected in tilapia caudal fin and co-expression of different opsin classes was found in individual melanophores and erythrophores at the single-cell level. We also showed these two major types of pigmented chromatophores (melanophores and erythrophores) in tilapia caudal fin demonstrated distinct photoresponses to incident light stimuli. Under illuminations, melanophores extended their processes and maintained dispersions by shuttling pigment granules (melanosomes). On the other hand, erythrophores aggregated and dispersed pigment granules (erythrosomes) at different spectral range. We suggest that this co-expression of different opsin classes is correlated to the dynamic photoresponses of these intrinsically photosensitive dermal chromatophores (ipDCs) in Nile tilapia. These findings are significant to our understanding of light-driven mechanisms within specific chromatophore classes, and advance our knowledge of how intrinsic dermal photosensitivity contributes to organismal survival and interaction.

Materials and Methods

Ethics Statement

The protocol used for the experimental fish was reviewed and approved by the Queen's University Animal Care Committee (Protocol NO: Hawryshyn-2010-004-R3-A1) and all procedures complied with the Canadian Council for Animal Care regulations.

Animal

Adult male Nile tilapia *Oreochromis niloticus* (Figure 1A; 46.7 ± 7.1 g body mass, 14.8 ± 0.6 cm standard length) were obtained from a local fish farm, Northern American Tilapia Inc. (Lindsay, Ontario, Canada). Fish were kept at a water temperature of 25°C under a 12 h:12 h L:D light cycle and the lighting condition was provided by full spectrum fluorescent lamps (see Figure 2; Full Spectrum Solutions, Inc., Jackson, MI, USA). Fish were anaesthetized by immersion in MS-222 (Syndel Laboratories Ltd., Qualicum Beach, BC, Canada) and sacrificed by cervical transection. Because circadian changes of opsin expression and pigmentary patterns have been reported in fish [44,45], dissections were always carried out from 11:00 am to 13:00 pm. All procedures complied with the Canadian Council for Animal Care regulations and the Queen's University Animal Care Committee.

Preparation of Split-fin Tissues and Isolation of Single Chromatophore

Chromatophores were isolated by methods described previously [46–48]. Briefly, the integumentary tissues were excised, rinsed with 70% ethanol as well as Ca^{2+}, Mg^{2+}-free, Dulbecco's phosphate-buffered saline (CMF-PBS: NaCl 136.9 mM, KCl 2.7 mM, Na_2HPO_4 8.1 mM, KH_2PO_4 1.5 mM; pH 7.2), and cut into pieces about 5-mm^2 in CMF-PBS. To remove the epidermis, tissues were incubated in EDTA-bicarbonate solution (pH 7.4) and stirred for 20 min. Following incubation with vigorous shaking in 0.25% collagenase type II (Sigma-Aldrich, St. Louis, MO, USA) for 30 min, chromatophores were isolated from split-fin tissues. The dissociated cells were filtered through 140-μm Nylon membrane filter (Millipore, Billerica, MA, USA)

Table 1. Primers used in RT-PCR assays.

Gene	Accession no. (GenBank)	Forward Primer (5′→3′) Reverse Primer (5′→3′)	Amplicon size (bp)
SWS1	AF191221	TCCACCTGTACGAGAACATCTCCAA GGTGTGCCAGCAAACAGGACAA	124
SWS2b	AF247120	CAAGAAGCTCCGGTCTCATC ATGCAGTTGGACCAAGGAAC	134
SWS2a	AF247116	CGCTCGGTAACTTTGCTTTC AGCACTGTAGGCCTTCTGGA	132
RH2b	DQ235681	CTGGTCACCGCTCAAAACAA TCAAAGGACCCAAGGAGAAATAG	148
RH2aβ	DQ235682	CACCATCACAATCACGTCTGCTAT CCAGGACAACAAGTGACCAGAG	122
RH2aα	DQ235683	CCATCACCATCACATCAGCTG CCAGGACAACAAGTGACCAGAG	120
LWS	AF247128	TCATCTCCTGGGAAAGATGG TCCAAATATGGGAGGAGCAC	134
RH1	AY775108	ATATGTTGGCTGCCCTATGC TGCTCCCTCCTCTTCTTCAA	216
OPN4	GR605566	ACTGCACTGAGCACCATCAC TAGATGACCGGAGCATTTCC	196
TMT	AF402774	CCGTCCAACTACTGCAAGGT CACGATCAGGCAGAAGACAAA	198
β-actin[a]	EF206796	TGCGTGACATCAAGGAGAAG CTCTCGTTCCCAATGGTGAT	136

[a] *β-actin* was used as a positive control in RT-PCR assays.

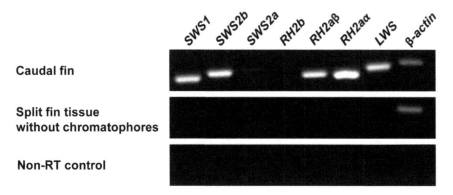

Figure 3. Cone opsin expression in tilapia caudal fin. Using RT-PCR, 7 tilapia cone opsins were detected in tilapia caudal fin but not in the split fin tissue without chromatophores. No detectable signal was found in the non-RT control sample. β-actin was used as a positive control in RT-PCR analysis.

and suspended in CMF-PBS. Specific classes of chromatophores were identified and selected under a dissecting microscope (Nikon Instruments Inc., Melville, NY, USA; SMZ1500) equipped with an epi-illumination system (Dolan-Jenner Industries, Boxborough, MA, USA) and polarizer (Nikon Instruments Inc., Melville, NY, USA; MNN40920).

RNA Extraction, and First Strand cDNA Synthesis

Total RNAs were extracted from caudal fins and single chromatophores (isolated from caudal fins) using Absolutely RNA Miniprep and Nanoprep Kits (Stratagene, La Jolla, CA, USA), respectively. Reverse transcriptions were performed by SuperScript III First-Strand Synthesis SuperMix for qRT-PCR (Invitrogen Canada, Burlington, ON, Canada), following the

Table 2. Cone opsin expression profile of tilapia melanophores.

Cell No	Cone opsin							Number of opsin classes detected per cell
	SWS1	SWS2b	SWS2a	RH2b	RH2aβ	RH2aα	LWS	
1	+	+						2
2	+							1
3	+			+				2
4	+						+	2
5				+		+	+	3
6				+				1
7	+				+		+	3
8				+			+	2
9	+			+				2
10	+			+				2
11	+						+	2
12				+				1
13	+							1
14	+	+		+				3
15	+					+		2
16	+		+					2
17	+							1
18	+		+					2
19	+							1
20			+		+		+	3
21	+			+	+			3
22	+			+	+			3
Total	17	2	3	10	4	2	6	

Table 3. Cone opsin expression profile of tilapia erythrophores.

Cell No	Cone opsin							Number of opsin classes detected per cell
	SWS1	SWS2b	SWS2a	RH2b	RH2aβ	RH2aα	LWS	
1	+	+	+	+	+			5
2	+							1
3	+							1
4				+				1
5	+			+	+			3
6	+				+			2
7	+			+	+			3
8					+			1
9	+			+	+			3
10	+							1
11	+			+	+			3
12						+		1
13	+			+				2
14	+							1
15	+			+				2
16	+	+	+	+		+		5
17	+		+	+		+	+	5
18	+			+		+		3
19	+		+	+		+		4
20	+	+	+	+	+	+		6
21	+	+	+	+	+	+		6
22	+	+	+		+	+		5
23	+				+		+	3
24	+							1
25			+		+			2
26				+				1
27				+				1
28	+			+			+	3
Total	22	5	8	16	12	8	3	

manufacturer's manual. To determine if genomic DNA contamination existed in the sample, reverse transcription for non-RT control was conducted with all reagents, except the reverse transcriptase.

RT-PCR Analysis

Primers for tilapia cone opsins were designed using Primer 3 software based on the sequences published on GenBank (refer to Table 1) and used in RT-PCR, and single-cell RT-PCR analysis. All of the primers were tested in PCR using GoTaq Flexi DNA Polymerase (Promega, WI, USA) in Eppendorf Mastercycler gradient (Eppendorf Canada, Mississauga, ON, Canada) under the same conditions [92°C, 2 min; 92°C, 30 s, 60°C, 30 s, 72°C, 30 s (40 cycles)] prior to being used in subsequent experiments. PCR products were sequenced (McGill University and Genome Quebec Innovation Centre, Montreal, Quebec, Canada) and verified by comparison with corresponding sequences on GenBank. Except cone opsins, we included *rhodopsin* (*RH1*), *melanopsin*

(homolog of zebrafish *OPN4a*), and *teleost multiple tissue* (homolog of pufferfish *TMT*) *opsin* in single-cell RT-PCR analysis and no expression was detected in individual chromatophores examined (data not shown; n = 7 for melanophores, and n = 12 for erythrophores).

To investigate cone opsin expression profiles in different types of chromatophores, single-cell RT-PCR was conducted under the aforementioned condition and a power analysis (statistical power = 0.9, $\alpha = 0.05$, two-tailed test) was administrated to determine the sample size of cells (n = 22 for melanophores, n = 28 for erythrophores).

Measurements of Chromatophore Photoresponses

Split-fin tissues containing chromatophores were incubated in the culture medium (mixture of Leibovits L15 medium, fetal calf serum, and water in a ratio 80:15:5, penicillin-G 100 U/ml, kanamycin 100 μg/ml) at 25 °C in a water-jacked CO_2 incubator in the dark for 2 days. After 2 days of culture, tissues were

Illumination

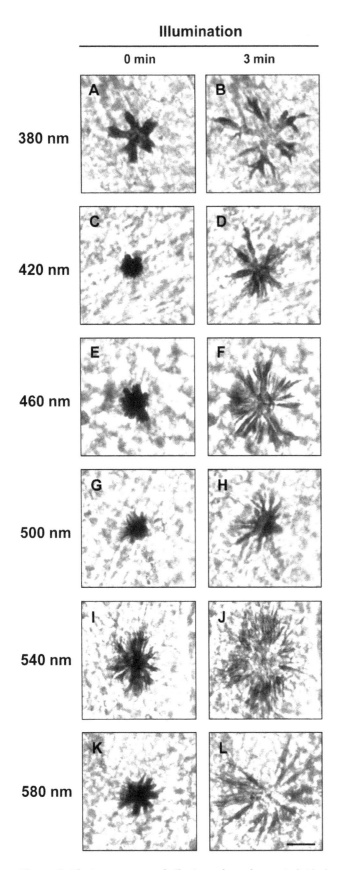

580 nm; n = 4 at each test wavelength) for 3 minutes. A,C,E,G,I,K: before illumination; B,D,F,H,J,L: after illumination. Scale bar: 40 μm.

immersed in PBS for 15 min before experiments. Light stimuli were generated by a 150 W xenon lamp system and a monochromator (Photon Technology International, London, ON, Canada). To measure photoresponses of melanophores and erythrophores, individual chromatophores were challenged with equal-quanta spectral irradiance (13.9 log photons $cm^{-2} s^{-1}$) and 10-nm FWHM values (Full Width at Half Maximum) at one of the stimulating wavelengths (380, 420, 460, 500, 540, and 580 nm) for 3 minutes. Images of chromatophore photoresponses were taken using a Qimaging Microimager II CCD camera and QCapture Suite V2.46 software (Qimaging, Burnaby, BC, Canada). All the experiments were conducted at 25°C in darkness and tissues were continuously perfused with PBS.

Results

Cone Opsin Gene Expression in Tilapia Caudal Fin

In sexually mature male tilapia, caudal fins typically bear pigmentation traits such as vertical bars (Figure 1B); these bars contain a high density of chromatophores. Cone opsin expression in tilapia caudal fin was investigated using RT-PCR with primer sets amplifying the target tilapia cone opsin genes (see Table 1). All cone opsin genes were detected in caudal fin but not in the split fin tissue without chromatophores (Figure 3). Therefore, we suggest cone opsin expression in tilapia caudal fin could be associated with the presence of chromatophores.

Cone Opsin Expression of Individual Melanophores and Erythrophores

To confirm cone opsin expression in individual chromatophore, we performed single-cell RT-PCR on melanophores and erythrophores. Opsin gene classes expressed in individual melanophores and erythrophores varied in number (Table 2 and 3). The results showed that co-expression of different classes of cone opsin genes was frequently found in chromatophores we examined (73% in melanophores and 64% in erythrophores). Among cells expressing only one opsin, melanophores expressed either *SWS1* or *RH2b* (Table 2); erythrophores expressed *SWS1* or one of the *RH2* group genes (*RH2b/RH2aβ/RH2aα*; Table 3). When expressing more than one class of opsins, most melanophores expressed *SWS1* with other class of opsin genes (14 out of 16) and erythrophores tended to co-express *SWS1* with at least one of the *RH2* group genes (*RH2b/RH2aβ/RH2aα*) (17 out of 18). The opsin expression profiles from melanophores and erythrophores suggested that *SWS1* and *RH2* genes are primarily expressed genes and may play important roles in chromatophore photosensitive functions.

Photoresponses of Melanophores and Erythrophores

In order to understand the relationship between opsin expression and the dynamic chromatophore photoresponse, we measured chromatophore responses at six wavelengths ranging from 380 to 580 nm. Both of melanophores and erythrophores were light-sensitive, but showed their photoresponses in different manners. Regardless of wavelengths, melanophores dispersed and tended to maintain their shape in the dispersion stage by shuttling pigment granules when receiving light stimulus (Figure 4; also see Movie S1). On the other hand, erythrophores exhibited wavelength-dependent photoresponses: aggregations were induced when short-wavelength light was applied (Figure 5A–D), while

Figure 4. Photoresponses of tilapia melanophores. Individual melanophores were challenged with equal-quanta spectral irradiance (13.9 log photons $cm^{-2} s^{-1}$) at one of the stimulating wavelengths (A,B: 380 nm; C,D: 420 nm; E,F: 460 nm; G,H: 500 nm; I,J: 540 nm; K,L:

Illumination

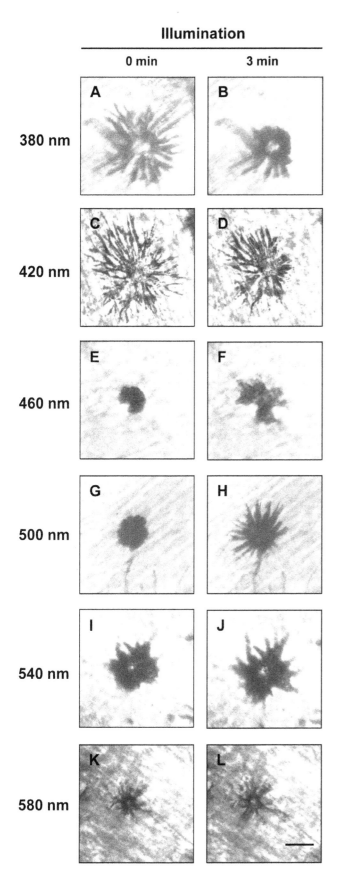

0 min	3 min
A	**B**
C	**D**
E	**F**
G	**H**
I	**J**
K	**L**

380 nm
420 nm
460 nm
500 nm
540 nm
580 nm

Figure 5. Photoresponses of tilapia erythrophores. Individual erythrophores were challenged with equal-quanta spectral irradiance (13.9 log photons cm^{-2} s^{-1}) at one of the stimulating wavelengths (A,B: 380 nm; C,D: 420 nm; E,F: 460 nm; G,H: 500 nm; I,J: 540 nm; K,L: 580 nm; n = 5 at each test wavelength) for 3 minutes. A,C,E,G,I,K: before illumination; B,D,F,H,J,L: after illumination. Scale bar: 20 μm.

dispersions occurred when cells were presented with middle- and long-wavelength light (Figure 5E–L).

Discussion

In previous studies, opsins detected in integumentary tissues are suggested to endow chromatophores with the ability to respond to light. However, the lack of direct evidence showing opsin expression at the single-cell level or the link of opsins and chromatophore photoresponses has hindered our understanding of the photosensitive mechanism within chromatophores. In the present study, cone opsin expression was detected in tilapia caudal fins where chromatophores are present. We further employed single-cell RT-PCR to demonstrate the specific opsin expression profiles in two chromatophore populations. We found *SWS1* is the prominent opsin expressed in both melanophores and erythrophores. In addition, the expression profile reveals that in most erythrophores, *SWS1* is co-expressed with at least one of the *RH2* group genes. When presented with light stimuli at different wavelengths, melanophores and erythrophores showed distinct responses. Under illuminations, melanophores dispersed and shuttled melanosomes to maintain cell shape in the dispersion stage while erythrophores acted in a wavelength-dependent manner. Although a variety of opsin repertoires are present in melanophores and erythrophores, some particular classes of opsins detected at higher frequencies may be responsible for their distinct photoresponses. More functional characterizations of ipDC photoresponses are required to clarify the involvement of opsins in this extraretinal photosensitive system.

In addition to inter- and intraspecific communication, there is also evidence that color changes are closely associated with habitat or background colors [18,49]. For example, when medaka *Oryzias latipes* are adapted to white background, their body colors become paler due to the aggregation, reduced cell size, or even melanophore apoptosis [19,50]. These results highlight the importance of environmental photic factors on body color patterns; however, how animals coordinate visual cues with their body color performances remains unclear. Previous work has suggested that chromatophores are able to directly respond to incident light, but this work lacked molecular evidence at the single-cell level to demonstrate that visual pigments are expressed in chromatophores. Studies using spectral analyses show that chromatophores in some vertebrates possibly express only one opsin, which regulates one corresponding photoresponse [36,37,51,52]. More recently, PCR data from tilapia skin tissue but not from single chromatophore, suggested tilapia erythrophores express two putative opsins, *RH2* and *LWS* [32]. In the present study, however, we detected *SWS1* and *RH2* group genes at higher frequency and *LWS* at lower frequency in opsin expression profile of erythrophores (see below and results of single-cell RT-PCR as well). Our results were obtained at the single cell level, which eliminated the possibility of misrepresented expression pattern from mixed cells, or contamination. Moreover, the previous study used primers sets restricted to two opsins, *RH2* and *LWS*. These explain the discrepancy between our work and previous results. Therefore, we suggest molecular work should be conducted at single-cell level to truly reflect the expression file of a specific chromatophore population.

Distinct photoresponse of tilapia melanophores and erythrophores suggest they may have different importance in physiological functions. Melanophores were suggested to prevent incident light from penetrating body surface and harming internal organs since melanin has been thought to provide effective photoprotection via absorption of UVA and UVB [53,54]. Nevertheless, based on the photoresponses measured herein, melanophores more likely serve as a broad-spectrum light filter. Instead, dynamic photoresponses of erythrophores can generate specific photic effects with other classes of chromatophores in a dermal chromatophore unit when confronting various light stimuli.

Recently, increasing evidence for extraretinal photoreceptors in varied locations across the body has been reported, and their visual pigments are suggested to aid in physiological regulations such as circadian rhythms, pupil size, and body coloration tuning [55]. Indeed, increasing evidence shows that some animals possess light-mediated behaviours due to their dermal sensitivity [56–58]. Moreover, evidence shows that *SWS1* opsin expression in pineal organ is detected during different life stages in a wide rage of teleosts [59]. Since high frequency of *SWS1* expression was found in cone opsin expression profiles of tilapia melanophores and erythrophores, ultraviolet photosensitivity may play important roles in some extraretinal photoreception. Although no expression of *RH1*, *OPN4*, and *TMT* opsins was detected in RT-PCR analysis, we cannot completely rule out the possibility that other types of opsins are involved in chromatophore photoresponses represented herein. The expression of non-visual opsins sensitive to UV and green light has been identified in photosensitive organs of different vertebrates [60–63]. For example, evidence has shown a non-mammalian type of UV-sensitive neuropsin (cOPN5L2) in birds [61] and green-sensitive vertebrate ancient (VA) opsins in zebrafish [62,63]. Because of the diversity of non-rod, non-cone visual pigments and their unclear functions in various organs, more molecular data and functional analysis are required to exclude their possible involvement in chromatophore photoresponses. Furthermore, the details of how photoreception and phototransduction take place within chromatophores are unknown so far. There is not yet direct evidence that opsins are expressed in cell membrane, cytoplasm, or intracellular granules. It is also unclear whether the opsins utilize the identical type of the light-sensitive chromophore in retina, although chromophores were isolated from extraretinal photoreceptors in some species [64–66]. It was suggested that non-image-forming photosensitive systems may adopt the chromophore selected by the visual system [65]. In order to understand if ipDCs compose a dermal photosensory system that independently detect a change in quantity and quality of light, expression studies alone are not enough and more cellular evidence and functional analysis are required, especially on under-characterized photosensitive systems.

Conclusions

Our data demonstrate opsin expression in tilapia integumentary tissue. More specifically, we identify and describe the fundamental molecular information of co-expression of opsins within individual chromatophores, which could be related to distinct photoresponses of different types of ipDCs. These observations indicate that diverse molecular mechanisms and photoreactive strategies may be employed by different classes of chromatophores, which play distinctive roles in pigment pattern formation and other physiological functions [53,67,68]. In the future, we will analyze chromatic performance of chromatophores to characterize the spectral sensitivity. This information will not only help us perform a comprehensive comparison of chromatic interactions of photoreceptors/opsins in retina and extraretinal tissues, but it will also provide an excellent chance to understand how animals coordinate multiple visual cues from different light-sensitive organs.

Supporting Information

Movie S1 Melanophore photoresponses under illumination. Melanophore photoresponses under illumination at 500 nm (13.9 log photons $cm^{-2} s^{-1}$) for 10 minutes. Under illuminations, melanophores tended to maintain their shape in the dispersion stage by shuttling pigment granules (melanosomes). Scale bars: 20 μm.

Acknowledgments

We thank Dr. Bob Montgomerie, Dr. Dongsheng Tu, Jun Liu, Shai Sabbah, Mark Hornsby, Shan Jiang, and Yi Niu for their valuable comments.

Author Contributions

Conceived and designed the experiments: SCC CWH. Performed the experiments: SCC. Analyzed the data: SCC CWH. Contributed reagents/materials/analysis tools: CWH. Wrote the paper: SCC. Critically read the manuscript: RMR CWH.

References

1. Auerswald L, Freier U, Lopata A, Meyer B (2008) Physiological and morphological colour change in Antarctic krill, *Euphausia superba*: a field study in the Lazarev Sea. J Exp Biol 211: 3850–3858.
2. Fuhrmann MM, Nygard H, Krapp RH, Berge J, Werner I (2011) The adaptive significance of chromatophores in the Arctic under-ice amphipod *Apherusa glacialis*. Polar Biol 34: 823–832.
3. de Velasco JB, Tattersall GJ (2008) The influence of hypoxia on the thermal sensitivity of skin colouration in the bearded dragon, *Pogona vitticeps*. J Comp Physiol B 178: 867–875.
4. Silbiger N, Munguia P (2008) Carapace color change in *Uca pugilator* as a response to temperature. J Exp Mar Biol Ecol 355: 41–46.
5. Chiou TH, Mäthger LM, Hanlon RT, Cronin TW (2007) Spectral and spatial properties of polarized light reflections from the arms of squid (*Loligo pealeii*) and cuttlefish (*Sepia officinalis* L.). J Exp Biol 210: 3624–3635.
6. Shashar N, Rutledge PS, Cronin TW (1996) Polarization vision in cuttlefish - A concealed communication channel? J Exp Biol 199: 2077–2084.
7. Chiao CC, Chubb C, Buresch KC, Barbosa A, Allen JJ, et al. (2010) Mottle camouflage patterns in cuttlefish: quantitative characterization and visual background stimuli that evoke them. J Exp Biol 213: 187–199.
8. Hanlon RT, Messenger JB (1988) Adaptive Coloration in Young Cuttlefish (*Sepia-Officinalis* L) - the Morphology and Development of Body Patterns and Their Relation to Behavior. Philos Trans R Soc B-Biol Sci 320: 437–487.
9. Skold HN, Amundsen T, Svensson PA, Mayer I, Bjelvenmark J, et al. (2008) Hormonal regulation of female nuptial coloration in a fish. Horm Behav 54: 549–556.
10. Svensson PA, Forsgren E, Amundsen T, Skold HN (2005) Chromatic interaction between egg pigmentation and skin chromatophores in the nuptial coloration of female two-spotted gobies. J Exp Biol 208: 4391–4397.
11. Muske LE, Fernald RD (1987) Control of a teleost social signal. I. Neural basis for differential expression of a color pattern. J Comp Physiol A 160: 89–97.
12. Pauers MJ, Kapfer JM, Fendos CE, Berg CS (2008) Aggressive biases towards similarly coloured males in Lake Malawi cichlid fishes. Biol Lett 4: 156–159.
13. Lanzing WJR, Bower CC (1974) Development of Color Patterns in Relation to Behavior in *Tilapia Mossambica* (Peters). J Fish Biol 6: 29–41.
14. Muske LE, Fernald RD (1987) Control of a teleost social signal. II. Anatomical and physiological specializations of chromatophores. J Comp Physiol A 160: 99–107.
15. Fujii R (1993) Cytophysiology of fish chromatophores. International Review of Cytology 143: 191–255.
16. Fujii R (2000) The regulation of motile activity in fish chromatophores. Pigment Cell Res 13: 300–319.
17. Goda M, Fujii R (1995) Blue chromatophores in two species of callionymid fish. Zool Sci 12: 811–813.
18. Hatamoto K, Shingyoji C (2008) Cyclical training enhances the melanophore responses of zebrafish to background colours. Pigment Cell Melanoma Res 21: 397–406.

19. Sugimoto M, Uchida N, Hatayama M (2000) Apoptosis in skin pigment cells of the medaka, *Oryzias latipes* (Teleostei), during long-term chromatic adaptation: the role of sympathetic innervation. Cell Tissue Res 301: 205–216.

20. Bagnara JT, Taylor JD, Hadley ME (1968) Dermal Chromatophore Unit. J Cell Biol 38: 67–79.

21. Grether GF, Kolluru GR, Nersissian K (2004) Individual colour patches as multicomponent signals. Biol Rev Camb Philos Soc 79: 583–610.

22. Kashina A, Rodionov V (2005) Intracellular organelle transport: few motors, many signals. Trends Cell Biol. 15: 396–398.

23. Rodionov V, Yi J, Kashina A, Oladipo A, Gross SP (2003) Switching between microtubule- and actin-based transport systems in melanophores is controlled by cAMP levels. Curr Biol 13: 1837–1847.

24. Burton D (2008) A physiological interpretation of pattern changes in a flatfish. J Fish Biol 73: 639–649.

25. Pye JD (1964) Nervous Control of Chromatophores in Teleost Fishes. I. Electrical Stimulation in Minnow (*Phoxinus Phoxinus* (L.)). J Exp Biol 41: 525–534.

26. Iga T, Takabatake I (1982) Action of melanophore-stimulating hormone on melanophores of the cyprinid fish *Zacco temmincki*. Comp Biochem Physiol C 73: 51–55.

27. Nagaishi H, Oshima N (1989) Neural control of motile activity of light-sensitive iridophores in the neon tetra. Pigment Cell Res 2: 485–492.

28. Oshima N, Makino M, Iwamuro S, Bern HA (1996) Pigment dispersion by prolactin in cultured xanthophores and erythrophores of some fish species. J Exp Zool 275: 45–52.

29. Oshima N, Nakamaru N, Araki S, Sugimoto M (2001) Comparative analyses of the pigment-aggregating and -dispersing actions of MCH on fish chromatophores. Comp Biochem Physiol C Toxicol Pharmacol 129: 75–84.

30. van der Salm AL, Metz JR, Bonga SE, Flik G (2005) Alpha-MSH, the melanocortin-1 receptor and background adaptation in the Mozambique tilapia, *Oreochromis mossambicus*. Gen Comp Endocrinol 144: 140–149.

31. Nery LE, Castrucci AM (1997) Pigment cell signalling for physiological color change. Comp Biochem Physiol A Physiol 118: 1135–1144.

32. Ban E, Kasai A, Sato M, Yokozeki A, Hisatomi O, et al. (2005) The signaling pathway in photoresponses that may be mediated by visual pigments in erythrophores of Nile tilapia. Pigment Cell Res 18: 360–369.

33. Kasai A, Oshima N (2006) Light-sensitive motile iridophores and visual pigments in the neon tetra, *Paracheirodon innesi*. Zool Sci 23: 815–819.

34. Oshima N, Yokozeki A (1999) Direct Control of Pigment Aggregation and Dispersion in Tilapia Erythrophores by Light. Zool Sci 16: 51–54.

35. Sato M, Ishikura R, Oshima N (2004) Direct effects of visible and UVA light on pigment migration in erythrophores of Nile tilapia. Pigment Cell Res 17: 519–524.

36. Naora H, Takabatake I, Iga T (1988) Spectral sensitivity of melanophores of a freshwater teleost, *Zacco temmincki*. Comp Biochem Physiol A 90: 147–149.

37. Negishi S (1985) Light response of cultured melanophores of a teleost adult fish, *Oryzias latipes*. J Exp Zool 236: 327–333.

38. Obika M, Meyer-Rochow VB (1990) Dermal and epidermal chromatophores of the Antarctic teleost *Trematomus bernacchii*. Pigment Cell Res 3: 33–37.

39. Oshima N, Nakata E, Ohta M, Kamagata S (1998) Light-induced pigment aggregation in xanthophores of the medaka, *Oryzias latipes*. Pigment Cell Res 11: 362–367.

40. Lythgoe JN, Shand J, Foster RG (1984) Visual pigment in fish iridocytes. Nature 308: 83–84.

41. Oshima N (2001) Direct reception of light by chromatophores of lower vertebrates. Pigment Cell Res 14: 312–319.

42. Yokoyama S (2000) Molecular evolution of vertebrate visual pigments. Prog. Retin. Eye Res. 19: 385–419.

43. Spady TC, Parry JWL, Robinson PR, Hunt DM, Bowmaker JK et al. (2006) Evolution of the cichlid visual palette through ontogenetic subfunctionalization of the opsin gene arrays. Mol Biol Evol 23: 1538–1547.

44. Halstenberg S, Lindgren KM, Samagh SPS, Nadal-Vicens M, Balt S, et al. (2005) Diurnal rhythm of cone opsin expression in the teleost fish *Haplochromis burtoni*. Visual Neurosci 22: 135–141.

45. Masagaki A, Fujii R (1999) Differential actions of melatonin on melanophores of the threeline pencilfish, *Nannostomus trifasciatus*. Zool Sci 16: 35–42.

46. Fujii R, Wakatabi H, Oshima N (1991) Inositol 1,4,5-Trisphosphate Signals the Motile Response of Fish Chromatophores. I. Aggregation of Pigment in the Tilapia Melanophore. J Exp Zool 259: 9–17.

47. Masada M, Matsumoto J, Akino M (1990) Biosynthetic pathways of pteridines and their association with phenotypic expression in vitro in normal and neoplastic pigment cells from goldfish. Pigment Cell Res 3: 61–70.

48. Oshima N, Suzuki M, Yamaji N, Fujii R (1988) Pigment Aggregation Is Triggered by an Increase in Free Calcium-Ions within Fish Chromatophores. Comp Biochem Physiol A Physiology 91: 27–32.

49. Sugimoto M, Yuki M, Miyakoshi T, Maruko K (2005) The influence of long-term chromatic adaptation on pigment cells and striped pigment patterns in the skin of the zebrafish, *Danio rerio*. J Exp Zool A 303: 430–440.

50. Sugimoto M, Oshima N (1995) Changes in adrenergic innervation to chromatophores during prolonged background adaptation in the medaka, *Oryzias latipes*. Pigment Cell Res 8: 37–45.

51. Daniolos A, Lerner AB, Lerner MR (1990) Action of light on frog pigment cells in culture. Pigment Cell Res 3: 38–43.

52. Moriya T, Miyashita Y, Arai J, Kusunoki S, Abe M, et al. (1996) Light-sensitive response in melanophores of *Xenopus laevis*: I. Spectral characteristics of melanophore response in isolated tail fin of *Xenopus* tadpole. J Exp Zool 276: 11–18.

53. Armstrong TN, Cronin TW, Bradley BP (2000) Microspectrophotometric analysis of intact chromatophores of the Japanese medaka, *Oryzias latipes*. Pigment Cell Res 13: 116–119.

54. Hofer R, Mokri C (2000) Photoprotection in tadpoles of the common frog, *Rana temporaria*. J Photochem Photobiol B-Biol 59: 48–53.

55. Shand J, Foster RG (1999) The extraretinal photoreceptors of non-mammalian vertebrates. In: Archer SN, Djamgoz MBA, Leow ER, Partridge JC, Vallerga S, editors. Adaptive Mechanisms in the Ecology of Vision: Kluwer Academic Publishers. 197–222.

56. Tosini G, Avery RA (1996) Dermal photoreceptors regulated basking behavior in the lizard *Podarcis muralis*. Physiol Behav 59: 195–198.

57. Ullrich-Luter EM, Dupont S, Arboleda E, Hausen H, Arnone MI (2011) Unique system of photoreceptors in sea urchin tube feet. Proc Natl Acad Sci USA 108: 8367–8372.

58. Yerramilli D, Johnsen S (2010) Spatial vision in the purple sea urchin *Strongylocentrotus purpuratus* (Echinoidea). J Exp Biol 213: 249–255.

59. Forsell J, Ekstrom P, Flamarique IN, Holmqvist B (2001) Expression of pineal ultraviolet- and green-like opsins in the pineal organ and retina of teleosts. J Exp Bio 204: 2517–2525.

60. Wada S, Kawano-Yamashita E, Koyanagi M, Terakita A (2012) Expression of UV-sensitive parapinopsin in the iguana parietal eyes and its implication in UV-sensitivity in vertebrate pineal-related organs. PLoS ONE 7: e39003.

61. Ohuchi H, Yamashita T, Tomonari S, Fujita-Yanagibayashi S, Sakai K, et al. (2012) A non-mammalian type opsin 5 functions dually in the photoreceptive and non-photoreceptive organs of birds. PLoS ONE 7: e31354.

62. Kojima D, Torii M, Fukada Y, Dowling JE (2008) Differential expression of duplicated VAL-opsin genes in the developing zebrafish. J Neurochem 104: 1364–1371.

63. Kojima D, Mano H, Fukada Y (2000) Vertebrate ancient-long opsin: a green-sensitive photoreceptive molecule present in zebrafish deep brain and retinal horizontal cells. J Neurosci 20: 2845–2851.

64. Foster RG, Garciafernandez JM, Provencio I, Degrip WJ (1993) Opsin Localization and Chromophore Retinoids Identified within the Basal Brain of the Lizard *Anolis-Carolinensis*. J Comp Physiol A 172: 33–45.

65. Provencio I, Foster RG (1993) Vitamin A2-based photopigments within the pineal gland of a fully terrestrial vertebrate. Neurosci Lett 155: 223–226.

66. Tabata M, Suzuki T, Niwa H (1985) Chromophores in the Extraretinal Photoreceptor (Pineal Organ) of Teleosts. Brain Res 338: 173–176.

67. Fletcher TC (1978) Defense mechanisms in fish. In: Malins DC, Sargent JR, editors. Biochemical and Biophysical Perspectives in Marine Biology. London: Academic Press. 189–217.

68. Kelsh RN, Harris ML, Colanesi S, Erickson CA (2009) Stripes and belly-spots-A review of pigment cell morphogenesis in vertebrates. Semin Cell Dev Biol 20: 90–104.

6

Molecular Genetic Analysis of Stomach Contents Reveals Wild Atlantic Cod Feeding on Piscine Reovirus (PRV) Infected Atlantic Salmon Originating from a Commercial Fish Farm

Kevin Alan Glover[1]*, Anne Grete Eide Sørvik[1], Egil Karlsbakk[1], Zhiwei Zhang[2], Øystein Skaala[1]

1 Institute of Marine Research, Bergen, Norway, 2 Jiangsu Institute of Marine Fisheries, NanTong City, P. R. China

Abstract

In March 2012, fishermen operating in a fjord in Northern Norway reported catching Atlantic cod, a native fish forming an economically important marine fishery in this region, with unusual prey in their stomachs. It was speculated that these could be Atlantic salmon, which is not typical prey for cod at this time of the year in the coastal zone. These observations were therefore reported to the Norwegian Directorate of Fisheries as a suspected interaction between a local fish farm and this commercial fishery. Statistical analyses of genetic data from 17 microsatellite markers genotyped on 36 partially-degraded prey, samples of salmon from a local fish farm, and samples from the nearest wild population permitted the following conclusions: 1. The prey were Atlantic salmon, 2. These salmon did not originate from the local wild population, and 3. The local farm was the most probable source of these prey. Additional tests demonstrated that 21 of the 36 prey were infected with piscine reovirus. While the potential link between piscine reovirus and the disease heart and skeletal muscle inflammation is still under scientific debate, this disease had caused mortality of large numbers of salmon in the farm in the month prior to the fishermen's observations. These analyses provide new insights into interactions between domesticated and wild fish.

Editor: Martin Krkosek, University of Toronto, Canada

Funding: This study was financed by the Norwegian Department of Fisheries. The funders had no role in study design, data collection and analysis, decision to publish, or preparation of the manuscript.

Competing Interests: The authors have declared that no competing interests exist.

* E-mail: kevin.glover@imr.no

Introduction

One of the most significant environmental challenges associated with the commercial culture of Atlantic salmon (*Salmo salar* L.) in marine net pens is containment. Within Norway, where statistics for the number of reported escapees are recorded by the Norwegian Directorate of Fisheries (NDF), the annual numbers of escapees has been in the hundreds of thousands for most years in the period 2000–2011 [1]. However, the true annual number of escapees has been estimated to be in the millions due to underreporting [2]. Farmed escapees can disperse over long distances [3,4], may enter rivers [5], and can display a range of ecological [6] and genetic interactions [7–12] with wild conspecifics. Thus, it is generally accepted that farmed escapees represent a potential threat to the integrity of native populations.

The application of molecular-genetic methods for wildlife conservation and fisheries management purposes, including forensic cases for law enforcement and regulation is expanding [13]. Typical wildlife forensic applications range from species identifications for morphologically unidentifiable tissues and samples, to population of origin identifications for individuals suspected to have been taken from locations where harvest is regulated or illegal [14], or even falsely claimed [15]. Analysis of stomach and faeces content from predators has also been

extensively conducted, and provided identification of prey items at the species [16–20], family [21], and even individual sample level [22].

The NDF are responsible for the development and implementation of aquaculture regulation in Norway. While escapement of fish from commercial aquaculture installations is not illegal in Norway, farmers are legally bound to report escapement from their farms. Despite this, underreporting represents a major challenge faced by the NDF. In response to this situation, genetic methods for the identification of escapees back to their farm of origin have been established and resulted in fines for companies found in breach of regulations [23,24].

In March 2012, local fishermen operating in a fjord in Northern Norway reported catching Atlantic cod (*Gadus morhua* L.), which forms an important commercial fishery in this region, with unusual prey fish in their stomachs. Most of these prey that were approximately 30–35 cm long, were partially or heavily degraded, and as such it was challenging to identify all of them morphologically (Fig. 1). Nevertheless, they did not look like herring (*Clupea harengus* L.) or smaller gadoid species which form an important part of the cod's diet in this region [25,26], and it was speculated by several fishermen that these could be Atlantic salmon. While Atlantic cod have been known to ingest Atlantic salmon smolts upon migration from freshwater into estuarine and marine environment [27,28], within a few weeks of entering the marine environment in the late spring and early summer, smolts have typically left fjord areas and migrate towards oceanic feeding grounds. As such, cod ingesting wild salmon of the observed size and time of year at this location was considered unusual by the local fishermen, and the situation therefore reported to the NDF as a suspected interaction between a local salmon farm and this commercial fishery. Here, we report the analysis of the prey in order to address the following questions: 1. What species are these prey, 2. If they are salmon, is it possible to identify them as wild or farmed (i.e., is this a rare natural phenomena or is it a human induced), and 3. If they are farmed Atlantic salmon, did they originate from a local farm?

Materials and Methods

Methodological approach

The present study was designed to address the three questions presented in the introduction. Diagnostic markers for identification of severely degraded Atlantic salmon and rainbow trout (*Oncorhynchus mykiss*) tissues have been recently developed [29]. However, the first attempt at identification of the prey was conducted with highly polymorphic microsatellite markers commonly implemented in Atlantic salmon population genetics projects. The reasoning for this was two-fold. First, in order to answer questions 2 and 3, an allele-frequency profile would be needed for each of the prey in order to match against the allele frequency profiles of farmed and wild salmon in the region. Second, a combination of past experiences with these microsatellite markers on partially degraded samples, together with inspection of the prey suggested that if they were indeed Atlantic salmon, it may be possible to successfully genotype the samples with these microsatellites.

Samples

This study is based in a fjord located in Northern Norway. For legal reasons, the exact locations of the cod captured in this study, and the local fish farm from which samples were taken, remain anonymous. Under supervision of the NDF, a total of 36 prey were sampled from cod stomachs by local fishermen (1–3 prey per cod stomach, all cod captured in the period March to April 2012). These cod were captured as part of a commercial harvest and were dead upon their stomachs being sampled. Thus, no specific permits were required for sampling the cod stomachs in this study. Both North east Arctic cod (NEAC) and Norwegian coastal cod (NCC) are known to form the basis of this commercial fishery at this time of the year in this region. However, no samples of nor data were recorded from these cod and as such it is not possible to exclude these fish of either type.

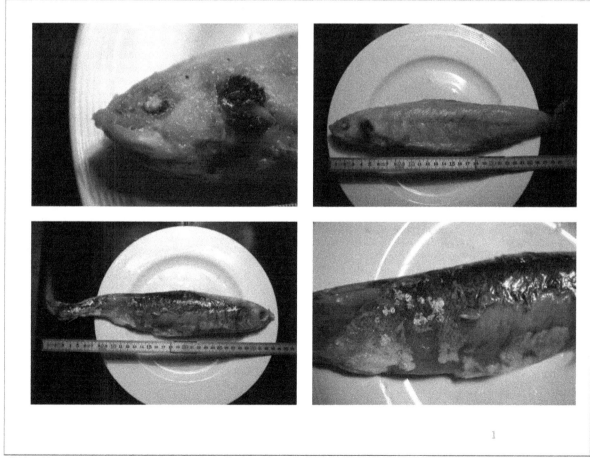

Friday, March 15, 2013

Figure 1. Examples of prey sampled from Atlantic cod stomachs. Most of the 36 prey were more severely digested than the specimens presented here and morphologically impossible to identify. However, not all prey were photographed.

Table 1. Summary statistics for samples from a local farm the group of escapees, and a local wild population.

Sample	N	Gene diversity			HWE		LD		Allelic diversity		Ne
		Ho	He	Fis	0.05	0.001	0.05	0.001	At	Ar	
Farm 1a	47	0.79	0.77	−0.026	0	0	17	1	156	151	43 (36–53)
Farm 1b	46	0.79	0.77	−0.027	1	0	12	0	157	152	125 (84–225)
Farm 1c	46	0.75	0.75	0.002	1	0	18	4	143	139	25 (21–30)
Prey-fish	37	0.76	0.75	−0.016	0	0	30	9	145	145	28 (24–35)
Wild	101	0.79	0.80	0.013	0	0	20	1	236	199	169 (135–222)

N = number of samples analysed, Ho and He = observed and expected heterozygosity, Fis = inbreeding coefficient, HWE = number of deviations from Hardy Weinberg equilibrium at two significance levels, LD = observed linkage disequilibrium at two significance levels, At = total number of alleles observed over 17 polymorphic loci, Ar = allelic richness based upon a re-sample size of 36–37 per locus/population combination and then totaled over all loci, Ne = effective population size as estimated by the LDNA method [47], with 95% confidence intervals in brackets and based upon including alleles down to and including those with frequencies of 0.005 in each population.

All cod were captured at one of six locations in the immediate vicinity of the only fish farm in the region containing salmon overlapping in size with the prey, or up to a maximum of 20 km further in the fjord. In addition to the 36 prey captured in cod stomachs, a single salmon post smolt, captured in the monitoring net located immediately beside the only salmon farm in this region, was sampled (this individual fish is hereafter referred to as "the escapee" and was similar in size to the prey). From all of these samples, two tissue samples per individual were taken for later genetic analysis.

Samples of salmon from the only farm in this fjord that contained fish overlapping in size with the prey were also collected. These fish were sampled by persons employed at the NDF. No specific permits were required to sample these fish, although the fish farmer gave access to their farm. The nearest alternative farm rearing fish overlapping in size with the prey was located over 100 km away (120 km away for the most distant captures of the cod) and not seen as a likely source, and therefore not sampled. From the local farm, a total of three samples, each consisting of approximately 47 fish, were taken from three separate cages. This represented the three genetic groups of fish on the farm, and is consistent with the sampling protocol for establishing a genetic baseline for identification of escapees back to their farm of origin [23,24].

A sample of wild Atlantic salmon, originating from the nearest river and in the immediate vicinity to where the cod with prey in their stomachs were captured was also included in this study. This wild salmon sample consisted of 101 adults captured by angling in the river in the seasons of 2007 and 2008. As these fish were captured and subsequently killed for consumption by sports

Table 2. Genetic relationships among the sets of samples as measured by pair-wise F$_{ST}$ (data in upper right diagonal), with associated P-values (data in lower left diagonal).

Sample	Farm1A	Farm 1B	Farm 1C	Prey	Wild
Farm1A		0.002	0.010	0.013	0.057
Farm 1B	0.144		0.006	0.006	0.052
Farm 1C	0.0008	0.0131		0.001	0.064
Prey	0.0008	0.0102	0.26		0.070
Wild	0.0001	0.0001	0.0001	0.0001	

fishermen, no permits for taking scale samples from these dead fish were required.

Disease status on the farm

Heart and skeletal muscle inflammation (HSMI) is an infectious disease [30] characterized by extensive inflammation and multifocal necrosis of myocytes in heart and red musculature [31]. A novel virus, piscine reovirus (PRV) has recently been detected in fish with HSMI. This virus is associated with the disease, shows elevated viral load in diseased fish, and is potentially responsible for the disease [32,33]. However, PRV infections are common in farmed salmon in Norway, and has also been documented in healthy fish including wild salmon [34]. Therefore, the role of PRV in HMSI remains under debate [34].

In the period January to February 2012 (i.e., a few weeks prior to the discovery of salmon-like prey in the stomachs of wild cod), the local farm reported losses of approximately 55000 fish (data from NDF farm biomass register). The causative disease was subsequently diagnosed as HMSI in February 2012. This diagnosis was based upon clinical analyses of fish from the farm by a local veterinary officer, and was subsequently confirmed by the Norwegian Veterinary Institute using histopathology (therefore, the presence or absence of PRV in these diseased fish is unknown).

Due to the background information regarding the disease status on the farm, samples from the prey captured in cod stomachs, and the single escapee, were analysed for the presence of PRV. This was on the basis that PRV could be present in the prey and the escapee if they originated from the farm where HMSI had caused mortality. PRV is also present in wild Norwegian salmon (also in fish not displaying HMSI), albeit at a lower frequency than in farmed escaped salmon (13.4% vs. 55.2% prevalence respectively) [34]. While this virus is typically identified in heart or head-kidney samples, due to the degraded state of the prey, only muscle samples were available for this test. Analyses were conducted by a Real Time PCR analysis company, PatoGen Analyse AS, accredited according to international standard ISO 17025. The samples were analyzed for PRV RNA at PatoGen in accordance with their in-house methods for Real Time PCR using an assay ('PRV-ST') targeting the L3 gene, sequenced previously [32]. The sequences of the forward and reverse primers for this assay are 5'-TCAACCACCTCCACACAAAAGA-3' and 5'-AACGAG TTGTGCGTGTGCC-3' respectively, and the probe VIC-5'-TTGGGATGTCGACGTTCT-3'. The standard curve based on tenfold dilutions in triplicates had a slope of −3.25 (R^2 = 0.998), and the Efficiency (E = $[10^{1/(-slope)}] - 1$) was 1.030. The cut-off C$_T$

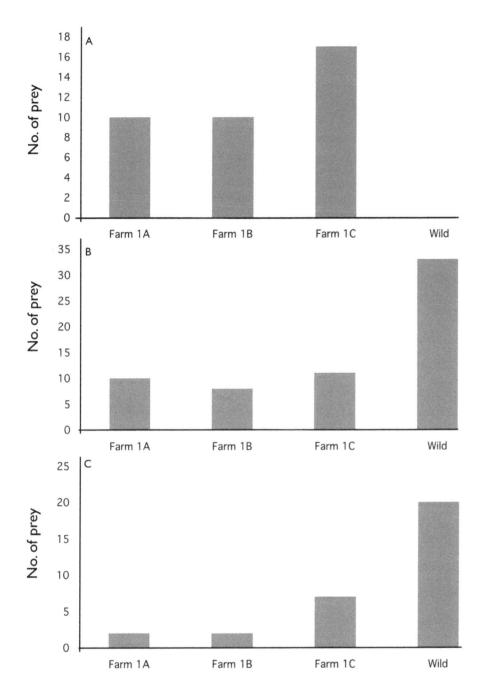

Figure 2. Genetic assignment of the prey to the samples collected from a local Norwegian farm and to the nearest wild Atlantic salmon population. A = direct assignment of prey to the genetically most similar sample, B = exclusion of prey from each sample in turn at α 0.01 threshold, C = exclusion of prey from each sample in turn at α 0.001 threshold. Note that individual prey can in theory be excluded from all or none of the samples, thus, exclusion does not sum to the exact number of prey in contrast to direct assignment which adds up to 37.

value was 37.0. The PRV-ST analysis was not accredited at the time of analysis and this work represents the first time these markers, produced by PatoGen AS, have been published.

Molecular genetic analyses

DNA extraction was conducted in 96-well format using a commercially available kit (Qiagen DNeasy®96 Blood & Tissue Kit). Each 96-well plate included two blank wells as negative controls. Routine genotyping control plays a standard role in genotyping in the laboratory at IMR [35,36]. Thus, each of the individual prey and the escapee were isolated twice to control

genotyping consistency. DNA quantity and quality was not measured.

All samples were subject to genotyping with a set of 18 microsatellites that are used in the laboratory for Atlantic salmon genetics projects. These loci were amplified in three multiplexes, using standard protocols for fresh tissues (full genotyping conditions available from authors upon request); *SSsp3016* (Genbank no. AY372820), *SSsp2210, SSspG7, SSsp2201, SSsp1605, SSsp2216* [37], *Ssa197, Ssa171, Ssa202* [38], *SsaD157, SsaD486, SsaD144* [39], *Ssa289, Ssa14* [40], *SsaF43* [41], *SsaOsl85* [42], *MHC I* [43] and *MHC II* [44]. PCR products were analysed on an

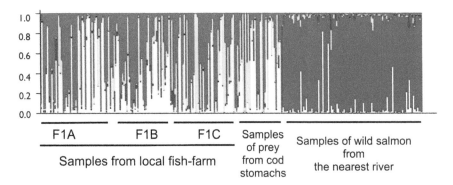

F1A   F1B   F1C   Samples   Samples of wild salmon
                                                       of prey   from
            Samples from local fish-farm   from cod   the nearest river
                                                    stomachs

Figure 3. Admixture analysis of salmon representing fish collected from the local Norwegian farm, prey captured in cod stomachs, and the nearest wild Atlantic salmon population. Results of admixture analysis are presented when the number of genetic clusters (i.e., *k*) is set to 4. Each genetic cluster is represented by a colour, and each individual's genetic assignment is represented by a vertical bar. Individuals may be admixed (i.e., mixtures of genetic clusters).

ABI 3730 Genetic Analyser and sized by a 500LIZ™ size-standard. The raw data was controlled manually twice before export for statistical analysis. No genotyping inconsistencies were observed among these re-analysed samples.

Statistical analysis

Once a DNA profile was successfully established for the individual prey, the single escapee, the local farm and wild salmon from a population in the region, several statistical tests, commonly implemented in population genetics studies, were conducted on these data. This was in order to primarily address three questions posed in the introduction. For these tests, the single escapee was pooled with the prey fish based upon pilot analysis documenting it to be genetically very similar to the prey (see results). Thus, for these analyses, the prey sample also included the single escapee.

First, the data were arranged in a population genetics program (MSA) [45], which was used to compute a range of summary statistics, and input files for other programs. Thereafter, the data was analysed in Genepop V3.3 [46] to compute gene diversities, Hardy Weinberg equilibrium, and linkage disequilibrium between pairs of loci within samples. The Fishers exact test (demorization 10 000; 100 batches; 5000 iterations) was implemented to test for statistical significance. The program LDNE [47] as used to compute the effective population size (*Ne*) for each of the samples. This program uses a one-sample approach to estimating *Ne* based upon the degree of LD observed within a sample.

Genetic identification of the prey was conducted by two different but complimentary methodological approaches. First, genetic assignment using the Rannala & Mountain method of computation [48] as implemented in the program GeneClass2 [49] was conducted. Here, the samples from the farms and of the wild fish were used as the pre-determined potential sources of the prey (i.e., the genetic baseline). Thereafter, direct genetic assignment was conducted. This method places each unknown fish (i.e., the individual prey fish) into the baseline sample that it resembles most. A limitation with direct assignment is that it assigns a potential source population to each of the unknown samples irrespective to the absolute degree of similarity. This may be acceptable in "closed systems" where all potential sources of the unknown samples are represented, however, in situations such as the present where not all potential sources are included in the baseline, it is important to get an estimation of the degree of similarity between the unknown sample(s) and each baseline sample. This is achieved by exclusion, and each individual is compared to each baseline sample, and a probability of belonging

(or more correctly, probability of not belonging) is computed. In the specific situation here, rejection from all baseline samples would suggest that the prey originated from a source not sampled.

The second approach to identifying the prey was to compute admixture (also referred to as Bayesian cluster analysis) using the program STRUCTURE 2.2 [50,51]. Individual admixture permits the identification and assignment of individual fish to genetic clusters (i.e., populations or genetic groups) without any "prior" regarding the population or location from which each individual sample originated. This permits, for example, identification of individuals that are of mixed genetic origin, and identification of individuals when mixed into samples predominantly of other genetic groups. The program was run using an admixture model with correlated allele frequencies and no prior. Runs consisted of a burn-in of 250 000 MCMC steps, followed by 250 000 steps. The program was run with all samples detailed included, with the number of populations set between *k* = 1–8 with 3 runs per *k*. The probability of the data was plotted, and the most appropriate *k* was determined at the point where the slope reached a plateau [50].

Results

Despite being partially digested (Fig. 1), microsatellite DNA profiles were successfully obtained from all 36 prey sampled from the cod stomachs, and the single escapee captured in the monitoring net placed outside the local farm. While some markers were not scored in some of the DNA isolations, when combining data from both isolates (after cross-checking to validate genotyping consistency), only two genotypes were missing from a total of 629 potential genotypes for the 37 fish analysed at 17 microsatellite loci (i.e., >99% genotyping coverage). This provided both conclusive evidence that the prey were indeed Atlantic salmon, and permitted the next step of their identification using a population genetics statistical approach.

Summary statistics for the combined sample of the prey (which included the single escapee captured in net), samples from the local farm and from the river demonstrate several trends (Table 1). All samples from the farm displayed less genetic diversity than the sample from the river, as measured by either the total number of alleles, or allelic richness which circumvents the problems of having different numbers of individuals representing each sample. Lower variation at polymorphic genetic markers is typical for farm samples in comparison with wild samples [52,53], and is linked to the fact that the fish sampled in a single cage often have a limited number of parents [54]. Almost all samples were in HWE, however, LD was observed in both the sample Farm 1C, and the sample of prey fish. Ne was very low in the sample of the prey and two of the samples from farms. In contrast, Ne was much higher in the sample of wild salmon and farm sample 1B. For all of these summary statistics, the prey resembled the farm samples very strongly, especially 1C, whereas they displayed very different parameters to the wild sample.

F_{ST} is an average measurement of genetic similarity between groups of samples or populations. Taken collectively, the prey were genetically strongly distinct to the wild salmon, and marginally different to the samples farm 1A and 1B (Table 2). These prey were genetically similar to the farm sample 1C. All farm samples were genetically distinct to the wild sample, supporting observations from the summary statistics presented above.

Self-assignment simulations including samples from the farm and the wild population demonstrated that overall, 70% of the fish in this set of samples would be correctly assigned to the sample from which they originated. Miss-assignment was caused almost exclusively by farmed fish being incorrectly placed into an alternative farm sample which reflects the overlapping genetic profile between these cages. None of the farmed fish were incorrectly assigned to the wild population, and only 3 of the 101 wild salmon were incorrectly assigned to any of the farmed samples (all were assigned to Farm 1B). Thus, these simulations demonstrate almost complete potential to identify whether the prey are more likely to have originated from this local farm (and thus a human-induced event) or the local wild population (and therefore an unusual natural event).

Direct assignment (which places the unknown sample, which is in this case was the 36 prey and the one salmon escapee captured in a net outside the farm, into the genetically most similar baseline sample) placed all of the prey and the single escapee into farm samples, and none into the wild sample (Fig. 2). Exclusion tests supported this, demonstrating that the majority of the prey and the single escapee could be conclusively excluded from the sample of wild salmon, whilst only 1 prey sample could be excluded from all of the farmed samples (and in that case the wild sample also).

Identification of the prey and the single escapee was also conducted using admixture analysis in the program Structure (Fig. 3). This program does not take into consideration any "priors" for the samples and each individual can represent a mixture of genetic groups or clusters. The probability of the data was plotted, and the most appropriate number of clusters k, was determined to be 4 (the point where the slope reached a plateau) [50]. Confirming results from other statistical tests presented above, admixture analysis demonstrated that there was a large genetic difference between the farmed salmon and the wild salmon in this data set, and importantly, that all of the prey, including the single salmon escapee, were closely associated with genetic clusters represented in the salmon from the local Norwegian farm, and not the local wild population. Data for other numbers of clusters (i.e., k set between 2–8) gave identical results (data not presented).

Real-time PCR analyses (PRV-ST assay) detected PRV virus RNA in 22 of the 37 prey samples. The positive samples represented 21 prey items from cod stomachs and the single salmon captured in the monitoring net placed outside the Norwegian farm. C_T values ranged from 27.8–35.3 (mean 33.0).

Discussion

Molecular-genetic tools to identify the aquaculture facility and in some cases even the specific cage of origin for Atlantic salmon [23,24], Atlantic cod [55,56] and farmed rainbow trout [57] escapees have been developed. However, the present study represents a new application of molecular-genetic methods in order to provide management authorities with the opportunity to monitor commercial aquaculture and its interaction with the natural environment. In addition, this study provides new insights into interactions between domesticated and wild fish.

Four main conclusions can be drawn from these analyses: 1. The partially digested and morphologically difficult to identify prey were revealed to be Atlantic salmon, 2. Based upon several independent genetic parameters, these salmon prey were identified as farmed and not from the local wild population, thus demonstrating this to be a human induced, as opposed to natural phenomena, 3. Despite partial digestion, the majority of the prey, including the single escapee, carried detectable levels of PRV. PRV is associated with the disease HSMI [32,33]. This disease had caused significant mortality of salmon on the local farm in the immediate time-period prior to the prey being captured in the wild cod, 4. The genetic profile of the salmon prey, and the single escapee, strongly matched the genetic profile of the fish in the local farm. Although genetic similarity is not unequivocal proof of origin [23], considering the nearest alternative farm that these individuals could have theoretically originated from was located over 100 km away in another fjord, these analyses provided the NDF with sufficient "circumstantial evidence" to initiate an investigation of the company owning this commercial aquaculture facility on the basis of potential mis-management.

The salmon farm in the study area was diagnosed with HSMI just weeks prior to the appearance of farmed salmon in the stomach of the local cod. Therefore, the prey recaptured from cod stomachs were examined for the presence of PRV, although virus levels may decline after an outbreak [58]. Despite partial digestion of the prey, and the fact that only muscle samples were available, PRV virus was still detected. Nevertheless, based upon the analyses conducted here, it is not possible to unequivocally resolve how the PRV infected farmed salmon entered the natural environment. They could be diseased dead fish deposited into the sea (which would represent an illegal practice in Norway) and thereafter ingested by the cod from the sea-bed, or they were escapees predated upon by the cod. Given that the farm had experienced significant mortality of fish (55000) through HMSI in the period immediately before the salmon were captured in the cod stomachs, indicate that the former explanation is the most likely.

Independent of how the fish entered the natural environment, this study demonstrates trophic transmission as a mechanism for interaction between salmon farming and wild populations. While Atlantic cod have been documented to predate upon wild Atlantic salmon smolts migrating from freshwater to the sea [27,28], Atlantic salmon is not typically predated upon by cod at the time of year in which the current study was conducted [25,26]. Furthermore, to our knowledge, this study represents the first documentation of Atlantic cod ingesting Atlantic salmon from a fish farm. Thus, it is possible that the cod investigated here, and forming part of the population in the study area at this time of year, have been exposed to PRV. The ability of PRV virus to be transmitted to new hosts via ingestion of infected prey is at present unknown, as is the susceptibility of Atlantic cod to the virus. However, PRV was not detected in 78 cod nor 850 other gadoids that were recently sampled in Norway and screened for this virus [59]. Since PRV infections are common in wild and farmed salmon in Norway, and also occur in wild sea trout (*Salmo trutta* L.) [34], it is likely that the common PRV type is specific to salmonids and that cod is not susceptible.

Analysis of animal stomach contents or faeces using molecular genetic methods has been widely applied to a range of taxa and biological questions [19,60]. These methods have primarily been conducted to identify prey items to a taxonomic classification, often species, usually involving the analysis of a single or low number of genes providing the required taxonomic capability for the potential prey species in question [17,18,20]. More recently, advances in next generation sequencing have permitted powerful additions to these approaches, leading to what is termed as DNA metabarcoding [61]. While the present study does not represent a technological advance for such molecular genetic methods, the application of microsatellite DNA analysis in diet analysis to provide identification beyond a taxonomic classification is novel. Here, it was possible to not only demonstrate that the prey was Atlantic salmon, but that the most probable source was a local farm and not a local wild salmon population. Finally, it was also possible to demonstrate that the prey carried a virus that has been associated with a disease that causes significant mortality of farmed salmon. Therefore, this study represents an extension of the biological questions that can be addressed via molecular genetic analysis of stomach contents. Other examples of diet analysis going beyond species identifications include analysis to the family level to demonstrate filial cannibalism in the wild [21], and predation mortality of Atlantic salmon of farmed, hybrid and wild parentage in a natural river system (Skaala unpublished). Also, identification of prey items to the individual level has been conducted, where microsatellite analysis of diet from greenland sharks (*Somniosus microcephalus*), together with a search of a DNA register for all minke whales (*Balaenoptera acutorostrata*) captured under commercial harvest in Norway [62] permitted connecting the whale and shark captures in both time and space to understand both their movement and diet habits [22].

For more than a decade, the annual reported escapement of salmon from Norwegian fish farms has been in the tens or hundreds of thousands [1]. This is likely to be an underestimate due to underreporting, and in the period 1998 to 2004, it is estimated that the mean annual number of escapees was 2.4 million [2], which is higher than the annual number of wild salmon returning to the Norwegian coastline to reproduce in the same period. While attention surrounding the impact of escapees has primarily been given to those alive as opposed to dead [7,9,63], the present study demonstrates that virus-infected farmed fish may be released to the environment by one method or another. The analyses in the present case highlight the potential to identify and track such events. Given the magnitude of escapement from commercial fish farms, this represents one of the most significant human-induced invasions of native populations by a species that has been subject to selective breeding. Therefore, this situation needs to be monitored for not only ecological [6] and genetic [7–12] interactions, but also disease interactions.

Acknowledgments

Tor Arne Helle, and other workers at the Norwegian Directorate of Fisheries are gratefully acknowledged for assistance in coordinating samples taken from fishermen, and gathering in background information. Morten Thuv is gratefully acknowledged for supplying the photographs upon which Fig. 1 is based.

Author Contributions

Conceived and designed the experiments: KAG EK ØS. Performed the experiments: KAG AGES ZZ. Analyzed the data: KAG EK. Contributed reagents/materials/analysis tools: KAG AGES ZZ EK. Wrote the paper: KAG AGES EK ZZ ØS.

References

1. Anonymous (2012) Oppdaterte rømmingstall. Available: http://www.fiskeridir.no/statistikk/akvakultur/oppdaterte-roemmingstall. Accessed: 2013 Mar 23. In Norwegian.
2. Saegrov H, Urdal K (2006) Escaped farmed salmon in the sea and rivers; numbers and origin. Rådgivende Biologer, Norway Report 947.
3. Skilbrei OT (2010) Adult recaptures of farmed Atlantic salmon post-smolts allowed to escape during summer. Aquaculture Environment Interactions 1: 147–153.
4. Hansen LP (2006) Migration and survival of farmed Atlantic salmon (*Salmo salar* L.) released from two Norwegian fish farms. Ices Journal of Marine Science 63: 1211–1217.
5. Fiske P, Lund RA, Hansen LP (2006) Relationships between the frequency of farmed Atlantic salmon, *Salmo salar* L., in wild salmon populations and fish farming activity in Norway, 1989–2004. Ices Journal of Marine Science 63: 1182–1189.
6. Jonsson B, Jonsson N (2006) Cultured Atlantic salmon in nature: a review of their ecology and interaction with wild fish. Ices Journal of Marine Science 63: 1162–1181.
7. Skaala O, Wennevik V, Glover KA (2006) Evidence of temporal genetic change in wild Atlantic salmon, *Salmo salar* L., populations affected by farm escapes. Ices Journal of Marine Science 63: 1224–1233.
8. Clifford SL, McGinnity P, Ferguson A (1998) Genetic changes in Atlantic salmon (*Salmo salar*) populations of northwest Irish rivers resulting from escapes of adult farm salmon. Canadian Journal of Fisheries and Aquatic Sciences 55: 358–363.
9. Glover KA, Quintela M, Wennevik V, Besnier F, Sørvik AGE, et al. (2012) Three decades of farmed escapees in the wild: A spatio-temporal analysis of population genetic structure throughout Norway. Plos One 7(8): e43129v.
10. Crozier WW (2000) Escaped farmed salmon, *Salmo salar* L., in the Glenarm River, Northern Ireland: genetic status of the wild population 7 years on. Fisheries Management and Ecology 7: 437–446.
11. Crozier WW (1993) Evidence of genetic interaction between escaped farmed salmon and wild Atlantic salmon (*Salmo salar* L) in a Northern Irish river. Aquaculture 113: 19–29.
12. Clifford SL, McGinnity P, Ferguson A (1998) Genetic changes in an Atlantic salmon population resulting from escaped juvenile farm salmon. Journal of Fish Biology 52: 118–127.
13. Ogden R (2010) Forensic science, genetics and wildlife biology: getting the right mix for a wildlife DNA forensics lab. Forensic Science Medicine and Pathology 6: 172–179.
14. Withler RE, Candy JR, Beacham TD, Miller KM (2004) Forensic DNA analysis of Pacific salmonid samples for species and stock identification. Environmental Biology of Fishes 69: 275–285.
15. Primmer CR, Koskinen MT, Piironen J (2000) The one that did not get away: individual assignment using microsatellite data detects a case of fishing competition fraud. Proceedings of the Royal Society B-Biological Sciences 267: 1699–1704.
16. Shehzad W, Riaz T, Nawaz MA, Miquel C, Poillot C, et al. (2012) Carnivore diet analysis based on next-generation sequencing: application to the leopard cat (Prionailurus bengalensis) in Pakistan. Molecular Ecology 21: 1951–1965.
17. Jarman SN, Gales NJ, Tierney M, Gill PC, Elliott NG (2002) A DNA-based method for identification of krill species and its application to analysing the diet of marine vertebrate predators. Molecular Ecology 11: 2679–2690.
18. Scribner KT, Bowman TD (1998) Microsatellites identify depredated waterfowl remains from glaucous gull stomachs. Molecular Ecology 7: 1401–1405.
19. Symondson WOC (2002) Molecular identification of prey in predator diets. Molecular Ecology 11: 627–641.
20. Matejusova I, Doig F, Middlemas SJ, Mackay S, Douglas A, et al. (2008) Using quantitative real-time PCR to detect salmonid prey in scats of grey Halichoerus grypus and harbour Phoca vitulina seals in Scotland – an experimental and field study. Journal of Applied Ecology 45: 632–640.
21. DeWoody JA, Fletcher DE, Wilkins SD, Avise JC (2001) Genetic documentation of filial cannibalism in nature. Proceedings of the National Academy of Sciences of the United States of America 98: 5090–5092.
22. Leclerc LM, Lydersen C, Haug T, Glover KA, Fisk AT, et al. (2011) Greenland sharks (Somniosus microcephalus) scavenge offal from minke (Balaenoptera acutorostrata) whaling operations in Svalbard (Norway). Polar Research 30: 7342.
23. Glover KA (2010) Forensic identification of fish farm escapees: the Norwegian experience. Aquaculture Environment Interactions 1: 1–10.
24. Glover KA, Skilbrei OT, Skaala Ø (2008) Genetic assignment identifies farm of origin for Atlantic salmon *Salmo salar* escapees in a Norwegian fjord. Ices Journal of Marine Science 65: 912–920.
25. Michalsen K, Johannesen E, Bogstad L (2008) Feeding of mature cod (Gadus morhua) on the spawning grounds in Lofoten. Ices Journal of Marine Science 65: 571–580.

26. Sivertsen SP, Pedersen T, Lindstrom U, Haug T (2006) Prey partitioning between cod (Gadus morhua) and minke whale (Balaenoptera acutorostrata) in the Barents Sea. Marine Biology Research 2: 89–99.

27. Hvidsten NA, Lund RA (1988) Predation on hatchery-reared and wild smolts of Atlantic salmon, Salmo salar L, in the estuary of river Orkla, Norway. Journal of Fish Biology 33: 121–126.

28. Hvidsten NA, Mokkelgjerd PI (1987) Predation on salmon smolts, Salmo salar L., in the estuary of the river Surna, Norway. Journal of Fish Biology 30: 273–280.

29. Dalvin S, Glover KA, Sorvik AGE, Seliussen BB, Taggart JB (2010) Forensic identification of severly degraded Atlantic salmon (Salmo salar L.) and rainbow trout (Oncohynchus mykiss) tissues Bmc Investigative Genetics 1: 12.

30. Kongtorp RT, Kjerstad A, Taksdal T, Guttvik A, Falk K (2004) Heart and skeletal muscle inflammation in Atlantic salmon, Salmo salar L.: a new infectious disease. Journal of Fish Diseases 27: 351–358.

31. Kongtorp RT, Taksdal T, Lyngoy A (2004) Pathology of heart and skeletal muscle inflammation (HSMI) in farmed Atlantic salmon Salmo salar. Diseases of Aquatic Organisms 59: 217–224.

32. Palacios G, Lovoll M, Tengs T, Hornig M, Hutchison S, et al. (2010) Heart and Skeletal Muscle Inflammation of Farmed Salmon Is Associated with Infection with a Novel Reovirus. Plos One 5.

33. Finstad OW, Falk K, Lovoll M, Evensen O, Rimstad E (2012) Immunohisto-chemical detection of piscine reovirus (PRV) in hearts of Atlantic salmon coincide with the course of heart and skeletal muscle inflammation (HSMI). Veterinary Research 43.

34. Garseth AH, Fritsvold C, Opheim M, Skjerve E, Biering E (2012) Piscine reovirus (PRV) in wild Atlantic salmon, Salmo salar L., and sea-trout, Salmo trutta L., in Norway. Journal of Fish Diseases. doi: 10.1111/j.1365-2761.2012.01450.x.

35. Glover KA, Hansen MM, Lien S, Als TD, Hoyheim B, et al. (2010) A comparison of SNP and STR loci for delineating population structure and performing individual genetic assignment. Bmc Genetics 11.

36. Haaland ØA, Glover KA, Seliussen BB, Skaug HJ (2011) Genotyping errors in a calibrated DNA -register: implications for identification of individuals. BMC Genetics 12: 36.

37. Paterson S, Piertney SB, Knox D, Gilbey J, Verspoor E (2004) Characterization and PCR multiplexing of novel highly variable tetranucleotide Atlantic salmon (Salmo salar L.) microsatellites. Molecular Ecology Notes 4: 160–162.

38. O'Reilly PT, Hamilton LC, McConnell SK, Wright JM (1996) Rapid analysis of genetic variation in Atlantic salmon (Salmo salar) by PCR multiplexing of dinucleotide and tetranucleotide microsatellites. Canadian Journal of Fisheries and Aquatic Sciences 53: 2292–2298.

39. King TL, Eackles MS, Letcher BH (2005) Microsatellite DNA markers for the study of Atlantic salmon (Salmo salar) kinship, population structure, and mixed-fishery analyses. Molecular Ecology Notes 5: 130–132.

40. McConnell SK, Oreilly P, Hamilton L, Wright JN, Bentzen P (1995) Polymorphic microsatellite loci from Atlantic salmon (Salmo salar) – genetic differentiation of North-American and European populations. Canadian Journal of Fisheries and Aquatic Sciences 52: 1863–1872.

41. Sanchez JA, Clabby C, Ramos D, Blanco G, Flavin F, et al. (1996) Protein and microsatellite single locus variability in Salmo salar L (Atlantic salmon). Heredity 77: 423–432.

42. Slettan A, Olsaker I, Lie O (1995) Atlantic salmon, Salmo salar, microsatellites at the SsOSL25, SsOSL85, SsOSL311, SsOSL417 loci. Animal Genetics 26: 281–282.

43. Grimholt U, Drablos F, Jorgensen SM, Hoyheim B, Stet RJM (2002) The major histocompatibility class I locus in Atlantic salmon (Salmo salar L.): polymorphism, linkage analysis and protein modelling. Immunogenetics 54: 570–581.

44. Stet RJM, de Vries B, Mudde K, Hermsen T, van Heerwaarden J, et al. (2002) Unique haplotypes of co-segregating major histocompatibility class II A and class II B alleles in Atlantic salmon (Salmo salar) give rise to diverse class II genotypes. Immunogenetics 54: 320–331.

45. Dieringer D, Schlotterer C (2003) MICROSATELLITE ANALYSER (MSA): a platform independent analysis tool for large microsatellite data sets. Molecular Ecology Notes 3: 167–169.

46. Raymond M, Rousset F (1995) GENEPOP (VERSION-1.2) – Population-genetics software for exact tests and ecumenicism. Journal of Heredity 86: 248–249.

47. Waples RS, Do C (2008) LDNE: a program for estimating effective population size from data on linkage disequilibrium. Molecular Ecology Resources 8: 753–756.

48. Rannala B, Mountain JL (1997) Detecting immigration by using multilocus genotypes. Proceedings of the National Academy of Sciences of the United States of America 94: 9197–9201.

49. Piry S, Alapetite A, Cornuet JM, Paetkau D, Baudouin L, et al. (2004) GENECLASS2: A software for genetic assignment and first-generation migrant detection. Journal of Heredity 95: 536–539.

50. Falush D, Stephens M, Pritchard JK (2003) Inference of population structure using multilocus genotype data: Linked loci and correlated allele frequencies. Genetics 164: 1567–1587.

51. Pritchard JK, Stephens M, Donnelly P (2000) Inference of population structure using multilocus genotype data. Genetics 155: 945–959.

52. Skaala Ø, Hoyheim B, Glover K, Dahle G (2004) Microsatellite analysis in domesticated and wild Atlantic salmon (Salmo salar L.): allelic diversity and identification of individuals. Aquaculture 240: 131–143.

53. Norris AT, Bradley DG, Cunningham EP (1999) Microsatellite genetic variation between and within farmed and wild Atlantic salmon (Salmo salar) populations. Aquaculture 180: 247–264.

54. Glover KA, Skaala Ø, Sovik AGE, Helle TA (2011) Genetic differentiation among Atlantic salmon reared in sea-cages reveals a non-random distribution of genetic material from a breeding programme to commercial production. Aquaculture Research 42: 1323–1331.

55. Glover KA, Dahle G, Jorstad KE (2011) Genetic identification of farmed and wild Atlantic cod, Gadus morhua, in coastal Norway. Ices Journal of Marine Science 68: 901–910.

56. Glover KA, Dahle G, Westgaard JI, Johansen T, Knutsen H, et al. (2010) Genetic diversity within and among Atlantic cod (Gadus morhua) farmed in marine cages: a proof-of-concept study for the identification of escapees. Animal Genetics 41: 515–522.

57. Glover KA (2008) Genetic characterisation of farmed rainbow trout in Norway: intra- and inter-strain variation reveals potential for identification of escapees. Bmc Genetics 9.

58. Lovoll M, Alarcon M, Jensen BB, Taksdal T, Kristoffersen AB, et al. (2012) Quantification of piscine reovirus (PRV) at different stages of Atlantic salmon Salmo salar production. Diseases of Aquatic Organisms 99: 7–12.

59. Wiik-Nielsen CR, Lovoll M, Sandlund N, Faller R, Wiik-Nielsen J, et al. (2012) First detection of piscine reovirus (PRV) in marine fish species. Diseases of Aquatic Organisms 97: 255–258.

60. Yoccoz NG (2012) The future of environmental DNA in ecology. Molecular Ecology 21: 2031–2038.

61. Pompanon F, Deagle BE, Symondson WOC, Brown DS, Jarman SN, et al. (2012) Who is eating who: diet assessment using next generation sequencing. Molecular Ecology 21: 1931–1950.

62. Glover KA, Haag T, Oien N, Walloe L, Lindblom L, et al. (2012) The Norwegian minke whale DNA register: a database monitoring commercial harvest and trade of whale products. Fish and Fisheries 13: 313–332.

63. Zhang Z, Glover KA, Wennevik V, Svåsand T, Sørvik AGE, et al. (2012) Genetic analysis of Atlantic salmon captured in a netting station reveals multiple escapement events from commercial fish farms. Fisheries Management and Ecology. doi: 10.1111/fme.12002.

7

Resilience of Coral-Associated Bacterial Communities Exposed to Fish Farm Effluent

Melissa Garren[1]*, **Laurie Raymundo**[2], **James Guest**[3], **C. Drew Harvell**[4], **Farooq Azam**[1]

1 Marine Biology Research Division, Scripps Institution of Oceanography, University of California San Diego, La Jolla, California, United States of America, **2** Marine Laboratory, University of Guam, Mangilao, Guam, **3** Marine Biology Laboratory, Department of Biological Sciences, National University of Singapore, Singapore, Singapore, **4** Department of Ecology & Evolutionary Biology, Cornell University, Ithaca, New York, United States of America

Abstract

Background: The coral holobiont includes the coral animal, algal symbionts, and associated microbial community. These microbes help maintain the holobiont homeostasis; thus, sustaining robust mutualistic microbial communities is a fundamental part of long-term coral reef survival. Coastal pollution is one major threat to reefs, and intensive fish farming is a rapidly growing source of this pollution.

Methodology & Principal Findings: We investigated the susceptibility and resilience of the bacterial communities associated with a common reef-building coral, *Porites cylindrica*, to coastal pollution by performing a clonally replicated transplantation experiment in Bolinao, Philippines adjacent to intensive fish farming. Ten fragments from each of four colonies (total of 40 fragments) were followed for 22 days across five sites: a well-flushed reference site (the original fragment source); two sites with low exposure to milkfish (*Chanos chanos*) aquaculture effluent; and two sites with high exposure. Elevated levels of dissolved organic carbon (DOC), chlorophyll *a*, total heterotrophic and autotrophic bacteria abundance, virus like particle (VLP) abundances, and culturable *Vibrio* abundance characterized the high effluent sites. Based on 16S rRNA clone libraries and denaturing gradient gel electrophoresis (DGGE) analysis, we observed rapid, dramatic changes in the coral-associated bacterial communities within five days of high effluent exposure. The community composition on fragments at these high effluent sites shifted towards known human and coral pathogens (i.e. *Arcobacter*, *Fusobacterium*, and *Desulfovibrio*) without the host corals showing signs of disease. The communities shifted back towards their original composition by day 22 without reduction in effluent levels.

Significance: This study reveals fish farms as a likely source of pathogens with the potential to proliferate on corals and an unexpected short-term resilience of coral-associated bacterial communities to eutrophication pressure. These data highlight a need for improved aquaculture practices that can achieve both sustainable industry goals and long-term coral reef survival.

Editor: Thomas Bell, University of Oxford, United Kingdom

Funding: This research was supported by grants from the GEF Coral Targeted Research Program, Gordon and Betty Moore Foundation, and National Science Foundation (NSF) grant #OCE06-48116 to Farooq Azam. James Guest was supported by the European Commission INCO-DEV REEFRES Project Number 510657. Melissa Garren was supported by an NSF Integrative Graduate Education and Research Training (IGERT) and an NSF Graduate Research Fellowship. The funders had no role in study design, data collection and analysis, decision to publish, or preparation of the manuscript.

Competing Interests: The authors have declared that no competing interests exist.

* E-mail: mgarren@ucsd.edu

Introduction

Reef-building corals are just one of many animals that have a mutualistic microbial community. However, the particular relationship between corals and their associated microbes may have a more direct linkage to ecosystem health compared with other species' symbiotic microbial community. The ecosystem-level influence of coral-associated microbial communities is rooted in the fundamental role that scleractinian (stony, reef-building) corals play in physically structuring the habitat and supporting reef organisms, and the roles coral-associated microbes have in maintaining holobiont health. In recent years, molecular methods have greatly expanded our ability to study coral-associated microbial communities. We now know some coral species-specific associations of microbes exist [1], that coral-associated communi-ties are distinct from water column-associated microbes [1–3], and that the bacterial communities of corals can shift under conditions of stress or disease [4–7]. The causative agents of many coral diseases remain unknown [8,9], as do the outcomes of interactions among coral-associated microbial communities and various environmental perturbations, such as nutrient enrichment.

Coral disease [10] and nutrient enrichment of coastal waters [11] actively contribute to the continued decline of coral reefs globally. As coastal development and coral disease incidence continue to rise [12], understanding the mechanisms linking them [13] and managing reefs for long-term survival become pressing management priorities.

Managing for resilience is a current priority in the field of coral reef management [14,15]. The goal is to develop and employ management strategies that increase the ability of reef ecosystems to withstand and recover from stress. Three cornerstones have been proposed to aid the empirical assessment of 'resilience': biodiversity, spatial heterogeneity, and connectivity [15]. The focus has been on broad reef-wide or region-wide assessments [16], but a potentially important and overlooked component to consider is the resilience capability of the individual corals that comprise the reefs. At this scale, we must consider coral-associated microbial communities [4,17]. The same three cornerstones of resilience can also be applied to the microbial communities within a single coral colony. Here we define resilience as the ability of the microbial communities to return to their original composition after disturbance by nutrient enriched waters despite the continued presence of this enrichment. The large body of work illuminating the role of coral-algal symbioses in large-scale bleaching patterns [18,19] is a good example of how understanding the biodiversity, spatial heterogeneity, and connectivity of single-celled dinoflagellates (zooxanthellae) provided critical links between the small-scale processes of coral-algal associations and region-wide bleaching patterns. The discovery of a vast genetic diversity of zooxanthellae led to the realization that not all algal symbionts of corals are equally tolerant to environmental stressors. The distribution of algal symbionts that were more susceptible to increased temperature stress explained not only the single colony-scale bleaching patterns observed in some corals [18], but also helped explain larger region-wide bleaching patterns. It has become clear that the zooxanthellae diversity and distribution patterns of a reef are worth considering when planning management strategies for a given region [20]. The parallel body of work for coral-associated bacteria, archaea, and viruses is much younger and smaller than that of zooxanthellae. We currently understand more about the biodiversity of the coral-associated microbes than we do about the spatial heterogeneity or connectivity of these communities [21]. Thinking about the microbial ecology of these microscale ecosystems in the context of the three resilience criteria will help illuminate the functional role these organisms play in coral health and disease.

This study builds on previous research that demonstrated that coral reef water-associated microbial communities were influenced by effluent from coastal milkfish (*Chanos chanos*) aquaculture in Bolinao, Philippines [17]. Since the corals that lived in the channel before the intensive aquaculture industry began have died out (E. Gomez, pers. comm.), we simulated the effect of the introduction of these fish pens on corals by transplanting live, clonally replicated corals from a relatively effluent-naïve site (Reference site) to sites routinely exposed to some amount of effluent (low exposure as sites Far-1 & 2; high exposure at sites Near-1 & 2). We investigated the biodiversity of coral-associated bacteria communities over 22 days in response to this chronic effluent exposure. This allowed us to investigate both the effect of spatial proximity to fish pens and temporal transitions in microbial communities. We tested the hypotheses that: 1) exposure to fish pen effluent would cause shifts in coral-associated bacterial community composition towards phylotypes not normally associated with healthy corals; 2) corals exposed to high levels of effluent would be colonized by fish pen-associated bacteria; and 3) there would be some degree of resilience in bacterial communities allowing corals to maintain or recover their original microbial communities despite the stress of effluent exposure. This study broadens our knowledge of the links among environmental conditions, bacterial community composition, and coral health. In the face of rising coastal development and changing environmental stimuli, understanding these connections will aid in developing effective and ecologically sound coral reef management strategies. A focus on microbial community shifts in response to organically enriched effluent will contribute to

understanding the connectivity of these communities on coral reefs, further advancing our understanding of how to manage for increased resilience.

Results

Fish Pen Effluent is Persistent, Spatially Extensive and Traceable in Real Time

Weekly mapping at 20 stations all around Santiago Island showed that the strongest chlorophyll *a* signals consistently occurred in the channel containing the majority of fish pens (Fig. 1). There are no major river discharges or other point-sources of nutrient enrichment at or up-current (south east) from the fish pens [22], thus the majority of organic matter input can be attributed to the fish pens. An elongated plume of chlorophyll *a* in the direction of the prevailing current was also consistent with the hypothesis that the fish pens were the source of enrichment in this system.

Bacterial abundance, frequency of dividing cells (FDC), cyanobacteria abundance, virus-like particle (VLP) abundance, concentrations of colony forming units (CFU) of *Vibrio* and kanamycin-resistant bacteria, dissolved organic carbon (DOC) concentration, and chlorophyll *a* concentrations at sites along a strong fish pen gradient were all higher during sampling in May and June 2008 than a previous sampling period in January 2007 ([17],Table 1). A pronounced wet season (May and June 2008) influenced the gradient of fish pen effluent distribution. These data also suggested that corals at the most well-flushed sites (such as the Reference site; Fig. 1) in our study area were no longer living in an oligotrophic system. The DOC, bacteria, VLP, and chlorophyll *a* concentrations were all indicative of a mesotrophic or eutrophic ecosystem. The relatively high percentage of total cells that were dividing at any given time point at any site (ranging from $8.6 \pm 3.1\%$ at Far-2 site to $17.0 \pm 1.1\%$ at the Fish Pens) indicated that the enriched DOC observed at all sites (ranging from $77.5 \pm 0.8\ \mu M$ at Far-1 to $189.8 \pm 1.0\ \mu M$ at Fish Pens) was readily utilizable to the microbial community to support growth. A truly oligotrophic environment normally exhibits DOC concentration of ~ 40 uM [23], which is much lower than our relatively well-flushed sites.

Coral-Associated Bacteria Communities Shift in Response to Effluent

One fragment out of the 40 transplants died at site Near-2 before the 5-day sampling point (T-5 days). All other fragments appeared visually healthy throughout the duration of the experiment. Large community shifts in the coral-associated bacteria samples were observed via denaturing gradient gel electrophoresis (DGGE) analysis five days after transplantation. Profiles were most radically different at the Near sites. However, changes were observed at all sites including the Reference, indicating a transplantation effect in addition to an effluent effect. By 22 days post-transplantation (T-22 days), all coral-associated microbial community profiles had shifted back towards their original profile patterns observed at T-0 (Fig. 2).

All corals transplanted to the Near sites at T-5 days showed, via DGGE profiles, a prominent band that matched the 16S rRNA sequence of the coral black band disease (BBD)-associated [24] bacterium *Desulfovibrio* sp. ([15]; Accession No. AY750147.1) with 100% similarity and 100% query coverage. The majority of the bands for this phylotype were present in coral fragments transplanted at sites Near-1 and Near-2 at T-5 days. This genus was not detected with genus-specific polymerase chain reaction (PCR) primers in the bacterial communities associated with the

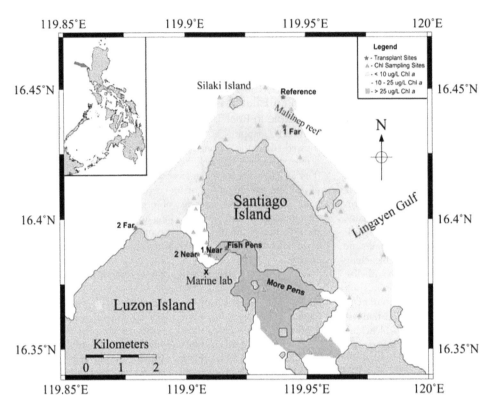

Figure 1. A map of the study sites and surface water *in vivo* chlorophyll *a* measurements averaged over 4 consecutive weekly samplings. Note that transplant site "Fish Pens" did not have corals placed there since no live coral currently exists at that site. It was the location of all water and sediment sampling for the experiment as a representative fish pen.

water column at any site and was only detectable in sediments from sites Near-2 and Fish Pens. Sequences related to the genus were not present in clone libraries for water or sediment bacterial communities. No coral fragment had detectable levels at T-0, but at T-5 days all corals had detectable levels at most sites, identified by genus-specific PCR analysis (Table 2). By T-22 days, only fragments at the Near sites had detectable levels of genus-specific PCR product remaining, the bands in DGGE were greatly diminished or absent, and no sequences were present in clone libraries from that time point.

Evidence suggested that water-associated bacteria from the Fish Pens had the ability to colonize corals exposed to effluent-rich waters. A phylotype most closely related to an uncultured *Roseobacter* (Accession No. EU627982.1) comprised enough of the water-associated bacteria community at the Fish Pens to be detectable via DGGE at T-0, but was completely absent in coral-associated bacteria communities. However, five days after transplantation, this band (verified through sequence analysis) became strongly visible in corals transplanted to Near sites but not in the fragments of the same colony that were transplanted to Far sites (Fig. 3). This same sequence was observed in January 2007 clone libraries from free-living bacteria of fish pen water [17], suggesting that this bacterium may be a persistent feature of fish-pen influenced seawater in Bolinao.

Clone libraries of coral fragments, water, and sediment revealed similar patterns to those seen via DGGE analysis. Very little overlap of phylotypes was observed among water, coral, and sediment bacteria communities. Of the 1172 sequences analyzed from all libraries, there were several noteworthy patterns. Eighteen sequences from the family Vibrionaceae were observed—14 of which came from corals at various site only during the T-5 days

time point, while the rest were observed in sediments from the Near sites. Nineteen sequences from the genus *Desulfovibrio* were observed in coral clone libraries– all but two occurrences were from T-5 days and 11 were from fragments at Near sites. Thirty sequences from an order of anaerobes, Clostridiales, were also present among the coral libraries—all but one occurrence were from T-5 days and all but six were from Near sites. Only one Clostridiales sequence was observed in a non-coral library: Far-1 sediment at T-0. Four occurrences of the anaerobic sulfite-oxidizing genus *Sulfitobacter* were observed in a coral at the Reference site (n = 1) at T-5 days, and at Near-1 (n = 3) at T-22 days. Two genera were coral-specific, appearing in coral libraries from all time points at all sites: *Alcanivorax* (n = 18) and *Halomonas* (n = 334). Two instances of sequences belonging to genera predominantly comprised of pathogens (human, porcine, bovine, feline, canine, and equine) occurred on corals at the Near sites five days post-transplantation. *Arcobacter*, which is generally associated with feces (human, porcine, and bovine) and has been found in sewage-contaminated waters, was present on a coral at site Near-2 while *Fusobacterium*, often associated with necrotic lesions and a variety of mammal diseases, was present on a coral at site Near-1. Sequences from the class Spirochaete, which have been found previously in diseased coral samples in association with BBD and White Plague-like syndromes [2,4,25], occurred only at T-5 days, mostly at Near sites (Fig. 4).

RDP-10 LibCompare analysis indicated that T-5 days libraries for fragments from different coral colonies at Near sites were more similar to each other than they were to their own original colonies at T-0 or T-22 days (p ≪ 0.05). In particular, the abundances of Bacteriodetes, Deltaproteobacteria, and Firmicutes were higher at T-5 days than any other time point (Fig. 4). Within a colony, the

Table 1. Comparison of Water Characteristics.

Site	Average values from May and June 2008 (T0, T-5days, T-22days)								Values from January 2007 (Garren et al. 2008)			
	Free-living Bacteria (cells x 10⁶/ml±SE)	Virus-like Particles (VLP x 10⁶/ml±SE)	Extracted Chl a (µg/L±SE)	Dissolved Organic Carbon (µM±SE)	Frequency of Dividing Cells (% total abundance dividing±SE)	Cyanobacteria Abundance (cells x 10⁴/ml±SE)	Vibrio (TCBS media) plate counts (CFU/ml±SE)*	Kanamycin resistant plate counts (25 µg/ml CFU/ml±SE)*	Free-living Bacteria Abundance (cells x 10⁶/ml±SE)	Virus-like Particles (VLP x 10⁶/ml±SE)	Extracted Chl a (µg/L±SE)	Dissolved Organic Carbon (µM±SE)
Ref	1.51 ± 0.03	8.0 ± 0.8	2.8 ± 1.2	93.4 ± 17.2	11.1 ± 2.9	11.3 ± 1.0	60 ± 0	398 ± 43				
Far-1	1.94 ± 0.05	14.2 ± 0.9	3.7 ± 0.1	77.5 ± 0.9	11.9 ± 2.0	7.5 ± 0.8	40 ± 12	1765 ± 1495	0.54 ± 0.03	10 ± 0.7	0.25 ± 0.03	69.7 ± 1.3
Far-2	1.76 ± 0.04	14.8 ± 1.0	4.1 ± 1.6	82.0 ± 4.6	10.5 ± 2.8	9.9 ± 1.2	53 ± 7	285 ± 95				
Near-1	7.43 ± 0.18	81.4 ± 4.3	63.4 ± 12.9	143.6 ± 1.4	12.8 ± 0.8	63.4 ± 4.6	160 ± 23	1008 ± 93	0.61 ± 0.06	70 ± 3	4.5 ± 0.2	141 ± 2.9
Near-2	5.71 ± 0.15	75.9 ± 3.6	60.2 ± 15.3	158.0 ± 2.5	12.4 ± 1.9	52.2 ± 5.9	220 ± 23	1218 ± 238				
Pens	10.25 ± 0.50	112.3 ± 4.7	99.8 ± 25.2	189.8 ± 1.0	16.7 ± 1.4	64.2 ± 8.0	340 ± 46	1413 ± 73	0.99 ± 0.03	61 ± 7	10.3 ± 0.2	162 ± 18.5

Water Characteristics from May/June 2008 compared to January 2007. (* signifies T0 data only).

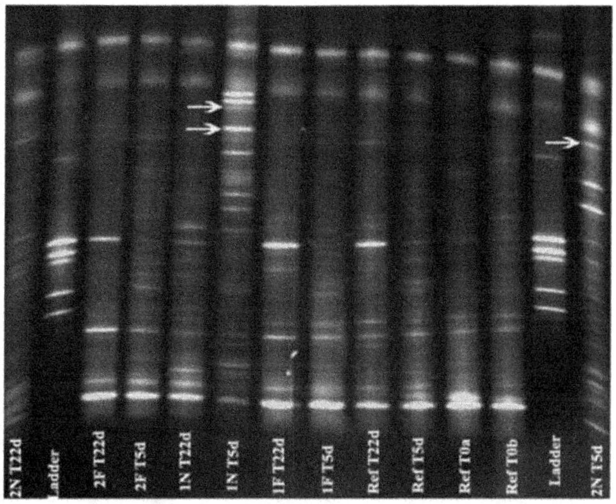

Figure 2. Denaturing gradient gel electrophoresis (DGGE) image of one coral colony (#1) at all sites (Far-1&2, Near-1&2, and Reference) across all time points (T-0, T-5 days, T-22 days). Arrows indicate *Desulfovibrio* bands that were sequence verified. There were two separate fragments sampled from each colony at T0. Both samples are shown on this gel as T0 a and b. Samples from the high effluent sites (Near-1&2) at T-5 days were the only visible *Desulfovibrio* bands.

phylotype distributions were fairly similar among Reference, Far-1 and Far-2 for all time points and Near-1 & 2 for T-0 and T-22 days. The other clone library pattern that revealed itself was an apparent increase in diversity at the high effluent sites at T-5 days. The number of classes represented at the Near sites increased at T-5 days, while the number decreased at the Reference and Far sites (Fig. 4). The Shannon-Wiener index of diversity also increased by 1.7954, from 1.5358 at T-0 to an average of 3.3312 at T-5 days, at the high effluent sites. In contrast, the values for corals at the low effluent sites increased by an average of 0.5140, and by 0.3636 at the reference site.

Discussion

There are several lines of evidence supporting our hypothesis that fish pen effluent induces changes in coral-associated bacterial

Table 2. *Desulfovibrio* Presence/Absence.

Coral & Time / Site	Colony 1		Colony 2		Colony 3		Colony 4	
	T-5d	T-22d	T-5d	T-22d	T-5d	T-22d	T-5d	T-22d
Ref	+	−	−	−	+	−	+	−
Far-1	−	−	+	−	+	−		
Far-2	+	−	+	−	+	−	+	
Near-1	+	+	+	+	+	+	+	+
Near-2	+	+	+	died	+	+	+	−

The presence (+) or absence (−) of *Desulfovibrio* as detected by genus-specific PCR amplification. No amplified product was detected in any colony at T0, so this time point is omitted from the table for simplicity. The T-22 days fragment for colony 2 died before T-5 days, and was the only mortality for the experiment. *Desulfovibrio* returned to undetectable levels for all colonies at all low-effluent sites by T-22 days (Ref, Far-1&2), yet remained detectable in most colonies at the high-effluent sites (Near-1&2).

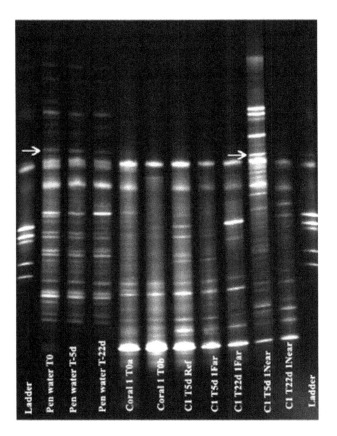

Figure 3. DGGE image showing *Roseobacter* **band (>97% sequence similarity) at all time points in the Fish Pen water, but only T-5 days at site Near-1 for coral 1.**

community composition. Both DGGE and clone library community composition profiles from all coral colonies at T-5 days at high effluent (Near) sites were more similar to other colonies transplanted to those sites than they were to the communities associated with the fragments from their own colonies transplanted to lower effluent (Far or Reference) sites or sampled from the high effluent sites at the later time point (T-22 days). Corals at high effluent sites were also the only samples with detectable levels of *Desulfovibrio* genus-specific PCR products at the final time point, by which time the same coral clones at lower effluent sites had recovered their original bacterial community composition. This suggests that high effluent exposure was either a source or a facilitator for change in the coral-associated bacterial communities.

The ability of a fish pen-associated *Roseobacter* to become a prominent member of the DGGE profiles from corals at high effluent sites during the first few days post-transplantation was consistent with our hypothesis that corals could become colonized by aquaculture-associated bacteria. Sequences typing pathogenic genera (*Arcobacter* and *Fusobacterium*) generally associated with humans that became associated with the corals were only present at high effluent sites. The fish pens are tended by people who stay in structures directly above the pens. These facilities lack any form of septic or plumbing systems. The presence of potential human pathogens associating with corals suggests that these fish pens represent more than just a source of fish-associated microbes onto the reefs, and that corals are sensitive to these inputs. Beyond the implication for corals, the human health implications of these pathogenic bacteria should not be ignored. These waters are also used for swimming and recreation, thus current aquaculture practices may require improvement for both human and coral health.

A noteworthy observation during this experiment was the restoration of the microbial communities by day 22 towards their

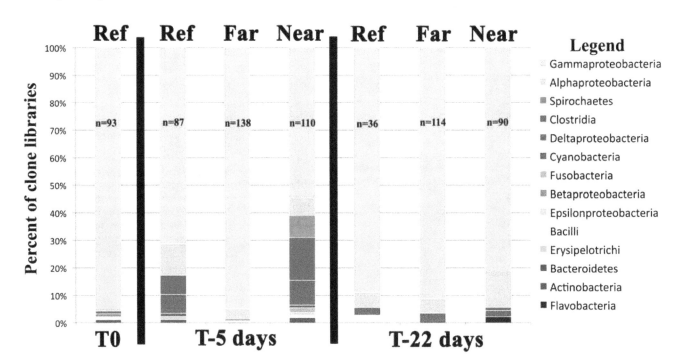

Figure 4. The distribution of phylotypes by class from 16S rRNA clone libraries. Libraries have been pooled by site (i.e. libraries from all fragments at a given time point from Far-1 and Far-2 are represented by "Far" and the same is true for "Near") and each individual time point (T0, T-5 days, T-22 days) is shown separately. The number of sequences represented is denoted as "n = ". Spirochaetes, previously seen only in corals infected with diseases, and Clostridia are both present even though fragments showed no visible signs of disease. Spirochaetes sequences are only present at T-5 days and predominantly at high effluent sites.

original T-0 composition across all sites, despite the surrounding water remaining essentially unchanged. Our original hypothesis was that the low effluent site corals would have the ability to maintain or restore their bacteria community composition, but we observed a much better recovery than anticipated in the high effluent site corals as well. This observation suggests that there is a preferred/stable coral-associated composition that the bacterial community tends toward, and generates hypotheses regarding mechanisms by which a conserved microbial community can be maintained. Whether the coral itself [26–29], its microbial associates [30,31], some combination of the two, or another component or set of components of the holobiont is responsible for restoring the bacterial community balance remains to be understood. However, the observation of the community changing in response to effluent exposure and subsequently recovering *without* effluent reduction suggests that the mechanisms for bacteria community control are strong. Illuminating these mechanisms will be a critical step towards understanding the role of bacterial communities in coral health and disease, and facilitating predictions of reef-wide responses to various stressors.

The fact that the coral-associated bacterial communities at all sites, including the Reference, changed to some degree at T-5 days demonstrated that there was an effect of transplantation. The appearance of *Desulfovibrio* bacteria, generally associated with coral black band disease (BBD), on all coral samples and at all sites at T-5 days was in agreement with our hypothesis that stress, in this case induced by transplantation, would allow corals to be opportunistically colonized by potentially harmful bacteria. Due to the geography and hydrography of the area, the two Near sites had to be located closer to each other than the two Far sites. Thus, one might expect the Near sites to be more similar to each other than the Far sites would be. However, the RDP LibCompare results analyzing clone libraries showed that microbial communities from all fragments at both Near sites at T-22 days were more similar to the communities associated with any fragment from the Far or Reference sites (from any time point) than they were to the communities associated with fragments from the very same colony sampled at T-5 days at that same Near site.

Unlike the *Roseobacter* that was transferred from the fish pen effluent to the coral, the source of *Desulfovibrio* bacteria seems more likely to be the coral itself. Given that *Desulfovibrio* phylotypes were not detectable in water or sediment samples from most sites, yet were detectable in corals at all sites during at least one time point—as it has been detected in corals from around the globe [2,32]— we hypothesize that this potentially detrimental phylotype was originally associated with the corals in undetectable numbers rather than with the surrounding environment. It seems likely that the stress of transplantation, through an unidentified mechanism, allowed these bacteria to proliferate rapidly. However, while transplantation likely initiated the proliferation, exposure to fish pen effluent seems the most likely cause for the continued proliferation of *Desulfovibrio*, as they were visible only at high effluent sites by the final time point. The same mechanisms controlling *Desulfovibrio* proliferation may also be responsible for the proliferation of Clostridia and Spirochaetes in this experiment. The two separate likely sources of newly colonizing bacteria, aquaculture effluent and the coral itself, highlights the potential for multiple pathways of microbial colonization to occur simultaneously. Investigating the ecological interactions among the coral animal, its associated microbiota, and water column-derived microbial invaders may shed light on synergistic roles played by host immunity [26–29,33] and microbe-microbe interactions [9,21,30,31] to protect the holobiont from undesired microbial colonization.

Though all but one of the 40 transplanted coral fragments survived the experiment, it is a worthwhile exercise to consider other potential trajectories this experiment could have taken and the possibilities for these fragments after the final time point. It seems that many of them, especially at the Near sites, may have been particularly vulnerable 5 days after transplantation given the relative increase of sequences affiliated with potential pathogens in molecular profiling results. A previous study showed decreased survival of juvenile corals at this site, and it is possible our corals could have followed a similar fate [22]. Had there been any additional stressor (such as increased water temperature or a major storm event), it is possible that we would have observed a drastically different survival rate. Also, the short time scale of this experiment cannot be extrapolated to imply that these fragments would have survived a long-term transplantation to these sites. The return of the bacteria communities towards their original structure seems to be a positive indication of more resilient capabilities than previously observed; however, the fact that some of the 5 day time point bacteria (such as members of *Desulfovibrio* and Clostridiales) were still detectable at the end of the experiment in corals at high effluent sites should not be overlooked. It is certainly possible that the persistence of these phylotypes could be an early indicator of future disease [34]. Also, the potential physiological costs –such as reduced growth, unsuccessful reproduction, or impaired ability to heal from wounds— of returning the bacteria communities to their original state after disturbance should be considered. There may be hidden costs of this response that we were unable to measure in this experiment.

Observing dramatic changes in community composition over the short time scale of 5 days calls attention to the sensitivity and susceptibility of these corals to physical stress and organic matter enrichment. The lack of visible signs of disease during this stress event raises the need to monitor and observe corals at smaller scales than current monitoring techniques allow. For instance, it may be informative to incorporate microscopy techniques into routine coral reef monitoring protocols. Studying the microbial ecology of sub-lethal and sub-visible effects of stress may provide some of the mechanistic links we need to understand and predict physiological responses of corals to various scenarios. These data provide insight into some of the microbial biodiversity that may be integral in resilience for these corals, and raise the question of whether or not increased diversity of associated bacteria is a desirable state for corals. Our data suggest that an increase in diversity could be indicative of, or correlated to, stress events. These data provide a broader understanding of potentially desirable and undesirable groups of bacteria through time, but they do not yet provide insight into the spatial heterogeneity at the colony scale or connectivity of these microbial consortia. To gain insight into those two resilience criteria, we must shift towards incorporating and developing more sophisticated visual methods of analysis. Some methods that could be incorporated in coral microbiological studies more routinely to this end include fluorescence *in situ* hybridization (FISH), confocal microscopy, and atomic force microscopy. Developing the proper tools to monitor these microscale microbial interactions *in situ* will increase our ability to understand and manage coral reefs in the face of a rapidly changing environment. On a positive note, the ability of the coral-associated bacterial communities to rebound and recover after this severe stress event, without any improvement in the water quality, suggests that corals may be able to survive short, severe stress events if they are not pushed beyond a threshold. The more we learn about individual coral resilience capabilities, the better we can develop management practices integrated across multiple spatial scales for the long-term success of reefs.

Methods

Study Sites

The study took place in the Bolinao, Pangasinan province, Philippines (16N, 119E). Milkfish (*Chanos chanos*) mariculture has been actively practiced in the area since 1995 [35]. The farms employ net pens measuring roughly 10 m×10 m×8 m with a stocking density of approximately 50,000 fish per pen and a pen density of 10 per hectare [22].

A site within the channel between Luzon and Santiago Island was chosen as a representative fish pen sampling site (referred to as "Fish Pens") for water and sediment collection. Two reef sites <1 km from the Fish Pens were selected as sites Near-1 and Near-2 (high effluent exposure) and two reef sites >5 km from the Fish Pens were selected as sites Far-1 and Far-2 (low effluent exposure). The Reference site was ~10 km from the Fish Pens on the outside of the Malilnep reef crest that is regularly flushed by the South China Sea (Fig. 1).

Experimental Design

Four colonies of *Porites cylindrica* were selected from the Reference Site. Twelve branches from each colony were gently removed using wire cutters. Two fragments from each colony were immediately flash frozen as the T-0 sampling. Two additional fragments were transported and affixed in place at each of the following transplantation sites: Reference, Far-1, Near-1, Far-2, Near-2 (note: corals were not transplanted directly at Fish Pens because live coral does not currently exist in that location). The fragments were affixed at each site in flexible plastic tubing with their source colony labeled with Dymo tape and zip-tied to plastic mesh tables installed 1 m above the substrate on rebar supports (Fig. S1). Three replicate tables were installed at each site at 2–3 m depth with the fragments from each colony randomized among the three tables. One fragment from each colony was collected from each site 5 days after transplantation and the second fragment was collected 22 days after transplantation. All fragments were monitored at each visit for visible signs of stress or disease. The experiment took place between May 19 and June 10, 2008.

Sample Collection

Four liter water samples were collected directly above the transplant tables at each site during each sampling period. They were kept cool and shaded until they were processed in the lab within 4 hr of collection. Triplicate sediment cores were collected with sterile 10 ml syringes from each site at T-5 days. Coral fragments were collected in sterile WhirlPaks (Nasco, USA), rinsed with sterile seawater, wrapped in aluminum foil, and flash frozen in liquid nitrogen.

Sample preservation for DNA

A particle fraction was operationally defined as 3 μm being the minimum particle size [36,37]. To preserve water samples for DNA extraction, 200 ml seawater were pre-filtered through a 3 μm-pore size filter (47 mm diameter, polycarbonate, Whatman) and the filtrate was put onto a 0.22 μm-pore size filter (47 mm diameter; Supor 200; Pall Corp.). The 0.22 μm filters were stored at −20°C in 250 μl of RNALater^{TM} (Ambion, USA). To preserve coral samples for DNA extraction, frozen coral fragments were thawed on ice, airbrushed in 2 ml sterile seawater to remove tissue and mucus along the length of the fragment except the portion that was in contact with the mounting tube, and stored at −20°C. Sediment cores were kept cold until they could be flash frozen in liquid nitrogen within 2 hr of collection and stored at −20°C.

Chlorophyll

Chlorophyll *a* concentrations were used as a proxy for phytoplankton biomass. Chlorophyll *a* was measured both *in vivo* and from extracted samples. *In vivo* measurements were made weekly using a hand-held fluorometer (Aquafluor, Turner Designs Inc., USA) for four weeks at 20 stations (Fig. 1) around Santiago Island to track the influence fish pen effluent. This hand-held *in vivo* fluorometer, coupled with a hand-held Global Positioning System (GPS) unit, allowed for real-time data collection regarding the depth and location of these hotspots. Extracted chlorophyll *a* samples were taken from a subset of these sites each week to create a standard curve that allowed the *in vivo* Relative Fluorescence Unit (RFU) readings to be translated into μg Chl *a*/L. Samples for extracted readings were also taken at each transplant sampling time point. For all extracted samples, seawater (50 ml) was filtered onto Whatman GF/F filters (25 mm) and stored in the dark at −20°C until processing. Samples were processed using the method described by Holm-Hansen et al. (1967). Briefly, filters were extracted in 5–10 ml methanol for two hours and fluorescence measured using a Turner Designs 700 fluorometer. Extracts were acidified and remeasured to determine total phaeophytin.

Dissolved organic carbon

Seawater aliquots (30 ml) were filtered through Whatman GF/F filters (25 mm diameter). The filtrates were acidified and analyzed for total organic carbon (TOC) content (Scripps Institution of Oceanography, Aluwihare Laboratory) on a Shimadzu TOC-V instrument fitted with an autosampler. Briefly, the concentration of each sample was calculated from an average of four 100 μL injections using a five-point potassium hydrogen phthalate standard curve and certified reference materials (courtesy of Dennis Hansell, Rosenstiel School of Marine and Atmospheric Science).

Bacteria and VLP abundances

Water samples were fixed with a 2% final concentration of 0.02 μm filter sterilized formaldehyde. The "particle" fraction was removed by pre-filtering through a 3 μm polycarbonate filter. Two to three milliliters of filtrate were put on a 0.22 μm polycarbonate filter to collect the "free-living" fraction of the sample. These samples were dried, wrapped in aluminum foil, and stored at −20°C. A quarter of each 0.22 μm filter was prepared for epifluorescence microscopy using Vectashield mounting medium containing DAPI (Vector Laboratories, USA), while the remainder of the filter was archived. Bacteria were counted in 20 haphazardly chosen fields of view (with one 100 μm×100 μm grid per field) at 1000×magnification on an Olympus IX-50 microscope. Ten haphazardly chosen fields were photographed at 1000x on a Nikon, Eclipse TE-2000U using NIS Elements software program to count the frequency of dividing cells (FDC) following the protocol described by Hagstrom et al. [38].

Plate counts were performed to quantify the number of culturable bacteria. Water samples (50–200 μl) from each site were spread onto Thiosulfate Citrate Bile Sucrose (TCBS) agar to select for *Vibrios* and Zobell Marine agar with 25 μg/ml Kanamycin to select for Kanamycin-resistant bacteria. Plates were incubated overnight at 30°C and colonies were counted 12 hr later.

Virus-like particles (VLPs) were quantified on 0.02 Anodisc filters (Whatman, USA) that had 500–1,000 μl of sample filtered through them. Filters were stained with 10x SYBR Gold (Invitrogen, USA) for 15 min, dried, mounted to slides using VectaShield (Vector Laboratories, USA). A Nikon Eclipse TE-

2000U and NIS Elements software program were used to image and count 20 haphazardly chosen fields per filter.

PCR amplification:

DNA was extracted from water filters, coral, and sediment samples using the UltraClean™ Soil Kit (MoBio).

To amplify community 16S rRNA genes for denaturing gradient gel electrophoresis (DGGE) analyses, the variable V3 region was targeted using primer 341f with a GC clamp (5′-CGCCCGCCGCGCGCGGCGGGCGGGGCG GGGGCAC-GGGGGGCCTACGGGAGGCAGCAG-3′) and primer 534r (5′-ATTACCGCGGCTGCTGG-3′) [39] following the amplification protocol described by Garren et al. [17]. The PCR products were separated by electrophoresis on a 1.0% agarose gel for confirmation of ~200 bp product and quantified using PicoGreen (Molecular Probes) following manufacturers instructions using a SpectraMax M2 plate reader (Molecular Devices, USA).

To amplify from all samples for clone libraries of community 16S rDNA genes, a nested PCR was performed using the universal primer 27F (5′-AGAGTTTGATCM TGGCTCAG-3′) and the Eubacterial-specific primer 1492R (5′-TACGGYTACCTT GT-TACGACTT-3′; [40]) in a 15 cycle amplification using a 55°C annealing temperature. One microliter of the product was used as the template for the second PCR reaction. Primers 341-forward (without GC clamp [39],) and 981-reverse (41) were used under the same conditions as for DGGE. Three microliters of the final PCR product was used as template for ligation and clone library construction (see below).

To assess the prevalence of this genus in samples that may have had abundances too low to detect via DGGE, we designed specific to probe environmental DNA samples. Genus specific primers were designed using NCBI's Primer-BLAST for a 576 base pair region of *Desulfovibrio spp.* to look for these phylotypes previously associated with Black-Band Disease (BBD) in corals. Primers 645F (CAAGCCCCCAACA CCTAGTA) and 1220R (TACCGTGG-CAACGATGAATA) were used in the following PCR protocol: an initial 94°C denaturing step for 5 min was followed by 34 cycles of amplification (45 sec denaturation at 94°C; 45 sec at 53.5°; 2 min extension at 72°C), and a final extension of 10 min at 72°C. Primer specificity was verified via cloning of PCR product.

Cloning

PCR products from half of the fragments analyzed via DGGE were cloned using Invitrogen's pCR4-TOPO for sequencing kit with Top-10 chemically competent cells following manufacturer's instructions. 48 or 96 colonies were picked for each cloning reaction and transferred into LB + Kanamycin media containing 10% glycerol for a 12 h incubation at 37°C, then submitted to a commercial sequencing service (Agencourt Genomic Services, MA, USA). Briefly, inserts were amplified using the M13 forward primer, sequencing reactions were performed using BigDye Terminator v3.1 (Applied Biosystems), and sequences were delineated using a PRISM™ 3730xl DNA Analyzer (Applied Biosystems).

Denaturing gradient gel electrophoresis (DGGE) analyses

PCR products (~200 bp) were separated by GC-content using a hot-bath DGGE system (CBS Scientific). One hundred nanograms of each PCR product were loaded onto 8.0% polyacrylamide gels in 0.5x TAE buffer (20 mM Tris, 10 mM sodium acetate, 0.5 mM Na$_2$EDTA, pH 8.2) with top-to-bottom denaturing gradients of 35–65% formamide and urea (100% denaturant being 40% [v/v] formamide and 7M urea). Electrophoresis was run at 55 V for 18 h at 60°C in 0.5x TAE. After electrophoresis, the gels were stained for 15 min in 0.5x SYBR Gold (Invitrogen) in TAE buffer. Gels were then imaged using a UVP Epi-chemi Darkroom with a charge coupled device (CCD) camera.

Community DGGE profiles were compared across sites and sample type using a standard ladder made from 100 ng each of two isolates collected from the fish pen site in January 2007 [17]. Bands of the same relative position among sites or sample type were identified. These select bands were excised from the gel and eluted using the protocol described by Long and Azam [42]. The bands were re-amplified and run on a new gel to confirm the position relative to a known standard. For those products of the correct position, the original excised band was again amplified (without GC-clamp) and the product used for cloning reactions (see Cloning).

Sequence analyses

Sequence data were trimmed, cleaned, and aligned using Sequencher 4.5. Final clean sequences were exported in FASTA format and imported into the Ribosome Database Project (RDP-10) online portal [43]. RDP was used to align sequences and classify sequences and to compare libraries. Where sequences from the various sample types clustered together, those sequences were hand aligned in Sequencher 4.5 and blasted in NCBI to determine putative identification. Sequences were submitted to GenBank (accession numbers GQ412750 - GQ413933). 99% similarity clusters were used to identify sequences affiliated with a specific organism to eliminate bias induced by Taq polymerase error [44]. Values for the Shannon-Weiner Index of diversity were calculated for clone libraries using FastGroupII program [45].

Acknowledgments

We are especially grateful to our hosts at the University of the Philippines, Dr. Edgardo Gomez, Dr. Wolfgang Reichardt, Ms. Joanne Tiquio, Ms. Christine Aguila, Dr. Mark Arboleda, and Ms. Miahnie Pueblos. Thank you to Ms. Courtney Couch, Mr. Mark Defley, and Mrs. Morgan Mouchka for their critical logistical and field support. We thank Drs. Marah Hardt and Krystal Rypien for insightful edits. Many thanks to Mr. John Goeltz and Mr. Trevor Davies.

Author Contributions

Conceived and designed the experiments: MG LR JG CDH FA. Performed the experiments: MG LR JG. Analyzed the data: MG. Contributed reagents/materials/analysis tools: CDH FA. Wrote the paper: MG. Edited the paper: LR CDH FA.

References

1. Rohwer F, Seguritan V, Azam F, Knowlton N (2002) Diversity and distribution of coral-associated bacteria. Marine Ecology-Progress Series 243: 1–10.

2. Frias-Lopez J, Zerkle AL, Bonheyo GT, Fouke BW (2002) Partitioning of bacterial communities between seawater and healthy, black band diseased, and dead coral surfaces. Applied and Environmental Microbiology 68: 2214–2228.

3. Rohwer F, Breitbart M, Jara J, Azam F, Knowlton N (2001) Diversity of bacteria associated with the Caribbean coral *Montastraea franksi*. Coral Reefs 20: 85–91.

4. Pantos O, Cooney RP, Le Tissier MDA, Barer MR, O'Donnell AG, et al. (2003) The bacterial ecology of a plague-like disease affecting the Caribbean coral *Montastrea annularis*. Environmental Microbiology 5: 370–382.

5. Ritchie KB, Smith GW (1995) Preferential carbon utilization by surface bacterial communities from water mass, normal, and white-band diseased *Acropora cervicornis*. Molecular Marine Biology and Biotechnology 4: 345–352.

6. Koren O, Rosenberg E (2006) Bacteria associated with mucus and tissues of the coral *Oculina patagonica* in summer and winter. Applied and Environmental Microbiology 72: 5254–5259.

7. Bourne D, Iida Y, Uthicke S, Smith-Keune C (2008) Changes in coral-associated microbial communities during a bleaching event. Isme Journal 2: 350–363.

8. Richardson LL (1998) Coral diseases: what is really known? Trends in Ecology & Evolution 13: 438–443.

9. Rosenberg E, Koren O, Reshef L, Efrony R, Zilber-Rosenberg I (2007) The role of microorganisms in coral health, disease and evolution. Nature Reviews Microbiology 5: 355–362.

10. Carpenter KE, et al. (2008) One-third of reef-building corals face elevated extinction risk from climate change and local impacts. Science 321: 560–563.

11. Fabricius KE (2005) Effects of terrestrial runoff on the ecology of corals and coral reefs: review and synthesis. Marine Pollution Bulletin 50: 125–146.

12. Remily ER, Richardson LL (2006) Ecological physiology of a coral pathogen and the coral reef environment. Microbial Ecology 51: 345–352.

13. Harvell D, Jordan-Dahlgren E, Merkel S, Rosenberg E, Raymundo L, et al. (2007) Coral disease, environmental drivers, and the balance between coral and microbial associates. Oceanography 20: 172–195.

14. Mumby PJ, Hastings A (2008) The impact of ecosystem connectivity on coral reef resilience. Journal of Applied Ecology 45: 854–862.

15. Nystrom M, Graham NAJ, Lokrantz J, Norstrom AV (2008) Capturing the cornerstones of coral reef resilience: linking theory to practice. Coral Reefs 27: 795–809.

16. Graham NAJ, Wilson SK, Jennings S, Polunin NVC, Robinson J, et al. (2007) Lag effects in the impacts of mass coral bleaching on coral reef fish, fisheries, and ecosystems. Conservation Biology 21: 1291–1300.

17. Garren M, Smriga S, Azam F (2008) Gradients of coastal fish farm effluents and their effect on coral reef microbes. Environmental Microbiology 10: 2299–2312.

18. Rowan R, Knowlton N, Baker A, Jara J (1997) Landscape ecology of algal symbionts creates variation in episodes of coral bleaching. Nature 388: 265–269.

19. Berkelmans R, van Oppen MJH (2006) The role of zooxanthellae in the thermal tolerance of corals: a 'nugget of hope' for coral reefs in an era of climate change. Proceedings of the Royal Society B-Biological Sciences 273: 2305–2312.

20. Knowlton N, Rohwer F (2003) Multispecies microbial mutualisms on coral reefs: The host as a habitat. American Naturalist 162: S51–S62.

21. Rosenberg E, Kellogg CA, Rohwer F (2007) Coral Microbiology. Oceanography 20: 146–154.

22. Villanueva RD, Yap HT, Montano MNE (2005) Survivorship of coral juveniles in a fish farm environment. Marine Pollution Bulletin 51: 580–589.

23. Torreton JP, Dufour P (1996) Bacterioplankton production determined by DNA synthesis, protein synthesis, and frequency of dividing cells Tuamotu atoll lagoons and surrounding ocean. Microbial Ecology 32: 185–202.

24. Viehman S, Mills DK, Meichel GW, Richardson LL (2006) Culture and identification of *Desulfovibrio* spp. from corals infected by black band disease on Dominican and Florida Keys reefs. Diseases of Aquatic Organisms 69: 119–127.

25. Sekar R, Kaczmarsky LT, Richardson LL (2008) Microbial community composition of black band disease on the coral host *Siderastrea siderea* from three regions of the wider Caribbean. Marine Ecology Progress Series 362: 85–98.

26. Gochfeld DJ, Aeby GS (2008) Antibacterial chemical defenses in Hawaiian corals provide possible protection from disease. Marine Ecology Progress Series 362: 119–128.

27. Mydlarz LD, Holthouse SF, Peters EC, Harvell CD (2008) Cellular Responses in Sea Fan Corals: Granular Amoebocytes React to Pathogen and Climate Stressors. PLoS One 3:Article No.: e1811.

28. Kvennefors ECE, Leggat W, Hoegh-Guldberg O, Degnan BM, Barnes AC (2008) An ancient and variable mannose-binding lectin from the coral *Acropora millepora* binds both pathogens and symbionts. Developmental and Comparative Immunology 32: 1582–1592.

29. Geffen Y, Rosenberg E (2005) Stress-induced rapid release of antibacterials by scleractinian corals. Marine Biology 146: 931–935.

30. Rypien KL, Ward JW, Azam F (In press) Antagonistic interactions among coral-associated bacteria. Environmental Microbiology, doi:10.111/j.1462-2920.2009.02027.x.

31. Ritchie KB (2006) Regulation of microbial populations by coral surface mucus and mucus-associated bacteria. Marine Ecology-Progress Series 322: 1–14.

32. Cooney RP, Pantos O, Le Tissier MDA, Barer MR, O'Donnell AG, et al. (2002) Characterization of the bacterial consortium associated with black band disease in coral using molecular microbiological techniques. Environmental Microbiology 4: 401–413.

33. Palmer CV, Mydlarz LD, Willis BL (2008) Evidence of an inflammatory-like response in non-normally pigmented tissues of two scleractinian corals. Proceedings of the Royal Society B-Biological Sciences 275: 2687–2693.

34. Raymundo LJH (2002) Effects of Conspecifics on Reciprocally-Transplanted Fragments of the Scleractinian Coral *Porites attenuata* Nemenzo in the Central Philippines. Coral Reefs 20: 263–272.

35. Holmer M, Marba N, Terrados J, Duarte CM, Fortes MD (2002) Impacts of milkfish (*Chanos chanos*) aquaculture on carbon and nutrient fluxes in the Bolinao area, Philippines. Marine Pollution Bulletin 44: 685–696.

36. Bidle KD, Fletcher M (1995) Comparison of free-living and particle-associated bacterial communities in the Cheasapeake Bay by stable low-molecular-weight RNA analysis. Applied and Environmental Microbiology 61: 944–952.

37. Crump BC, Armbrust EV, Baross JA (1999) Phylogenetic analysis of particle-attached and free-living bacterial communities in the Columbia river, its estuary, and the adjacent coastal ocean. Applied and Environmental Microbiology 65: 3192–3204.

38. Hagstrom A, Larsson U, Horstedt P, Normark S (1979) Frequency of dividing cells, a new approach to the determination of bacterial-growth rates in aquatic environments. Applied and Environmental Microbiology 37: 805–812.

39. Muyzer G, Dewaal EC, Uitterlinden AG (1993) Profiling of complex microbial populations by denaturing gradient gel electrophoresis of polymerase chain reaction amplified genes coding for 16S ribosomal RNA. Applied and Environmental Microbiology 59: 695–700.

40. Weisburg WG, Barns SM, Pelletier DA, Lane DJ (1991) 16S ribosomal DNA amplification for phylogenetic study. Journal of Bacteriology 173: 697–703.

41. Lane DJ, Pace B, Olsen GJ, Stahl DA, Sogin ML, et al. (1985) Rapid determination of 16S ribosomal RNA sequences for phylogenetic analyses. Proceedings of the National Academy of Sciences of the United States of America 82: 6955–6959.

42. Long RA, Azam F (2001) Microscale patchiness of bacterioplankton assemblage richness in seawater. Aquatic Microbial Ecology 26: 103–113.

43. Wang Q, Garrity GM, Tiedje JM, Cole JR (2007) Naive Bayesian classifier for rapid assignment of rRNA sequences into the new bacterial taxonomy. Applied and Environmental Microbiology 73: 5261–5267.

44. Acinas SG, Sarma-Rupavtarm R, Klepac-Ceraj V, Polz MF (2005) PCR-induced sequence artifacts and bias: Insights from comparison of two 16S rRNA clone libraries constructed from the same sample. Applied and Environmental Microbiology 71: 8966–8969.

45. Yu YN, Breitbart M, McNairnie P, Rohwer F (2006) FastGroupII: A web-based bioinformatics platform for analyses of large 16S rDNA libraries. Bmc Bioinformatics 7.

8

Tracing Asian Seabass Individuals to Single Fish Farms Using Microsatellites

Gen Hua Yue*, Jun Hong Xia, Peng Liu, Feng Liu, Fei Sun, Grace Lin

Molecular Population Genetics Group, Temasek Life Sciences Laboratory, 1 Research Link, National University of Singapore, Singapore, Singapore

Abstract

Traceability through physical labels is well established, but it is not highly reliable as physical labels can be easily changed or lost. Application of DNA markers to the traceability of food plays an increasingly important role for consumer protection and confidence building. In this study, we tested the efficiency of 16 polymorphic microsatellites and their combinations for tracing 368 fish to four populations where they originated. Using the maximum likelihood and Bayesian methods, three most efficient microsatellites were required to assign over 95% of fish to the correct populations. Selection of markers based on the assignment score estimated with the software WHICHLOCI was most effective in choosing markers for individual assignment, followed by the selection based on the allele number of individual markers. By combining rapid DNA extraction, and high-throughput genotyping of selected microsatellites, it is possible to conduct routine genetic traceability with high accuracy in Asian seabass.

Editor: Carlos Garcia de Leaniz, Swansea University, United Kingdom

Funding: This study is supported by the National Research Foundation (NRF), Singapore. The funders had no role in study design, data collection and analysis, decision to publish, or preparation of the manuscript.

Competing Interests: The authors have declared that no competing interests exist.

* E-mail: genhua@tll.org.sg

Introduction

Food traceability is becoming increasingly important in ensuring food-safety in the agrifood industry [1,2,3,4,5]. As a number of traceability concepts and technologies are available [4], consideration needs to be given to the reliability and precision of traceability systems. Previous experience has shown that conventional tagging and labelling systems are prone to high error rates and may not have sufficient reliability and precision [1,6]. DNA technology can overcome these existing problems in traditional labelling systems by tracking animals and their products through their DNA. This enables the tracking of any food products through the supply chain back to the source animals, and offers high reliability and precision of traceability [5,7]. The precision and reliability of a DNA-based traceability system depend on the number and type of DNA markers used in the system [5]. DNA-based traceability requires collection of samples for extracting DNA, and specialized facilities to detect the DNA [2]. Although RAPD and AFLP markers have been used in genetic traceability of food products [8,9], genotyping markers is usually tedious, and results of RAPD and AFLP analyses are not highly reproducible [10]. Microsatellites [11], which are short (2–6 bp) tandemly repetitive DNA sequences, are the markers of choice for traceability [2] because of their high abundance, high polymorphism and ease of scoring by PCR. Automatic genotyping of PCR products amplified with fluorescently labelled primers, and automated DNA sequencers considerably increases efficiency and precision of genotyping microsatellites and decreases the cost for genotyping [12]. Recently, single nucleotide polymorphism (SNPs) has been tested for genetic traceability [13]. A study showed that identification of highly informative SNP loci from large panels could provide a powerful approach to delineate genetic relationships at the individual and population levels [13]. However, for most aquaculture species, the number of SNPs is limited [14].

Aquaculture is the fastest increasing sector in agriculture. According to FAO's recent statistics [15], international trade of aquaculture fish products has reached a record high. Aquaculture plays a very important role in the economy of Asian countries [15]. The future and sustainable development of aquaculture will be progressively more market driven, and will rely heavily on its capacity to meet consumers' expectations. Therefore, the establishment of a comprehensive traceability system within the aquaculture industry is becoming increasingly important [1,3]. Methods for tracing escapes to single fish farms using microsatellites were reported in salmon and rainbow trout [16,17,18,19]. A simulation study showed that at least 15 microsatellites were required to reach 95% correct assignment decisions [20]. The number of markers required differs from species to species, and depends on the many factors such as genetic diversity and population structure [20,21]. Therefore, it is essential to examine the power of DNA markers for genetic traceability for each species. It is known that the quality of farmed fish, which can vary greatly between farms, is mainly influenced by the quality of farming environment (e.g. water quality), feeds used, feeding regimes and the culture methods implemented [22,23]. Thus, there is a growing need to develop highly reliable, rapid and cost-effective molecular tools for discriminating among farmed fish cultured in different systems and/or in different geographical locations [3,7].

The Asian seabass (*Lates calcarifer*) is a highly valued aquaculture fish species. It has been farmed in Southeast Asia and Australia since 1980s [24], and recently some countries in Europe, such as Germany, France, as well as the USA, have started to culture this species [24]. In this species, a large number of DNA markers have been characterized [25,26,27], mapped to linkage maps [28,29] and applied to study population structure [30,31], to map QTL [28,32] and to conduct parentage analysis for selective breeding [33]. However, no DNA markers have been used in genetic traceability in Asian seabass. The objective of this study is to develop a cost-effective and precise DNA-based tracking system for Asian seabass by evaluating the efficiency of each of the 16 selected microsatellites and their combinations, and by selecting the most powerful markers for individual assignment in four populations of Asian seabass. The selected microsatellites, in combination with the rapid and cost-effective method of DNA extraction developed previously [34], as well as the automatic genotyping of PCR products with DNA sequencers, will enable the routine genetic traceability with high accuracy in Asian seabass.

Materials and Methods

Fish samples

Fin clips of all Asian seabass breeding fishes (i.e. spawners) from three local farms (MAC, FARM-1 and FARM-2) were collected and stored in 75% ethanol. The MAC, FARM-1, and FARM-2 contained 104 (51 males and 53 females), 148 (66 males and 82 females) and 40 (20 males and 20 females) spawners respectively. The spawners of MAC were caught from the wild at sea near Singapore. For sampling spawners of Asian seabass, no specific permits were required for the described field studies. The spawners of FARM-1 originated from Indonesia, while the spawners of the FARM-2 were a mixture of different origins, including a few individuals from the western part of Australia. In addition, we obtained fin clip samples of 62 wild Asian seabass from the western part of Australia from 2003–2005. However, the exact sampling locations were not known. From the three local farms (MAC, FARM-1 and FARM-2), we collected the fin clips of 88, 96 and 96 juveniles (three months post hatch) respectively, and stored them in 75% ethanol. We also obtained 88 fin clips from offspring from a hatchery located in Darwin, Australia in 2006, and stored them in 75% ethanol. The spawners in the hatchery located in Darwin originated from the wild of the western part of Australia.

DNA extraction and genotyping of microsatellites

DNA of each sample was isolated on 96-well plates using a very rapid and cost effective method developed in our laboratory previously [34].

Sixteen microsatellites (Lca002, Lca016, Lca020, Lca021, Lca040, Lca050 Lca057, Lca058, Lca062, Lca063, Lca064, Lca069, Lca070, Lca074, Lca086 and Lca098) were selected from a set of 27 markers that were characterized previously [35]. One primer of each pair was labelled with a fluorescent dye (either Fam or Hex or Ned) at the 5′ end of the primer. The PCR reaction for each sample consisted of 10 ng of genomic DNA, 0.5 units of *Taq* polymerase (Finnzymes, Vantaa, Finland), 1x PCR buffer containing 1.5 mM MgCl$_2$, 0.2 µM dNTPs, and 50 nM of each primer. PCR was conducted on PTC-100 PCR machines (MJ Research, CA, USA) under the following conditions: 2 min denaturation at 94°C; 35 cycles of 30 s at 94°C, 30 s at 55°C and 30 s at 72°C and a final extension at 72°C for 10 min. PCR products were analyzed on an ABI3730xl DNA sequencer (Applied Biosystems, Foster City, USA). Fragment sizes were analyzed against the ROX-500 standard using GeneMapper 4.1 (Applied Biosystems, Foster City, USA). Genotypes were exported to Excel table for later data analysis.

Statistical analysis

Allele number (A), expected (He) and observed (Ho) heterozygosity, and fixation index (f) of each microsatellite were estimated using the software Genetic Data Analysis (GDA) [36]. Allelic richness, a parameter for allelic diversity independent of sample size, was estimated with the software FSTAT [37] for each population. Polymorphism information content (PIC) and the exclusion probabilities (NE-1) of each microsatellite were calculated with the software CERVUS 3.0 [38]. F_{ST} between populations, statistical significance of population pairwise F_{ST} and molecular variance (AMOVA) were analyzed using the software ARLEQUIN 3.1 [39]. In AMOVA, the total variance is partitioned into separate components, each of which describes the proportion of the total variance at distinct hierarchical levels (within population and among populations). Ratios of the variance components can then be used to define population structure. Significance was tested by comparing observed values to null distribution generated by permutation using 10,000 replicates.

Two software WHICHRUN [40] and GENECLASS [41] were used for individual assignment. WHICHRUN performed individual assignment to populations based on a maximum likelihood method. The GENECLASS software [41] can implement three individual assignment tests: Bayesian-based approach, assignment based on reference population allele frequencies and assignment based on genetic distance. The base line used for the assignment in GENECLASS was the spawners from the four populations. As the Bayesian-based approach is commonly used, we used this method for our current analysis.

The software WHICHLOCI [42] was used to examine the assignment score for each marker and their combinations, and to select the most powerful microsatellites for individual assignment. In order to examine the power of single markers for individual assignment, we tested different methods for selecting markers: (1) based on the allele number of single markers, (2) based on the expected heterozygosity, and (3) based on the assignment score estimated using the software WHICHLOCI. We combined 2–16 markers (from the marker with the highest to the lowest allele number, expected heterozygosity and assignment score, respectively) to test the power of combinations of single markers for assignment. The base populations were the spawners from the four populations, whereas the juveniles were the individuals to be assigned.

Results

Polymorphisms and assignment power of 16 microsatellites

Sixteen microsatellites were individually examined for their polymorphisms and power for individual assignment in four Asian seabass populations consisting of 354 spawners. The polymorphisms and power for individual assignment of the 16 markers are shown in Table 1. The allele number of these markers ranged from 6 for Lca069 to 26 for Lca086, with an average allele number of 13.3. The average observed heterozygosity was 0.70, while the expected heterozygosity was 0.78. At the locus Lca086, the majority (Ho = 0.91) of the 354 individuals were heterozygous, whereas at the locus Lca050, only 47% (Ho = 0.47) of the 354 spawners were heterozygous. The polymorphism information content (PIC) varied from 0.41 for Lca050 to 0.91 for Lca058. The fixation index of 16 markers averaged at 0.08, with a range from 0.00 for Lca070 to 0.32 for Lca063. The non-exclusion probability (NE-I) ranged from 0.34 for Lca050 to 0.01 for Lca086. Assignment scores of individual markers were estimated with the

Table 1. Genetic variation and assignment power of 16 microsatellite loci in the four studied populations of Asian seabass.

Locus	Allele no.	Ho	He	PIC	NE-I	Score*	f
LCA002	12	0.65	0.73	0.70	0.11	5.1	0.11
LCA016	14	0.69	0.83	0.82	0.04	10.8	0.18
LCA020	13	0.72	0.80	0.77	0.07	6.1	0.10
LCA021	7	0.87	0.83	0.80	0.05	5.0	−0.06
LCA040	9	0.63	0.78	0.75	0.08	8.1	0.20
LCA057	13	0.74	0.79	0.76	0.07	6.6	0.06
LCA058	22	0.82	0.91	0.90	0.02	6.9	0.09
LCA062	17	0.73	0.89	0.88	0.02	8.0	0.18
LCA063	9	0.49	0.73	0.69	0.12	8.5	0.32
LCA064	12	0.72	0.81	0.78	0.06	5.5	0.10
LCA069	6	0.70	0.78	0.74	0.08	6.6	0.09
LCA074	11	0.58	0.64	0.60	0.17	2.8	0.09
LCA086	26	0.91	0.91	0.91	0.01	7.7	0.01
LCA098	19	0.69	0.77	0.75	0.07	7.8	0.10
Lca070	10	0.76	0.76	0.72	0.10	4.5	0.00
Lca050	13	0.47	0.45	0.41	0.342	4.2	0.03
Mean	13.3	0.70	0.78	0.77	1.2E-19	-	0.08

Ho: observed heterozygosity; He: expected heterozygosity; PIC: polymorphism information content; NE-1: Non-exclusion probability (identity); Score*: Assignment score calculated using the software WHICHLOCI [42] and f: Fixation index.

software WHICHLOCI. Lca016 showed the highest assignment score (10.8), while Lca050 displayed the lowest score (4.2) (Table 1).

Genetic diversity and relationships among the four populations

The mean number of alleles, observed and expected heterozygosity, and fixation index of 16 microsatellites in each of the four populations are presented in Table 2. The Australian population and FARM-1 showed more alleles than the other two local farms in Singapore.

F_{ST} analysis showed that the Australian population was significantly (F_{ST} = 0.11–0.13, $P<0.01$) different from the three populations in Singapore. The differentiation of the three local populations was small (F_{ST} = 0.048–0.067), but statistically

Table 2. Number of samples (N), mean number of alleles (A), mean allelic richness (Ar), observed (Ho) and expected (He) heterozygosity, and fixation index (f) estimated with 16 microsatellites in four populations of Asian seabass.

Population	N	A	Ar	Ho	He	f
AUS	62	9.75	8.91	0.60	0.74	0.18
MAC	104	8.50	7.54	0.72	0.73	0.01
FARM-1	148	9.75	7.78	0.70	0.72	0.03
FARM-2	40	7.19	7.19	0.72	0.74	0.03

significant ($P<0.05$). AMOVA also revealed that the Australia population was significantly different from the three local farms (among population variation = 8.37%, $P<0.01$).

Methods of selecting markers and the number of markers required for assigning individuals to single propulsions

We examined the power of marker combinations for individual assignment by selecting markers based on three parameters (i.e. assignment score, allele number and expected heterozygosity). Using both ML and BS methods, we had the lowest correct assignment of the juveniles from FARM-1. Therefore, in the following results, we showed the percentage of correct assignment for all juveniles (Overall) and juveniles from FARM-1.

Using only three most powerful markers (Lca016, Lca062 and Lca021) selected based on their assignment scores, 97.8% and 96.7% of juveniles could be correctly assigned to their original populations using the ML and BS methods, respectively (Figure 1). With all the 16 markers, all 368 juveniles could be correctly assigned to their original populations.

Based on the allele number of individual markers, using five most polymorphic markers (Lca086, Lca098, Lca058, Lca062 and Lca016) together with the ML and BS methods, the percentage of overall correct assignments reached over 95% (Figure 2). Most of the individuals which could not be correctly assigned were from FARM-1.

Based on the expected heterozygosity of individual markers, the percentage of correct assignment reached 95% by using up to 15 markers (Figure 3).

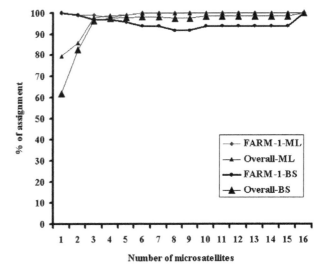

Figure 1. The correct assignment percentage of individuals corresponding to the number of loci selected on the basis of the ranking of score (from high to low) obtained by WHICHLOCI. The marker order was Lca016, Lca063, Lca040, Lca062, Lca098, Lca086, Lca058, Lca057, Lca069, Lca020, Lca064, Lca002, Lca021, Lca070, Lca050 and Lca074. The individual assignment was performed using the Bayesian (BS) and maximum likelihood (ML) methods.

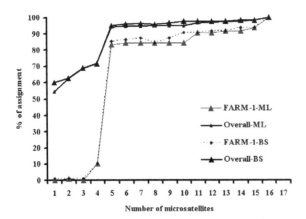

Figure 2. The percentage of correct assignment of individuals corresponding to the number of loci selected on the basis of the ranking of allele number of the markers from high to low. The marker order was Lca086, Lca058, Lca098, Lca062, Lca016, Lca050, Lca057, Lca020, Lca064, Lca002, Lca074, Lca070, Lca063, Lca040, Lca021 and Lca069. The assignment was conducted using the Bayesian (BS) and maximum likelihood (ML) methods.

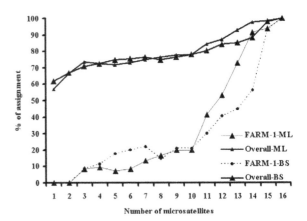

Figure 3. The percentage of correct assignment of individuals corresponding to the number of loci selected on the basis of expected heterozygosity (from high to low expected heterozygosity). The marker order was Lca086, Lca058, Lca062, Lca021, Lca016, Lca064, Lca057, Lca020, Lca069, Lca040, Lca098, Lca070, Lca063, Lca074, Lca002 and Lca050. The individual assignment was carried out using the Bayesian (BS) and maximum likelihood (ML) methods.

Discussion

Microsatellites as markers for genetic traceability in Asian seabass

Traceability is critically important in ensuring food safety and building consumers' confidence [7]. Although physical labels have been used for traceability for a very long time, they are still not very reliable, as physical labelling can be easily lost and changed [1]. Traceability using DNA markers is more reliable [4], and has been used in the livestock industry [2]. In the seafood industry, genetic traceability using DNA markers is just in its infancy [1], although genetic tracing had been used in aquaculture, such as tracking escapees of salmon and rainbow trout [16,19]. In this study, we tested the power of 16 microsatellites and their combinations in assigning individuals to their original populations. In general, the 16 microsatellites were highly polymorphic in the four populations. Using only the three most efficient microsatellites selected based on the assignment score, over 95% of 368 individuals could be correctly assigned to their original populations. These data suggest that these 16 microsatellites could be used for future routine individual/product assignment of Asian seabass and that the selection of most efficient markers for parentage assignment is essential.

Population structure influenced the power of DNA markers for traceability

The genetic differentiation between the Australian and Asian populations was much bigger than that among the three Asian populations, which is in agreement with the results of previous studies on genetic relationships of Asian seabass populations [30,31]. It is known that population structure and relationships influenced the power of DNA markers for traceability [20]. In this study, the population FARM-1 was the most diverse, containing the majority of alleles. The juveniles originating from this FARM-1 were assigned to other populations when only a few microsatellites were used, suggesting that when selecting DNA markers for individual assignment, population structure and relationships must be taken into account. It is advisable to conduct some pilot studies to examine the efficiency of DNA individual markers to select the

most efficient DNA markers for routine genetic traceability in pupations where genetic traceability to be conducted.

Other factors influencing the power of DNA markers for traceability

Besides the structure of populations of spawners, other factors such as the allele number of markers, heterozygosity of markers, and assignment score of individual markers, number of markers [21,43] and other factors (e.g. statistical methods) may also influence the efficiency of individual DNA markers for individual assignment,. In this study, the two statistical methods showed similar results. Therefore, we only analysed the effects of the allele number, heterozygosity and assignment score of individual markers on the efficiency of individual assignment. To reach over 95% of correct assignment, based on the assignment scores estimated with the software WHICHLOCI, only three markers with the highest score were required, whereas based on allele number and expected heterozygosity, at least 6 and 15 microsatellites were required, respectively. These data suggest that selection of markers based on the assignment score estimated with the software WHICHLOCI is most effective for individual assignment. Therefore, in practice, to accomplish high efficiency of assignment of individuals, it is essential to conduct some small-scale feasibility studies to examine the power of markers for individual assignment using the software WHICHLOCI.

Conclusions

We have tested the efficiency of 16 microsatellites and their combinations in assigning individuals to populations of Asian seabass where they originated. Three most effective microsatellites were required to assign over 95% of fish to the correct populations. Selection of markers based on the assignment score estimated with the software WHICHLOCI was most effective in choosing markers for individual assignment, followed by the selection based on the allele number of individual markers. Therefore, for routine genetic traceability, it is essential to conduct some small-scale feasibility studies to select the most efficient DNA markers. By combining the rapid DNA extraction method developed previously by us [34] and automatic genotyping of selected microsatellites using sequencers, it is possible to conduct routine genetic traceability with high accuracy in Asian seabass.

Acknowledgments

We would like to thank MAC, AVA for cooperation on this project and our former lab members Loongcheung Lo, Zheyuan Zhu and Felicia Feng for their help in sample collection.

References

1. Hastein T, Hill B, Berthe F, Lightner D (2001) Traceability of aquatic animals. Rev Sci Tech Oie 20: 564–583.
2. Dalvit C, De Marchi M, Cassandro M (2007) Genetic traceability of livestock products: A review. Meat Sci 77: 437–449.
3. Guerard F, Sellos D, Le Ga Y (2005) Fish and shellfish upgrading, traceability. In: Gal YL, Ulber R, editors. Marine Biotechnology I. NY: Springer. pp. 127–163.
4. Opara LU (2003) Traceability in agriculture and food supply chain: a review of basic concepts, technological implications, and future prospects. J Food Agri Envir 1: 101–106.
5. Rodríguez-Ramírez R, González-Córdova AF, Vallejo-Cordoba B (2011) Review: Authentication and traceability of foods from animal origin by polymerase chain reaction-based capillary electrophoresis. Ana Chim Acta 685: 120–126.
6. Loftus R (2005) Traceability of biotech-derived animals: application of DNA technology. Rev Sci Tech Oie 24: 231–242.
7. Ogden R (2008) Fisheries forensics: the use of DNA tools for improving compliance, traceability and enforcement in the fishing industry. Fish Fish 9: 462–472.
8. Cirillo A, Del Gaudio S, Di Bernardo G, Galderisi U, Cascino A, et al. (2009) Molecular characterization of Italian rice cultivars. Eur J Food Res Tech 228: 875–881.
9. Maldini M, Marzano FN, Fortes GG, Papa R, Gandolfi G (2006) Fish and seafood traceability based on AFLP markers: Elaboration of a species database. Aquaculture 261: 487–494.
10. Liu ZJ, Cordes JF (2004) DNA marker technologies and their applications in aquaculture genetics. Aquaculture 238: 1–37.
11. Weber JL, May PE (1989) Abundant class of human DNA polymorphisms which can be typed using the polymerase chain-reaction. Am J Hum Genet 44: 388–396.
12. Yue GH, Beeckmann P, Bartenschlager H, Moser G, Geldermann H (1999) Rapid and precise genotyping of porcine microsatellites. Electrophoresis 20: 3358–3363.
13. Glover KA, Hansen MM, Lien S, Als TD, Høyheim B, et al. (2010) A comparison of SNP and STR loci for delineating population structure and performing individual genetic assignment. BMC Genet 11: 2.
14. Liu Z (2011) Next generation sequencing and whole genome selection in aquaculture. New York: Wiley.
15. FAO (2010) The State of World Fisheries and Aquaculture – 2010 (SOFIA). Roma: FAO.
16. Glover KA, Skilbrei OT, Skaala Ø (2008) Genetic assignment identifies farm of origin for Atlantic salmon Salmo salar escapees in a Norwegian fjord. ICES J Mar Sci 65: 912–920.
17. Glover KA, Hansen MM, Skaala Ø (2009) Identifying the source of farmed escaped Atlantic salmon (Salmo salar): Bayesian clustering analysis increases accuracy of assignment. Aquaculture 290: 37–46.
18. Glover KA (2010) Forensic identification of fish farm escapees: the Norwegian experience. Aquaculture Environment Interactions 1: 1–10.
19. Glover KA (2008) Genetic characterisation of farmed rainbow trout in Norway: intra- and inter-strain variation reveals potential for identification of escapees. BMC Genet 9: 87.
20. Hayes B, Sonesson AK, Gjerde B (2005) Evaluation of three strategies using DNA markers for traceability in aquaculture species. Aquaculture 250: 70–81.
21. Hansen MM, Kenchington E, Nielsen EE (2001) Assigning individual fish to populations using microsatellite DNA markers. Fish Fish 2: 93–112.
22. Sargent JR, Bell JG, McGhee F, McEvoy J, Webster JL (2001) The nutritional value of fish. In: Kestin SC, Warriss PD, editors. Farmed Fish Quality. Oxford, United Kingdom: Fishing News Books, Blackwll Science Ltd. pp. 3–12.
23. Shearer KD (2001) The effect of diet composition and feeding regime on the proximate composition of farmed fishes. In: Kestin SC, Warriss PD, editors. Farmed Fish Quality. Oxford, United Kingdom: Fishing News Books, Blackwll Science Ltd. pp. 31–41.
24. Fishbase (2012) Available: http://www.fishbase.org/search.php.Accessed 2012 Jan. 28.
25. Yue G, Li Y, Orban L (2001) Characterization of microsatellites in the IGF-2 and GH genes of Asian seabass (Lates calcarifer). Mar Biotechnol 3: 1–3.
26. Yue GH, Li Y, Chao TM, Chou R, Orban L (2002) Novel microsatellites from Asian sea bass (Lates calcarifer) and their application to broodstock analysis. Mar Biotechnol 4: 503–511.
27. Zhu ZY, Wang CM, Lo LC, Feng F, Lin G, et al. (2006) Isolation, characterization, and linkage analyses of 74 novel microsatellites in Barramundi (Lates calcarifer). Genome 49: 969–976.
28. Wang CM, Lo LC, Zhu ZY, Pang HY, Liu HM, et al. (2011) Mapping QTL for an adaptive trait: the length of caudal fin in Lates calcarifer. Mar Biotechnol 13: 74–82.
29. Wang CM, Zhu ZY, Lo LC, Feng F, Lin G, et al. (2007) A microsatellite linkage map of barramundi, Lates calcarifer. Genetics 175: 907–915.
30. Yue GH, Zhu ZY, Lo LC, Wang CM, Lin G, et al. (2009) Genetic variation and population structure of Asian seabass (Lates calcarifer) in the Asia-Pacific region. Aquaculture 293: 22–28.
31. Zhu ZY, Lin G, Lo LC, Xu YX, Feng F, et al. (2006) Genetic analyses of Asian seabass stocks using novel polymorphic microsatellites. Aquaculture 256: 167–173.
32. Wang CM, Lo LC, Zhu ZY, Yue GH (2006) A genome scan for quantitative trait loci affecting growth-related traits in an F1 family of Asian seabass (Lates calcarifer). BMC Genomics 7: 274.
33. Wang CM, Lo LC, Zhu ZY, Lin G, Feng F, et al. (2008) Estimating reproductive success of brooders and heritability of growth traits in Asian seabass using microsatellites. Aqua Res 39: 1612–1619.
34. Yue GH, Orban L (2005) A simple and affordable method for high-throughput DNA extraction from animal tissues for polymerase chain reaction. Electrophoresis 26: 3081–3083.
35. Zhu ZY, Wang CM, Lo LC, Lin G, Feng F, et al. (2010) A standard panel of microsatellites for Asian seabass (Lates calcarifer). Anim Genet 41: 208–212.
36. Lewis PO, Zaykin D (2000) Genetic Data Analysis: Available: http://hydrodictyon.eeb.uconn.edu/people/plewis/software.php.Accessed on 2012 Jun 28.
37. Goudet J (1995) FSATA (vers.1.2): a computer program to calculate F-statistics. J Herd 85: 485–486.
38. Kalinowski ST, Taper ML, Marshall TC (2007) Revising how the computer program CERVUS accommodates genotyping error increases success in paternity assignment. Mol Ecol 16: 1099–1106.
39. Excoffier L, Laval G, Schneider S (2005) Arlequin ver. 3.0: An integrated software package for population genetics data analysis. Evolut Bioinfo 1: 47–50.
40. Banks M, Eichert W (2000) Computer note. whichrun (version 3.2): a computer program for population assignment of individuals based on multilocus genotype data. J Herd 91: 87–89.
41. Piry S, Alapetite A, Cornuet JM, Paetkau D, Baudouin L, et al. (2004) GENECLASS2: a software for genetic assignment and first-generation migrant detection. J Herd 95: 536–539.
42. Banks MA, Eichert W, Olsen JB (2003) Which genetic loci have greater population assignment power? Bioinformatics 19: 1436–1438.
43. Fan B, Chen Y, Moran C, Zhao S, Liu B, et al. (2005) Individual-breed assignment analysis in swine populations by using microsatellite markers. Asian Austral J Anim Sci 18: 1529.

Author Contributions

Conceived and designed the experiments: GHY. Performed the experiments: GHY JHX PL FL FS GL. Analyzed the data: GHY. Contributed reagents/materials/analysis tools: GHY. Wrote the paper: GHY.

9

Verification of Intraovum Transmission of a Microsporidium of Vertebrates: *Pseudoloma neurophilia* Infecting the Zebrafish, *Danio rerio*

Justin L. Sanders[1]*, Virginia Watral[1], Keri Clarkson[2], Michael L. Kent[1,2]

1 Department of Microbiology, Oregon State University, Corvallis, Oregon, United States of America, **2** Department of Biomedical Sciences, Oregon State University, Corvallis, Oregon, United States of America

Abstract

Direct transmission from parents to offspring, referred to as vertical transmission, occurs within essentially all major groups of pathogens. Several microsporidia (Phylum Microsporidia) that infect arthropods employ this mode of transmission, and various lines of evidence have suggested this might occur with certain fish microsporidia. The microsporidium, *Pseudoloma neurophilia*, is a common pathogen of the laboratory zebrafish, *Danio rerio*. We previously verified that this parasite is easily transmitted horizontally, but previous studies also indicated that maternal transmission occurs. We report here direct observation of *Pseudoloma neurophilia* in the progeny of infected zebrafish that were reared in isolation, including microscopic visualization of the parasite in all major stages of development. Histological examination of larval fish reared in isolation from a group spawn showed microsporidian spores in the resorbing yolk sac of a fish. Infections were also observed in three of 36 juvenile fish. Eggs from a second group spawn of 30 infected fish were examined using a stereomicroscope and the infection was observed from 4 to 48 hours post-fertilization in two embryos. Intraovum infections were detected in embryos from 4 of 27 pairs of infected fish that were spawned based on qPCR detection of *P. neurophilia* DNA. The prevalence of intraovum infections from the four spawns containing infected embryos was low (~1%) based on calculation of prevalence using a maximum likelihood analysis for pooled samples. Parasite DNA was detected in the water following spawning of 11 of the infected pairs, suggesting there was also potential for extravovum transmission in these spawning events. Our study represents the first direct observation of vertical transmission within a developing embryo of a microsporidian parasite in a vertebrate. The low prevalence of vertical transmission in embryos is consistent with observations of some other fish pathogens that are also readily transmitted by both vertical and horizontal routes.

Editor: Dan Zilberstein, Technion-Israel Institute of Technology Haifa 32000 Israel, Israel

Funding: This study was supported by grants from the National Institutes of Health (NIH NCRR 5R24RR017386-02 and NIH NCRR P40 RR12546-03S1). The funders had no role in the study design, data collection and analysis, decision to publish or preparation of the manuscript.

Competing Interests: The authors have declared that no competing interests exist.

* E-mail: Justin.Sanders@oregonstate.edu

Introduction

Pathogens employ a range of different mechanisms to infect new hosts. Vertical or maternal transmission is characterized by pathogens being transmitted to progeny from parents, usually through an infected female host. This mode of transmission is employed by some obligate parasites, which cannot complete their life cycles without a host. The methods by which pathogens are transmitted can exert a powerful influence on virtually all aspects of the biology of the host, either directly or indirectly. Indeed, it was argued that pathogens and their effects on hosts are responsible for nearly all aspects of biological organization [1]. The population structure of the host species has an important influence on mode of transmission employed by pathogens [1,2]. For example, in populations where individuals are spatially separated, minimizing the opportunity for horizontal transfer, selection would favor maternal, or vertical, transmission, ensuring propagation of the parasite.

The mode of transmission also has a direct influence on the virulence of a parasite with selection favoring levels of parasite reproduction, and thus virulence, that provide the highest level of fitness for the parasite. Pathogens which are transmitted vertically are generally less virulent that those which are transmitted horizontally as there is selective pressure for the survival of the infected female to live to reproductive age and pass the infection on to progeny [3]. As a result, theses parasites generally do not proliferate to high numbers in the host, minimizing disease and thus host mortality. The virulence-related characteristics of vertically transmitted pathogens are generally limited to those that increase the number of susceptible hosts, including feminization of male hosts [4].

Microsporidia are obligate intracellular pathogens with species infecting virtually all animal phyla. This distinctive group has undergone numerous taxonomic revisions since first being described [5]. Originally assigned to the Schizomycetes, a group containing yeast-like fungi [6], the Microsporidia were subsequently moved to other groups such as the Sporozoa [7] and Archaezoa [8]. Their assignment to the Archaezoa, a group of protists considered to be "primitive" due to the absence of some typical eukaryotic features, was based mainly on their apparent lack of mitochondria and was further supported by early molecular phylogenetic analyses based on small subunit ribosomal RNA gene sequences. This was, however, later found to be in error as subsequent molecular and ultrastructural analyses have shown the presence of relictual mitochondria [9] and more sophisticated phylogenetic analyses that account for rate variation and the presence of numerous fungal-type products, notably trehalose and chitin, placed the Microsporidia once again among the Fungi. Their exact placement among the Fungi (i.e., as early-branching or sister to the Fungi) is currently debated [10].

Microsporidia have been shown to employ a diverse range of transmission strategies. Several microsporidian species are transmitted horizontally, generally by ingestion of spores from either water contaminated by feces or tissue from dead infected hosts[11]. Maternal transmission of microsporidian parasites has been noted for several species infecting crustaceans [12] and insects [13]. Whereas some microsporidian species are only vertically transmitted [14], many are transmitted both horizontally and vertically. Maternal transmission occurs by two general mechanisms, which in arthropods are characterized as transovarial (within the egg) and transovum (parasite is shed outside of the egg during egg laying/spawning). These terms were developed to describe vertically transmitted microsporidia of insects, and are somewhat confusing when applied to vertebrates. Hence, we refer to these two types of transmission as intraovum and extraovum, respectively.

Regarding vertebrates, the earliest recorded evidence of vertical transmission comes from Hunt et al [15] who observed *Encephalitozoon cuniculi* infections in gnotobiotic rabbits, strongly suggesting transplacental transmission of this parasite. Also, several lines of indirect evidence support vertical transmission of microsporidia in fishes. The strongest evidence of vertical transmission in fish comes from *Ovipleistophora ovariae*, a microsporidium that infects the golden shiner, *Notemigonus crysoleucas*, and has been observed to infect only female fish. In this species, spores are found almost exclusively within the ovaries and developing oocytes [16]. While large amounts of *O. ovariae* DNA has been detected within surface-decontaminated, spawned eggs and developing larval fish [17], the parasite has not been directly observed in the progeny of infected fish. *Ovipleistophora mirandellae* also infects the oocytes and eggs of cyprinid fishes [18] and thus has been suggested [19] to undergo intraovum transmission. *Loma salmonae* is a microsporidian pathogen of salmon and trout, and has been observed in the ovigerous stroma, but not eggs [20]. The translocation of *Loma salmonae* to fish farms in Chile which had been populated using only surface decontaminated, fertilized eggs is also suggestive of vertical transmission [20,21].

The zebrafish, *Danio rerio*, is an important laboratory model for toxicology, developmental biology, cancer, and infectious disease research. It has a well-characterized immune system [22] and the complete genome has been sequenced with the genetic map showing an overall highly conserved synteny with the human genome [23]. A microsporidium, *Pseudoloma neurophilia*, is responsible for chronic infections of zebrafish [24,25] and is common in zebrafish research colonies [26]. The high prevalence of a microsporidium in laboratory zebrafish populations provides an ideal opportunity to learn more about the transmission characteristics of this parasite. As the name implies, *Pseudoloma neurophilia* chronically infects neural tissue and develops into mature spores mainly in the hindbrain, spinal cord, motor nerve ganglia and spinal nerve roots. The parasite enters the host via the intestinal epithelium and infects extraintestinal skeletal muscle myocytes in early stages of the infection [27]. Infections by the parasite result in a spectrum of disease ranging from minimal to no clinical presentation to acute mortality. We have shown that *P. neurophilia* is horizontally transmissible by bath exposure and cohabitation

with live fish, ostensibly *per os*. There is growing evidence of vertical transmission for this parasite as follows: 1. The presence of microsporidian spores visualized by histology in the ovaries and developing oocytes of infected adult female fish [28], 2. *P. neurophilia* DNA detected by PCR in spawn water and eggs from infected adults [29], 3. High numbers of *P. neurophilia* spores in ovaries of infected females [25], 4. Repeated detection of *P. neurophilia* in the progeny of two lines of zebrafish screened as part of a protocol to develop a specific pathogen free colony leading to the exclusion of these lines from the colony [30].

Despite the significant indirect evidence, observation of the intra-ovum transmission of a microsporidian parasite in a vertebrate has not been verified. By utilizing a combination of a qPCR method [29] and microscopy on a number of spawning fish and progeny, we provide direct evidence of the intraovum transmission of *P. neurophilia*. Additionally, we provide data supporting the hypothesis that extraovum maternal transmission is also an important route of transmission.

Results

Group spawns

One of 24 egg pools was positive for *P. neurophilia* DNA. Histological examination of 7 dpf larval fish revealed the presence of microsporidian spores in the epidermis (Figure 1A) and resorbing yolk sac (Figure 1B) of one of 112 fish examined. Histological examination of juveniles from this spawn, reared in isolation, at 8 wk post fertilization revealed spores in various tissues in 3 of 36 fish examined, such as the lamina propria of the intestine (Figure 1C), the ovigerous stroma (Figure 1D), and the inner ear.

Eggs from a second group spawn of 30 adults from a population of infected fish were examined using a stereomicroscope with a transmitted light source. Distinctive opaque regions were seen in two embryos at approximately 4 hours post-fertilization (Figure 2A). One embryo was maintained at 28°C and examined again at 24 and 48 hours post fertilization during which time it appeared to develop normally (Figures 2B, 2C). The other embryo was sacrificed and examined by wet mount microscopy. Upon examination using higher magnification, these regions were found to consist of large aggregates of refractile spores with characteristic polar vacuoles, consistent in shape and size to *Pseudoloma neurophilia* (Figure 2D). The embryo was then homogenized in sterile water and the spores quantified by hemocytometer. Thirty thousand spores were present in this particular embryo.

Prevalence of *P. neurophilia* within oocytes and detection in spawning water

Pseudoloma neurophilia DNA was detected within eggs from 4, and in the spawn water of 11, of 27 paired spawns (Table 1). The mean prevalence of *P. neurophilia* within spawned eggs from these 4 positive spawns was calculated to be 0.9% (CI 0.06–3.45%).

Histological examination of spawning pairs

Spores were observed in the spinal cords and brains of 43 out of 54 spawning adult fish that were examined. The parasite was observed in the ovigerous stroma of 7 of 27 females, and within developing follicles of 4 of these 27 fish (Table 1). Of the 4 females with eggs that tested positive by qPCR after spawning, spores were observed in the ovigerous stroma of two and within a developing oocyte of one. No spores were observed in the testes of any male fish examined nor in any tissues of fish in which spores were not also observed in the spinal cord.

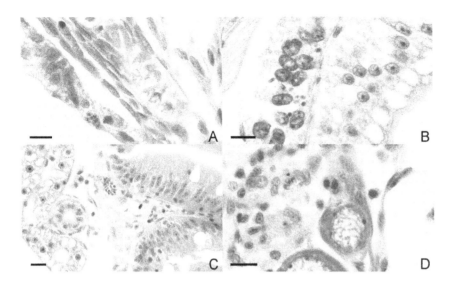

Figure 1. Spores of *Pseudoloma neurophilia* in Luna-stained histological sections of progeny of infected zebrafish, *Danio rerio*. A. Spores (red) in the epidermis of a 7 d post-fertilization (pf) larval zebrafish. **B.** Spores in the resorbing yolk-sac of the same 7 dpf larval zebrafish. **C.** Spore aggregate beneath the intestinal epithelium of an 8 wk pf juvenile fish. **D.** Spores in the ovigerous stroma adjacent to developing follicles in an 8 wk pf fish. Bar =10 µm.

Dose response of larval fish to *P. neurophilia*

Larval fish in two trials became infected when exposed to 300 or 500 spores/ml, but not at lower concentrations (Table 2). Whereas there was variability between replicates and trials, owing primarily to the high numbers of mortalities observed during the duration of the experiment, there was a trend for a dose response.

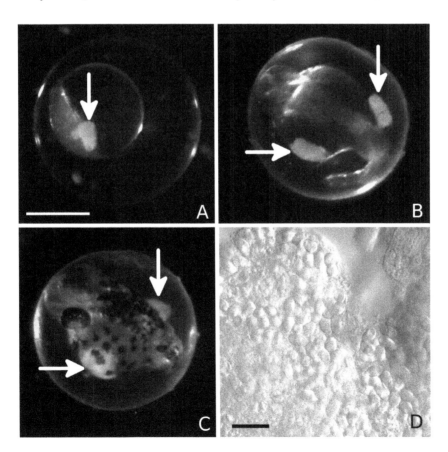

Figure 2. Spores of *Pseudoloma neurophilia* in developing embryo of zebrafish, *Danio rerio*. A. Aggregated spores (arrow) in a 4 hpf embryo. Bar =0.5 mm. **B.** Two foci of spores (arrows) visible in the same embryo at 24 hpf. **C.** Spores (arrows) in the same embryo at 48 hpf. **D.** Differential interference contrast micrograph of spores from an embryo. Bar =10 µm.

Table 1. Detection of *Pseudoloma neurophilia* in egg pools obtained from paired spawning of *Danio rerio* adults and tested by qPCR and in spawning adult fish by histology.

Spawn	Total eggs	Pool Size	Number positive pools/Total pools	Number of spores detected in positive egg pools	Estimated percent prevalence (95% confidence intervals)	Spores detected (water)	Histology - Male	Histology - Female
1	64	10	0/7	-	-	12159	ND	S
2	445	30	0/15	-	-	ND	ND	ND
3	543	30	0/19	-	-	16788	ND	S, F
4	463	30	1/16	132560	0.2 (0.01–1.0)	ND	ND	S
5	151	30	2/5	430847, 568	1.3 (0.2–4.0)	1418	ND	S
6	107	20	0/6	-	-	4417	S, K	S
7	223	20	0/12	-	-	1858	S	S
8	185	30	0/10	-	-	12642	S	S, O
9	135	20	0/7	-	-	ND	ND	S
10	159	20	0/8	-	-	1180	S	ND
11	295	30	0/11	-	-	1385	S	S, F
12	323	30	0/12	-	-	977	S	S
13	72	30	0/3	-	-	1549	S, K	S, K
14	249	30	0/9	-	-	ND	S	ND
15	247	30	0/9	-	-	703	S	S, I
16	370	20	0/18	-	-	ND	S, K	S
17	232	15	0/15	-	-	ND	S	ND
18	223	30	0/8	-	-	ND	S	S
19	285	30	0/10	-	-	ND	ND	S
20	193	30	0/7	-	-	ND	S, K	S, P
21	177	30	0/6	-	-	ND	S, K	S, O
22	76	15	1/6	998	1.5 (0.01–6.3)	ND	S, K	S, O
23	90	30	0/4	-	-	ND	S, K	S
24	188	30	1/7	22901	0.6 (0.03–2.5)	ND	S, K	S, K, O
25	178	20	0/9	-	-	ND	S, P	S, P
26	154	15	0/10	-	-	ND	S, K	S, K
27	398	30	0/14	-	-	ND	S	S, O

Estimation of prevalence of *P. neurophilia* in individual populations of eggs obtained by paired spawning of zebrafish. Prevalence is calculated using the maximum likelihood estimation method of Williams and Moffit [59]. ND = None detected, S= spinal, F= follicle, I= intestinal epithelium, K= kidney, O= ovigerous stroma, P = pancreas.

Table 2. Dose response of larval zebrafish exposed to low numbers of *Pseudoloma neurophilia*.

	500 sp/ml	300 sp/ml	100 sp/ml	50 sp/ml
Trial 1				
A	5/11	1/14	0/8	0/6
B	2/10	0/9	0/5	0/3
C	6/20	1/11	0/5	0/6
Total percent infected	**31.7**	**5.9**	**0.0**	**0.0**
Trial 2				
A	0/3	1/2	0/4	0/1
B	4/6	1/4	0/4	0/5
C	2/5	1/2	0/1	0/3
Total percent infected	**42.9**	**37.5**	**0.0**	**0.0**

Two trials were performed in which three replicates of larval zebrafish were exposed to low concentrations of *P. neurophilia* spores. Fish surviving to seven day post exposure were euthanized and examined by microscopy for the presence of *P. neurophilia* spores.

Discussion

A combination of indirect (qPCR) and direct (microscopy) methods demonstrated intraovum transmission of the microsporidian parasite, *Pseudoloma neurophilia*, in its zebrafish host. Evidence for the vertical transmission of fish pathogens has often been indirect or circumstantial, especially for those that can be transmitted both horizontally and vertically. Determination of whether vertical transmission occurs, and by which mechanism (i.e., intraovum versus extraovum) can be complicated by several factors including low overall prevalence of vertical transmission, variation of occurrence both within and between clutches of eggs, and the ability of many of these pathogens to survive for long periods in the water leading to extraovum transmission. Here we verify for the first time intraovum transmission of microsporidium infecting a vertebrate host.

There are several similarities and some differences between *P. neurophilia* and *Ovipleistophora ovariae*. The latter microsporidium was first described in ovarian infections of the golden shiner, *Notemigonus crysoleucas*, another cyprinid fish that is an important bait fish raised in aquaculture in the United States. It is found with high prevalence in females from commercial fish farms and has been shown to greatly reduce fecundity as the fish ages. *Ovipleistophora ovariae* is found primarily in the ovigerous stroma and within intermediate to fully mature oocytes. However, it has also been observed in liver and kidney tissue from a few infected females [31]. It has not been observed in tissue from male fish, however DNA from the parasite has been detected in testes [32]. Horizontal transmission of *O. ovariae* has been demonstrated by feeding fry and fingerlings spores adsorbed to feed [33]. By treating eggs from infected shiners with RNase Away, as done in our study, Phelps and Goodwin (2008) detected high numbers of copies of *O. ovariae* DNA in all pools of eggs tested, providing strong evidence of intraovum transmission of the parasite. Fry hatched from the same clutch of eggs also showed high levels of *O. ovariae* DNA at 48 hours post hatch. However, no parasites were observed in these fish by microscopy. The parasite almost completely replaces the egg interior with most infected eggs, and thus Summerfelt (1972) suggested that embryos from these eggs

are not viable, but rather serve as a source of infection to siblings. In this case, therefore, *O. ovariae* is actually transmitted by extraovum maternal transmission because the infected egg itself does not directly result in an infected fish.

In contrast to *O. ovariae*, relatively few zebrafish eggs and embryos are infected by *P. neurophilia*. We were able to observe spores of *Pseudoloma neurophilia* in one 7 day post fertilization fish (i.e. 5 day post-hatch) by histology, and this low prevalence correlates with that found by screening eggs by qPCR (~1%). While we are unable to differentiate between merogonic or spore stages of *P. neurophilia* using the qPCR assay, the observation of mature spores within a developing embryo and developing oocytes confirms the presence of this stage in at least some cases of intraovum transmission. This does not preclude the presence of presporogonic stages within oocytes, which are more difficult to visualize by standard histological methods. The other member of the genus *Ovipleistophora*, *O. mirandellae*, also infects ovaries, testis, and eggs of several fishes [19,34]. It has been suggested to be vertically transmitted, but we are not aware of empirical data supporting this hypothesis.

There are other examples of fish pathogens that, like *P. neurophilia*, are transmitted both horizontally and vertically, and show a low prevalence of infected eggs. It is not necessary to have a high prevalence of infected eggs and embryos within a clutch to infect the next generation as vertical transmission is followed by robust horizontal transmission within the F1 siblings. Two bacterial and one viral pathogen of salmonid fishes use these modes of transmission. *Renibacterium salmoninarum*, for example, is an obligate Gram positive bacterial pathogen that has been spread around the world with egg shipments [35]. It was present within surface disinfected eggs from experimentally infected rainbow trout *Onchorynchus mykiss* at a prevalence of 1.7% [36], and from two heavily infected coho salmon *Oncorhynchus kisutch* from separate studies at prevalences of 5.8% [37] and 11.6% [38]. Similarly, 13% of newly spawned eggs from infected steelhead trout, *Onchorynchus mykiss*, were found to be infected with *Flavobacterium psychrophilum* [39], a serious pathogen in salmon and trout hatcheries [40]. The small size of these bacteria allow for their passage into the egg after it is released from the female through the egg micropyle prior to fertilization, whereas the larger size of microsporidian spores such as *P. neurophilia* (approximately 3.5 μm by 5 μm) prevents it from entering through the approximately 1.7–2.5 μm diameter micropyle of the developed zebrafish egg [41], requiring it to be present in the oocyte prior to spawning. This could account for the somewhat higher prevalence of intraovum transmission observed with these bacteria.

Infectious hematopoietic necrosis virus (IHNV), a virus infecting salmonids, was responsible for the failure of successful culture of sockeye salmon, *Oncorhynchus nerka*, in Alaska prior to 1981 [42]. The implementation of a risk management approach consisting of the use of virus-free water supplies, rigorous disinfection, and compartmentalization of eggs and fry to contain virus outbreaks when they occur was unsuccessful in completely resolving IHNV epizootics [43]. One possibility for this failure was suggested to be that intraovum vertical transmission of the virus occurs, resulting in a few subclinically infected progeny. There has been evidence to support the intraovum transmission of IHNV. Mulcahy and Pascho [44] described the isolation of IHN virus from fry and eggs held in the laboratory under virus-free conditions in two occurrences. They found that only a small proportion of eggs and fry tested contained IHN from one female with a high titer present in the body cavity.

The low prevalence of intraovum transmission of *P. neurophilia* observed is adequate to maintain the infection between generations, given the large numbers of larval siblings in a single spawn and the high rates of horizontal transmission. Indeed, observing the infection by histology, we detected *P. neurophilia* in only one 7 dpf larva, but within 8 weeks increased to 11%, including spores developing within the ovigerous stroma of one fish. Following initial infection, the parasite employs several routes of transmission [25]. Waterborne transmission occurs by co-habitation with infected fish and by ingestion of infected tissue. Zebrafish spawn every few weeks, releasing spores and infected eggs into the water which are then fed upon by tanks mates. Also, feeding on infected carcasses is an efficient route of transmission. Zebrafish do not generally attack other fish in an aquarium or school, but quickly scavenge dead fish.

In addition to intraovum transmission, *Pseudoloma neurophilia*, as well as IHN and *Renibacterium salmoninarum*, employs a second maternal transmission mechanism; extraovum transmission. High levels of *P. neurophilia* in the ovaries and in spawn water have been reported previously [25,29], and with our paired spawn analysis the spawn water contained the parasite more often than eggs. Also, the dose response experiment agreed with the previously studies [45] showing that larvae, which begin feeding about 4–5 d after fertilization, are very susceptible. Again this is similar to *Renibacterium salmoninarum*, which is often found at high prevalence and concentration in the ovarian fluid from infected female salmon [46]. Whereas very few eggs were internally infected by this bacterium, it was detected on the surface of 38% of the eggs [36]. This strategy of extraovum transmission likely occurs with the IHN virus as well, and extends to infecting the next generation beyond the infected parents own progeny in a given river or stream [47,48]. Traxler et al. [47] showed no correlation between infection levels in ovarian fluid compared to eggs, indicating that extraovum transmission is the major route of vertical transmission for this viral pathogen.

The mode of vertical transmission we report has practical implications. Extraovum transmission in fish can be mitigated by the use of rigorous surface decontamination of eggs whereas intraovum transmission (i.e., pathogens present within the egg) would render any surface decontamination method ineffective. This approach has been used to avoid extraovum viruses and bacteria following spawning with salmonids [49,50], but is ineffective for *P. neurophilia* [45]. The high prevalence of *P. neurophilia* in laboratory zebrafish colonies, the risk of intraovum transmission, and the lack of disinfectants that will kill spores but not eggs [45] led to the development of a zebrafish colony which is specific pathogen free (SPF) for *P. neurophilia* [30], accomplished by rigorous screening of brood fish and progeny for the pathogen with our PCR tests [29,51].

The population of *P. neurophilia* used in this study has been maintained for >4 years by horizontal passaging in zebrafish held in aquaria (i.e., exposing fish to infected spinal material, cohabitation with infected fish). This may have resulted in the selection of parasite strains which are most efficiently transmitted horizontally, resulting in the very low prevalence of intraovum transmission observed in the current study. Intraovum transmission does occur nonetheless, and the occurrence of this method of transmission could likely be experimentally manipulated. Several species of Microsporidia are transmitted both horizontally and vertically [4,52]. Even among species which are transmitted solely transovarially, transmission efficiency has been observed to be variable between broods [52].

The zebrafish has been used to study host-pathogen interactions of numerous bacterial and viral pathogens [53–55]. *Pseudoloma neurophilia* infections of zebrafish would provide a good model to investigate the evolution of vertical transmission as it employs both horizontal and vertical transmission.

We summarize the transmission of *P. neurophilia* in Figure 3: The primary mode of transmission appears to be horizontal, with many numbers of *P. neurophilia* spores being released into the water from ovaries of females during spawning (Fig. 3a), which occurs on a routine basis every few weeks. Feces and urine are also a source of infection from live fish. This is supported by the presence of *P. neurophilia* spores developing in the renal tubules of (Fig. 3b). Zebrafish are relatively docile fish, but quickly scavenge dead tank mates, providing another route of horizontal transmission.

Alternatively, vertical transmission, either intraovum (Fig. 3c) or extraovum (Fig. 3d) can occur via the female host. Spores shed by females in the ovarian fluid during spawning can then be ingested by the developing larvae, leading to new infections. Intraovum transmission can lead to death of the infected embryos (Fig. 3e) or directly to viable, infected hosts (Fig. 3f) which can subsequently infect other hosts by horizontal transmission and later, by vertical transmission. Whereas the occurrence of intraovum transmission appears to be rare, it is nonetheless important for maintenance of the parasite in laboratory zebrafish colonies especially considering that infected embryos appear to be able to develop normally, at least to the post-hatch larval stage.

Methods

All experimental protocols for the procedures using zebrafish were approved by the Oregon State University Institutional Animal Care and Use Committee (Proposal Number: 4148).

Dose response in larval fish

Larval zebrafish have previously been shown to be highly susceptible to *Pseudoloma neurophilia*, with an estimated 310 spores/fish exposure resulting in 2 of 6 fish becoming infected at 7 days post exposure [45]. In order to further assess the potential for extraovum transmission of *P. neurophilia* from spawned ovarian fluid to progeny, 5 dpf larval zebrafish were exposed to low numbers of the parasite and examined by wet mount for the presence of microsporidian spores developing within tissues. Two separate dose-response trials were conducted. Zebrafish embryos (AB line) were obtained from the *P. neurophilia* specific pathogen free zebrafish colony housed at the Sinnhuber Aquatic Research Laboratory at Oregon State University [30]. Embryos were placed in sterilized 100 ml glass beakers (a total of 6 beakers) in groups of 25 embryos/beaker and held in 25 ml of autoclaved system water at 28°C. At 4 dpf, larval fish were fed 500 µl of a concentrated paramecium suspension and four groups were inoculated with a suspension of *P.neurophilia* spores obtained from infected adult fish at a concentration of 300 spores/ml. At 7 days post exposure, all fish were euthanized by an overdose of Finquel (MS-222, Argent Laboratories, Redmond, WA, USA) and examined microscopically by wet mount. The presence of spores was recorded for individual fish.

Infections in progeny from group spawns

Whereas the microsporidium has been observed in eggs at various stages of development and detected by PCR in eggs after spawning [29], it has not been visualized in embryos prior to hatching. Therefore, to obtain large numbers of potentially infected eggs, we harvested eggs that were obtained using group spawning from a known population of infected fish.

One population of presumed infected fish was spawned (approximately 75 fish) in a 10 L tank with a nylon mesh false bottom insert. Fish were held at 28°C overnight. Eggs were

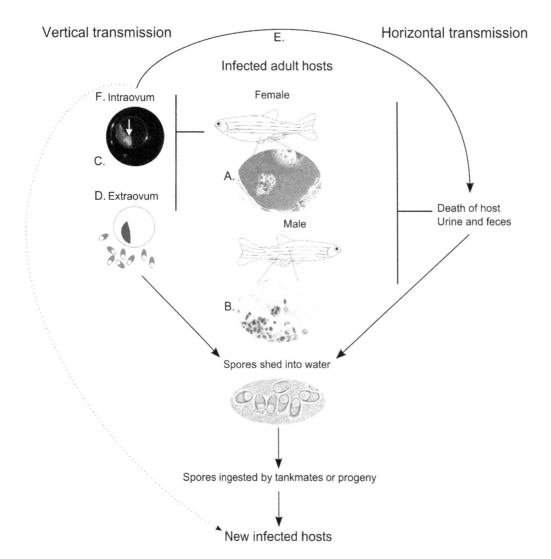

Vertical transmission   E.   Horizontal transmission

Infected adult hosts

F. Intraovum   Female

C.

D. Extraovum   Death of host
                                          Urine and feces

                          Male

B.

Spores shed into water

Spores ingested by tankmates or progeny

New infected hosts

Figure 3. Modes of transmission of *Pseudoloma neurophilia* **in the zebrafish,** *Danio rerio.* **A.** Luna-stained histological section showing *P. neurophilia* spores (red) within a secondary oocyte. Sexually mature female fish have been shown to harbor the parasite in both ovigerous stromal tissue and within various developmental stages of oocytes. **B.** Luna-stained histological section of kidney from an adult male zebrafish with spores present (red) within the epithelium of a renal tubule. The presence of spores in these structures is proposed to be one method by which spores can be released into the environment by live fish. **C.** *P. neurophilia* spores (arrow) present within a developing embryo. **D.** The presence of high numbers of *P. neurophilia* spores in spawn water and the high susceptibility of larval fish can result in infected progeny. **E.** Intraovum transmission of *P. neurophilia* is proposed to result in either the death of the developing embryo or larvae with the subsequent release of spores into the water infecting tank mates or **F.** live, infected animals that then go on to transmit the parasite horizontally and, later, vertically.

collected (approximately 2000), rinsed with sterile system water, placed in 90 mm glass petri dishes (approximately 200 eggs/dish) and held at 28°C in sterile embryo media [56]. At 2.5 dpf, 24 pools of ten eggs each were collected and placed in a 1.5 ml microcentrifuge tube. DNA was extracted from the egg pools and then tested by qPCR using the method described in Sanders & Kent (2011).

At 7 dpf, 112 fish were euthanized by an overdose of Finquel and fixed overnight in Dietrich's fixative. Fish were then placed in a 4X6 agarose array [57], topped with molten agarose and processed for histology. Fifteen serial 5 μm sections were cut from each array, stained with the Luna stain [58], and examined by light microscopy. At 8 weeks old, the remaining fish were euthanized and fixed in Dietrich's fixative. Thirty-six fish were processed for histology, stained with the Luna stain and examined as described above.

A second population of 30 infected adult fish was spawned. Eggs were examined for the presence of opaque regions potentially indicative of aggregates of microsporidian spores using a stereomicroscope with a transmitted light source.

Infections in eggs and spawn water from fish spawned in pairs

In order to determine the prevalence of *P. neurophilia* present within eggs after spawning, we applied the method described by Phelps and Goodwin (2008) to determine the prevalence of *Ovipleistophora ovariae* within golden shiner eggs. Adult zebrafish that were previously exposed to *P. neurophilia* by feeding infected material were spawned in pairs in 1 L of water in a tank with a false-bottom insert. Successfully spawning pairs were euthanized by an overdose of Finquel and fixed in Dietrich's fixative for subsequent processing and histological examination. Single sagittal sections were cut from these adult fish, stained with the Luna stain, and examined for the presence of *P. neurophilia* spores. Water from the spawning tanks was collected and filtered through a 1.2 μm nitrocellulose filter and processed as previously described [29]. Eggs were collected and placed in pools ranging from 10–30 eggs, depending on the total number of eggs collected from each clutch, in 1.5 ml tubes. The commercial DNA/RNA decontaminating product RNase AWAY (Molecular BioProducts, Inc., San Diego, California, USA) was added at full concentration to pools of eggs in order to destroy any *P. neurophilia* spores or DNA present on the exterior of the eggs. After 10 minutes, the RNase AWAY was aspirated, the eggs were rinsed twice with DNA-free water and DNA was extracted from the egg pools as previously described [29] with the exception that sonication was performed in a Bioruptor sonicating bath (Diagenode, Denville, NJ, USA) at high power for 14 minutes (30 s on, 30 s off at 4°C) prior to digestion with lysis buffer.

Phelps and Goodwin [17] determined that the RNase AWAY treatment removed DNA of *Ovipleistophora ovariae* spores outside the eggs of golden shiners. To verify that this method is efficacious for *P. neurophilia* spores outside the eggs of zebrafish, we conducted the following. Eggs were obtained from a population of *P. neurophilia*-free zebrafish and placed in pools of 10 eggs in 1.5 ml tubes. Each tube was spiked with 1,000 *P. neurophilia* spores and half were treated with RNase AWAY and processed as described above with the remainder untreated. *Pseudoloma neurophilia* DNA was detected in the spiked, untreated eggs whereas no *P. neurophilia* DNA was detected by qPCR in spiked egg samples that had been treated with RNase AWAY. In order to determine whether this treatment was affecting DNA present within the egg, qPCR was performed using primers which amplify a fragment of the zebrafish *pou5f1* gene [26] on eggs which had been treated with RNase AWAY and eggs which had not. Crossing threshold (Cq) values were analyzed using Welch's t test in the statistical software package R and found

to not differ significantly between RNase AWAY-treated and untreated eggs (p>0.1).

DNA extracted from egg pools and spawn water filters was analyzed by qPCR on an ABI 7500 sequence detection system using the method previously described [29]. For quantification of parasite in samples testing positive, two standard curves were obtained by spiking spawn water filters and egg pools obtained from a group spawn of known *Pseudoloma neurophilia*-free zebrafish with 100,000 *P. neurophilia* spores. Serial two-fold dilutions were made using DNA extracted from these spiked samples and the quantity of parasite was determined by plotting the Cq value of samples against the standard curve.

Estimation of *P. neurophilia* prevalence from pooled samples obtained from paired spawns was calculated using the statistical software environment R and the functions "llprevr" and "dprev" developed by Williams and Moffitt [59] and is available online (http://www.webpages.uidaho.edu/chrisw/research/prevalence/). This method calculates pathogen prevalence using a maximum likelihood estimator based on the results of tests performed on samples consisting of variable pool sizes and provides 95% confidence intervals.

Acknowledgments

We thank the Oregon State University Veterinary Diagnostic Laboratory for histological slide preparation and S. Giovannoni for helpful comments and review of this manuscript.

Author Contributions

Conceived and designed the experiments: JLS VW MLK. Performed the experiments: JLS VW KC. Analyzed the data: JLS VW KC MLK. Contributed reagents/materials/analysis tools: JLS VW MLK. Wrote the paper: JLS MLK.

References

1. Herre EA (1993) Population structure and the evolution of virulence in nematode parasites of fig wasps. Science 259: 1442–1445. doi:10.1126/science.259.5100.1442.

2. Ebert D, Herre EA (1996) The evolution of parasitic diseases. Parasitol Today 12: 96–101. doi:10.1016/0169-4758(96)80668-5.

3. Ewald PW (1987) Transmission modes and evolution of the parasitism-mutualism continuum. Ann N Y Acad Sci 503: 295–306.

4. Dunn A, Terry R, Smith J (2001) Transovarial transmission in the microsporidia. Adv Parasitol 48: 57–100.

5. Corradi N, Keeling PJ (2009) Microsporidia: a journey through radical taxonomical revisions. Fungal Biol Rev 23: 1–8. doi:10.1016/j.fbr.2009.05.001.

6. Nageli KW (1857) Uber die neue krankheit der seidenraupe und verwandte organismen. Bot Z 15: 760–761.

7. Balbiani G (1882) Sur les microsporidies ou sporospermies des articules. C R Acad Sci 95: 1168–1171.

8. Cavalier-Smith T (1983) A 6-kingdom classification and a unified phylogeny. In: Schenck H, Schwemmler W, Endocytobiology. II. Berlin: Walter de Gruyter. pp. 1027–1034.

9. Williams B, Hirt R, Lucocq J, Embley T (2002) A mitochondrial remnant in the microsporidian *Trachipleistophora hominis*. Nature 418: 865–869. doi:10.1038/nature00990.1.

10. Williams B, Keeling P (2011) 2 Microsporidia–Highly Reduced and Derived Relatives of Fungi. In: Pöggeler S, Wöstemeyer J, Evolution of Fungi and Fungal-Like Organisms, The Mycota XIV. Berlin, Heidelberg: Springer Berlin Heidelberg. pp. 25–36. doi:10.1007/978-3-642-19974-5_2.

11. Canning EU, Lom J, Dyková I (1986) The microsporidia of vertebrates. London: Academic Press.

12. Terry RS, Smith JE, Sharpe RG, Rigaud T, Littlewood DTJ, et al. (2004) Widespread vertical transmission and associated host sex-ratio distortion within the eukaryotic phylum Microspora. Proc Biol Sci 271: 1783–1789. doi:10.1098/rspb.2004.2793.

13. Becnel JJ, Andreadis TG (1999) Microsporidia in Insects. In: Wittner M, Weiss L, The Microsporidia and microsporidiosis. ASM Press. pp. 447–501.

14. Terry RS, Smith JE, Bouchon D, Rigaud T, Duncanson P, et al. (1999) Ultrastructural characterisation and molecular taxonomic identification of

Nosema granulosis n. sp., a transovarially transmitted feminising (TTF) microsporidium. J Eukaryot Microbiol 46: 492–499.

15. Hunt RD, King NW, Foster HL (1972) Encephalitozoonosis: Evidence for vertical transmission. J Infect Dis 126: 212–214.

16. Summerfelt R (1964) A new microsporidian parasite from the golden shiner, *Notemigonus crysoleucas*. Trans Am Fish Soc 93: 6–10. doi:10.1577/1548-8659(1964)93[6:ANMPFT]2.0.CO;2.

17. Phelps NBD, Goodwin AE (2008) Vertical transmission of *Ovipleistophora ovariae* (Microspora) within the eggs of the golden shiner. J Aquat Anim Heal 20: 45–53. doi:10.1577/H07-029.1.

18. Vaney C, Conte A (1901) Sur une nouvelle Microsporidie, *Pleistophora mirandellae*, parasite de l'ovaire d'Alburnus mirandella Blanch. C R Acad Sci Paris 133: 644–646.

19. Maurand J, Loubes C, Gasc C, Pelletier J, Barral J (1988) *Pleistophora mirandellae* Vaney & Conte, 1901, a microsporidian parasite in cyprinid fish of rivers in Herault: taxonomy and histopathology. J Fish Dis 11: 251–258. doi:10.1111/j.1365-2761.1988.tb00546.x.

20. Docker M, Devlin R, Richard J, Khattra J, Kent ML (1997) Sensitive and specific polymerase chain reaction assay for detection of *Loma salmonae* (Microsporea). Dis Aquat Organ 29: 41–48. doi:10.3354/dao029041.

21. Brown A, Kent ML, Adamson M (2010) Low genetic variation in the salmon and trout parasite *Loma salmonae* (Microsporidia) supports marine transmission and clarifies species boundaries. Dis Aquat Organ 91: 35–46. doi:10.3354/dao02246.

22. Traver D, Herbomel P, Patton EE, Murphey RD, Yoder JA, et al. (2003) The zebrafish as a model organism to study development of the immune system. Adv Immunol 81: 253–330.

23. Postlethwait JH, Yan YL, Gates MA, Horne S, Amores A, et al. (1998) Vertebrate genome evolution and the zebrafish gene map. Nat Genet 18: 345–349. doi:10.1038/ng0498-345.

24. Matthews JL, Brown AM, Larison K, Bishop-Stewart JK, Rogers P, et al. (2001) *Pseudoloma neurophilia* n. g., n. sp., a new microsporidium from the central nervous system of the zebrafish (*Danio rerio*). J Eukaryot Microbiol 48: 227–233. doi:10.1111/j.1550-7408.2001.tb00307.x.

25. Sanders JL, Watral V, Kent ML (2012) Microsporidiosis in zebrafish research facilities. ILAR J Natl Res Counc Inst Lab Anim Resour 53: 106–113. doi:10.1093/ilar.53.2.106.

26. Murray K, Dreska M, Nasiadka A, Rinne M, Matthews JL, et al. (2011) Transmission, diagnosis, and recommendations for control of Pseudoloma neurophilia infections in laboratory zebrafish (Danio rerio) facilities. Comp Med 61: 322–329.

27. Cali A, Kent ML, Sanders JL, Pau C, Takvorian PM (2012) Development, ultrastructural pathology, and taxonomic revision of the Microsporidial genus, *Pseudoloma* and its type species *Pseudoloma neurophilia*, in skeletal muscle and nervous tissue of experimentally infected zebrafish Danio rerio. J Eukaryot Microbiol 59: 40–48. doi:10.1111/j.1550-7408.2011.00591.x.

28. Kent ML, Bishop-Stewart JK (2003) Transmission and tissue distribution of *Pseudoloma neurophilia* (Microsporidia) of zebrafish, *Danio rerio* (Hamilton). J Fish Dis 26: 423–426. doi:10.1046/j.1365-2761.2003.00467.x.

29. Sanders JL, Kent ML (2011) Development of a sensitive assay for the detection of *Pseudoloma neurophilia* in the eggs and viscera of the zebrafish Danio rerio. Dis Aquat Organ 96: 145–156. doi:10.3354/dao02375.

30. Kent ML, Buchner C, Watral VG, Sanders JL, Ladu J, et al. (2011) Development and maintenance of a specific pathogen-free (SPF) zebrafish research facility for *Pseudoloma neurophilia*. Dis Aquat Organ 95: 73–79. doi:10.3354/dao02333.

31. Summerfelt RC, Warner MC (1970) Geographical distribution and host-parasite relationships of *Plistophora ovariae* (Microsporida, Nosematidae) in *Notemigonus crysoleucas*. J Wildl Dis 6: 457–465.

32. Phelps NBD, Goodwin AE (2007) Validation of a quantitative PCR diagnostic method for detection of the microsporidian *Ovipleistophora ovariae* in the cyprinid fish *Notemigonus crysoleucas*. Dis Aquat Organ 76: 215–221. doi:10.3354/dao076215.

33. Summerfelt RC (1972) Studies on the transmission of *Plistophora ovariae*, an ovary parasite of the golden shiner. Final Report: Project 4-66-R. National Marine Fisheries Service. pp. 1–19.

34. Pekkarinen M, Lom J, Nilsen F (2002) *Ovipleistophora* gen. n., a new genus for *Pleistophora mirandellae*-like microsporidia. Dis Aquat Organ 48: 133–142.

35. Fryer JL, Sanders JE (1981) Bacterial kidney disease of salmonid fish. Annu Rev Microbiol 35: 273–298. doi:10.1146/annurev.mi.35.100181.001421.

36. Bruno D, Munro A (1986) Observations on *Renibacterium salmoninarum* and the salmonid egg. Dis Aquat Organ 1: 83–87.

37. Evelyn T (1984) The salmonid egg as a vector of the kidney disease bacterium *Renibacterium salmoninarum*. In: ACUIGRUP, Fish diseases, fourth COPRAQ session. Madrid: EDITORA ATP. pp. 111–117.

38. Evelyn TPT, Ketcheson JE, Prosperi-Porta L (1984) Further evidence for the presence of Renibacterium salmoninarum in salmonid eggs and for the failure of povidone-iodine to reduce the intra-ovum infection rate in water-hardened eggs. J Fish Dis 7: 173–182. doi:10.1111/j.1365-2761.1984.tb00921.x.

39. Brown L, Cox W, Levine R (1997) Evidence that the causal agent of bacterial cold-water disease Flavobacterium psychrophilum is transmitted within salmonid eggs. Dis Aquat Organ 29: 213–218.

40. Holt R, Rohovec J, Fryer J (1993) Bacterial cold-water disease. In: Inglis V, Roberts R, Bromage N, Bacterial Diseases of Fish. Boston, MA: Blackwell Scientific Publications. pp. 3–22.

41. Hart NH, Donovan M (1983) Fine structure of the chorion and site of sperm entry in the egg of *Brachydanio*. J Exp Zool 227: 277–296. doi:10.1002/jez.1402270212.

42. Meyers T, Thomas J (1990) Infectious hematopoietic necrosis virus: trends in prevalence and the risk management approach in Alaskan sockeye salmon culture. J Aquat Anim Ldots: 37–41.

43. Meyers T (1998) Healthy juvenile sockeye salmon reared in virus-free hatchery water return as adults infected with infectious hematopoietic necrosis virus (IHNV): a case report and. J Aquat Anim Heal. doi:10.1577/1548-8667(1998)0102.0.CO.

44. Mulcahy D, Pascho R (1985) Vertical transmission of infectious haematopoietic necrosis virus in sockeye salmon, *Oncorhynchus nerka* (Walbaum): isolation of virus from dead eggs and fry. J Fish Dis.

45. Ferguson J, Watral V, Schwindt A, Kent M (2007) Spores of two fish microsporidia (*Pseudoloma neurophilia* and *Glugea anomala*) are highly resistant to chlorine. Dis Aquat Organ 76: 205–214.

46. Pascho RJ, Chase D, McKibben CL (1998) Comparison of the membrane-filtration fluorescent antibody test, the enzyme-linked immunosorbent assay, and the polymerase chain reaction to detect *Renibacterium salmoninarum* in salmonid ovarian fluid. J Vet Diagn Invest 10: 60–66. doi:10.1177/104063879801000111.

47. Traxler G, Roome J, Lauda K, LaPatra S (1997) Appearance of infectious hematopoietic necrosis virus (IHNV) and neutralizing antibodies in sockeye salmon *Onchorynchus nerka* during their migration and maturation period. Dis Aquat Organ 28: 31–38. doi:10.3354/dao028031.

48. LaPatra S, Groberg W, Rohovec J, Fryer J (1991) Delayed fertilization of steelhead (*Oncorhynchus mykiss*) ova to investigate vertical transmission of infectious hematopoietic necrosis virus. Proceedings Second International Symposium on Viruses of Lower Vertebrates. Corvallis: Oregon State University. pp. 261–268.

49. Stead S, Laird L (2002) The handbook of salmon farming. New York NY: Springer.

50. Kent ML, Kieser D (2003) Avoiding the introduction of exotic pathogens with atlantic salmon, *Salmo salar*, reared in British Columbia. In: Cheng-Sheng L, PJ O, Biosecurity in Aquaculture Production Systems: Exclusion of Pathogens and Other Undesirables. Baton Rouge, Louisiana, 70803: The World Aquaculture Society. pp. 43–50.

51. Whipps CM, Kent ML (2006) Polymerase chain reaction detection of *Pseudoloma neurophilia*, a common microsporidian of zebrafish (*Danio rerio*) reared in research laboratories. J Am Assoc Lab Anim Sci JAALAS 45: 36–39.

52. Smith J, Dunn A (1991) Transovarial transmission. Parasitol Today 7: 146–147.

53. Van der Sar AM, Appelmelk BJ, Vandenbroucke-Grauls CMJE, Bitter W (2004) A star with stripes: zebrafish as an infection model. Trends Microbiol 12: 451–457. doi:10.1016/j.tim.2004.08.001.

54. Sullivan C, Kim C (2008) Zebrafish as a model for infectious disease and immune function. Fish Shellfish Immunol 25: 341–350. doi:10.1016/j.fsi.2008.05.005.

55. Phelps HA, Neely MN (2005) Evolution of the zebrafish model: From development to immunity and infectious disease. Zebrafish 2: 87–103. doi:10.1089/zeb.2005.2.87.

56. Westerfield M (2007) The zebrafish book: a guide for the laboratory use of zebrafish (*Danio rerio*). 5th ed. Eugene, OR: University of Oregon Press.

57. Sabaliauskas NA, Foutz CA, Mest JR, Budgeon LR, Sidor AT, et al. (2006) High-throughput zebrafish histology. Methods San Diego Calif 39: 246–254. doi:10.1016/j.ymeth.2006.03.001.

58. Peterson TS, Spitsbergen JM, Feist SW, Kent ML (2011) Luna stain, an improved selective stain for detection of microsporidian spores in histologic sections. Dis Aquat Organ 95: 175–180. doi:10.3354/dao02346.

59. Williams CJ, Moffitt CM (2005) Estimation of pathogen prevalence in pooled samples using maximum likelihood methods and open-source software. J Aquat Anim Heal 17: 386–391. doi:10.1577/H04-066.1.

10

Aquaculture Can Promote the Presence and Spread of Antibiotic-Resistant Enterococci in Marine Sediments

Andrea Di Cesare[1]*, **Gian Marco Luna**[2], **Carla Vignaroli**[1], **Sonia Pasquaroli**[1], **Sara Tota**[1], **Paolo Paroncini**[1], **Francesca Biavasco**[1]

1 Department of Life and Environmental Sciences, Polytechnic University of Marche, Ancona, Italy, **2** Institute of Marine Sciences, National Research Council, Venezia, Italy

Abstract

Aquaculture is an expanding activity worldwide. However its rapid growth can affect the aquatic environment through release of large amounts of chemicals, including antibiotics. Moreover, the presence of organic matter and bacteria of different origin can favor gene transfer and recombination. Whereas the consequences of such activities on environmental microbiota are well explored, little is known of their effects on allochthonous and potentially pathogenic bacteria, such as enterococci. Sediments from three sampling stations (two inside and one outside) collected in a fish farm in the Adriatic Sea were examined for enterococcal abundance and antibiotic resistance traits using the membrane filter technique and an improved quantitative PCR. Strains were tested for susceptibility to tetracycline, erythromycin, ampicillin and gentamicin; samples were directly screened for selected tetracycline [*tet*(M), *tet*(L), *tet*(O)] and macrolide [*erm*(A), *erm*(B) and *mef*] resistance genes by newly-developed multiplex PCRs. The abundance of benthic enterococci was higher inside than outside the farm. All isolates were susceptible to the four antimicrobials tested, although direct PCR evidenced *tet*(M) and *tet*(L) in sediment samples from all stations. Direct multiplex PCR of sediment samples cultured in rich broth supplemented with antibiotic (tetracycline, erythromycin, ampicillin or gentamicin) highlighted changes in resistance gene profiles, with amplification of previously undetected *tet*(O), *erm*(B) and *mef* genes and an increase in benthic enterococcal abundance after incubation in the presence of ampicillin and gentamicin. Despite being limited to a single farm, these data indicate that aquaculture may influence the abundance and spread of benthic enterococci and that farm sediments can be reservoirs of dormant antibiotic-resistant bacteria, including enterococci, which can rapidly revive in presence of new inputs of organic matter. This reservoir may constitute an underestimated health risk and deserves further investigation.

Editor: Melanie R. Mormile, Missouri University of Science and Technology, United States of America

Funding: This work was supported by the Italian Ministry of Research and Education (contract PRIN 2008–I31J10000050001FYXAXL_003) (http://www.miur.it/). The funders had no role in study design, data collection and analysis, decision to publish, or preparation of the manuscript.

Competing Interests: The authors have declared that no competing interests exist.

* E-mail: andrix.di.cesare@alice.it

Introduction

The spread of antibiotic-resistant microorganisms in the environment is widely recognized as an important public health issue, and there is concern on the future ability to treat infectious diseases. Contaminated seawater and sediments can become reservoirs of virulent and antibiotic-resistant strains of fecal bacteria [1,2,3], including enterococci [4], which are capable of transmitting resistance genes to other bacteria by horizontal gene transfer mechanisms, thus contributing to dissemination of resistance genes into the marine environment.

The presence of resistant bacteria raises particular concern at fish-farm sites, where a large use of antibiotics has been made in recent years [2]. Resistant bacteria can reach aquaculture sites also *via* agricultural and urban wastewaters; these contain the typical intestinal flora and pathogens of animals and humans, which are usually resistant to antibiotics [5]. These emerging contaminants can accumulate in the underlying sediments, where they interact with the benthic microbial communities [6]. Even in absence of continuous antimicrobial administration, resistant microorganisms can persist in protected reservoirs such as sediments or fish gut [4,7]. Sediments are a particularly favorable environment for benthic allochthonous bacteria since they provide nutrients and protection from biotic and abiotic stress, allowing their long-term persistence in a culturable state or even their re-growth [8,9,10].

Enterococci are part of the human and animal intestinal microflora and are used as fecal indicator bacteria (FIB) for monitoring recreational waters and for assessing potential risks for human health [11,12]. They have been recognized as major agents of nosocomial infections [13,14] whose treatment is often complicated by antibiotic resistance (AR), either intrinsic and acquired [15]. Acquired AR is mainly due to integration of external genetic material mediated by transposon or plasmid transfer [16,17].

A greater understanding of the stress-resistance ability of *Enterococcus* species, virulence traits and AR is required for a full appreciation of the complexity of *Enterococcus* species in causing human disease [18]. While a number of papers have documented the presence, fate and reservoirs of enterococci in coastal marine systems and other aquatic environments [11,19,20,21,22], little information is available on the distribution of resistant enterococci and their determinants at aquaculture sites [4,23].

In this study, sediment samples were analyzed to investigate the impact of fish aquaculture on the spread and abundance of tetracycline-, macrolide-, β-lactam- and aminoglycoside-resistant benthic enterococci. Both culture-dependent and molecular tools were used to quantify enterococcal abundance and directly search for resistance genes. *In vitro* enrichment assays in the presence of antibiotics were also carried out to investigate the possible consequences of their release into the marine environment, with emphasis on the abundance of benthic enterococci and the profile of resistance genes, which are potentially transmissible between both autochthonous and allochthonous bacterial species.

Materials and Methods

Ethics Statement

All necessary permits were obtained for the described field studies. The approval for sediment sampling was obtained from the owner of the private aquaculture facility, who wishes to remain anonymous. The sampling activities were not performed in a protected area and they did not involve invertebrates, plant species, corals or fish.

Site, Sediment Sampling and Environmental Variables

Sediments were collected in June 2011 at a fish farm in Varano lagoon (central Italy; Figure 1). The farm consists of several ponds receiving water from the lagoon through a canal. Samples were collected from 3 stations in the largest pond (latitude 41° 54' 33.37" N; longitude 15° 45' 8.91" E), which measured 54×21×2 m and hosted about 14,000 seabream and seabass (data provided by the owner). Station (St.) 1 was located in an area of the pond used for feed administration; St. 2 was still in the pond but far from the feeding area; and St. 3 was upstream, in the water supply canal connecting the pond to the lagoon, and was thus unaffected by farming activities (control station). The water temperature was 29°C. The farm owner denied all antibiotic use in the pond, either for therapeutic or for growth promotion purposes, and reported using exclusively non-medicated feed (Hendrix, Verona, Italy). Sediments were collected using sterile Plexiglas corers, placed in sterile containers, and stored in the dark until delivery to the laboratory (max 5 h). Sub-samples were used for cultural and molecular microbiological analyses and to determine grain size, (by the sieving technique) and total organic matter, which was determined as the difference between dry weight (60°C, 48 h) of the sediment and the weight of the residue after combustion for 2 h at 450°C [8].

Enterococcus spp. Isolation and Enumeration

The membrane filter (MF) technique was used for the enumeration of culturable *Enterococcus* spp. Briefly, 30 g of sediment from each station was suspended in 300 ml of saline solution, vigorously shaken and sonicated to detach bacteria as described previously [1]. The supernatant was pre-filtered through a 30 μm membrane; 10 ml aliquots of the suspension and 1/10 dilutions were filtered (0.2 μm pore size), and filters were placed on Slanetz-Bartley plates (SB; Oxoid, Basingstoke, UK) and incubated for 48 h at 37°C. Grown colonies were counted and the abundance of *Enterococcus* spp. was expressed as CFU/g of wet sediment. Selected colonies were further amplified on SB plates and incubated for 48 h at 42°C. To establish if they belonged to the genus *Enterococcus* amplified cultures were tested for growth at 42°C in the presence of 6.5% NaCl.

Antibiotic Susceptibility Testing

Minimum Inhibitory Concentrations (MIC) were determined by broth microdilution according to CLSI guidelines [24]; the results were interpreted according to CLSI M100-S21 (2011). *Enterococcus faecalis* ATCC 29212 was used as the control strain.

Tetracycline (TET), erythromycin (ERY), ampicillin (AMP) and gentamycin (CN) were purchased from Sigma-Aldrich (Saint Louis, MO, USA).

In vitro Enrichment by Sediment Incubation in Rich Broth Supplemented with Antibiotics

Aliquots of sediment from the 3 stations were incubated in rich medium in the presence of one of the four antibiotics. Four sterile bottles per station were prepared, each containing 5 g of sediment and 50 ml Brain Hearth Infusion (BHI) broth (Oxoid) added with TET (10 μg/ml), ERY (20 μg/ml); AMP (20 μg/ml) or CN (250 μg/ml). Antibiotic concentrations were those generally used to select antibiotic-resistant isolates [4]. After 24 h incubation at 37°C DNA was extracted from sediment and broth.

DNA Extraction and Purification

DNA was extracted using different protocols depending on sample type. The commercial kits Ultra Clean Mega Soil DNA Isolation (MoBio, Carlsbad, CA, USA) and Fast DNA SPIN Kit for Soil (Q•BIO Gene, Fountain Parkway Solon, OH, USA) without (10 g aliquots) and with antibiotic enrichment (0.5 g aliquots) were used for farm sediments, whereas the procedure described by Hynes et al. (1992) was used for the antibiotic-enriched broth cultures [25]. DNA extracts to be used undiluted were further purified with the Wizard SV Gel and PCR Clean-Up System (Promega, Madison, WI, USA) to remove PCR inhibitors.

PCR Detection of Resistance Genes

Two Multiplex-PCR assays were developed to detect simultaneously *tet*(M), *tet*(L) and *tet*(O) and *erm*(B), *erm*(A) and *mef*, respectively. Three new primer pairs were designed to detect *tet*(O), *erm*(A) and *erm*(B) genes. For each target gene, several sequences deposited in the NCBI database were converted into FASTA format using the *NCBI Genome Workbench* software (http://www.ncbi.nlm.nih.gov/tools/gbench/), aligned using the *Clustal-XII* software (http://www.clustal.org/), and the primers were designed on the conserved regions using the *NetPrimer* software (http://www.premierbiosoft.com/netprimer/index.html). Those showing the highest specificity by BLAST analysis were selected. PCR assays were performed in a final volume of 50 μl containing 5 μl of DNA (diluted 100 times or undiluted and purified) using a T Personal thermal cycler (Biometra, Göttingen, Germany). The PCR cycling program was as follows: 95°C for 10 min, followed by 35 cycles at 94°C for 30 s, 53°C [*tet*(M), *tet*(L), *tet*(O)] or 54°C [*erm*(B), *erm*(A), *mef*] for 30 s, 72°C for 90 s and final extension at 72°C for 7 min. Each mix contained 600 μM dNTPs, 6 mM MgCl$_2$, 1× Buffer, 0.5 μM of each primer [1 μM of those targeting *tet*(M)], and 1.25 U hot-start Taq DNA polymerase (AmpliTaq Gold, Applied Biosystem, Foster City, CA, USA). The resistance genes *blaZ* and *aac (6')-Ie aph (2")-Ia* were sought as previously described [26]. The control strains and primer pairs used in PCR assays are reported in Tables 1 and 2, respectively.

Enumeration of Enterococci by Real Time Quantitative PCR (qPCR)

A Real Time Quantitative PCR (qPCR) assay was used to determine *Enterococcus* spp. abundance. The standard calibration curve was generated using a purified 23S rDNA amplicon obtained by a PCR reaction performed using DNA from *E. faecalis* ATCC 29212 and primers ECST748F and ENC854R, as previously described [8]. The 23S amplicon was purified by Gene Elute PCR Clean-up (Sigma-Aldrich) and quantified using an ND-1000 Nanodrop (Thermo Scientific, Wilmington, NC, USA).

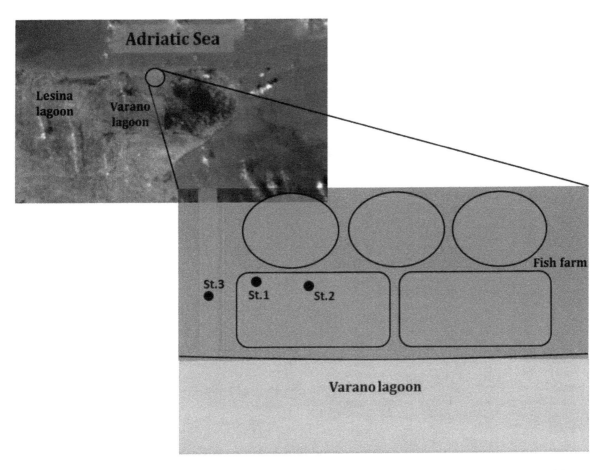

Figure 1. Location of the fish farm and of the sampling stations. The map is from http://earthobservatory.nasa.gov/, image courtesy Jesse Allen.

qPCRs were performed using the iCycler iQ-5 (Biorad, Hercules, OR, USA) in a 25 µl volume containing 2.5 µl of sample DNA, 0.2 µM of each (ECST748F and ENC854R), 12.5 µl of iQ^{TM} SYBR® Green Supermix (Biorad), and Milli Q water (Millipore, Billerica, MA, USA) to reach the final volume. The amplification reaction was as follows: 95°C for 3 min, followed by 35 cycles at 95°C for 15 s, 60°C for 30 s and 72°C for 15 s. Melt curve analysis was carried out from 59°C to 95°C, with increments of 0.5°C/10 s. Suitable dilutions (i.e. containing from 10^{-6} to 10^{-9} ng of DNA) of 23S rDNA of the purified amplicon of *E. faecalis* ATCC 29212 were used for construction of the standard curve. Similar PCR reactions using DNA from sediment samples, either diluted and undiluted to account for potential qPCR inhibition [8,27], were run together. These analyses consistently showed that undiluted DNA extracts were inhibited, as demonstrated by a threshold cycle (Ct) delay between qPCR results on this DNA extract and serial 10-fold dilutions (1:10 and 1:100). While the expected Ct difference between 10-fold dilutions in the absence of inhibition is 3.32, in our samples it was typically between 1 and 2 cycles less than expected without inhibition (data not shown). Each reaction was performed in triplicate. Reproducibility of the qPCR reaction was assessed by determining intra- and interassay repeatability of the standard curve. The coefficient of variation (CV) to evaluate intra-assay repeatability was calculated on the basis of the Ct value, by testing in triplicate the 4 dilutions containing from 10^{-6} to 10^{-9} ng DNA of the target gene. The CV for interassay reproducibility was calculated based on the Ct value of the 4 dilutions in four different analysis sessions. The Limit of Detection (LOD) was determined [28].

Data Analysis

The abundance of *Enterococcus* spp. cells was calculated on the basis of qPCR results as follows: assuming that 1 base pair (bp) of double-stranded DNA is equal to 660 Da (1 Da = 1.66 $\times 10^{-15}$ ng in the metric system), 1 bp is equal to 1.095×10^{-12} ng. Since amplicon size is 91 bp, one copy of the amplicon corresponds to 0.0996×10^{-9} ng DNA. Considering that each enterococcal cell contains 4 copies of 23S rDNA [29], each cell contains

Table 1. Control strains used in PCR assays.

Bacterial strain	Resistance gene(s)	Reference or source
E. faecium CM 4·2E	*tet*(M),*tet*(L),*erm*(B)	[4,44]
E. faecalis PM 2·2T	*tet*(M), *tet*(L)	[4,44]
E. faecium CF 2·1E	*tet*(M),*tet*(L),*erm*(B)	[4,44]
S. aureus MU 50	*erm*(A)	ATCC[a]
*S. aureus*29213	*blaZ*	ATCC[a]
S. pyogenes 7008	*tet*(O),*mef*	[45]
E. faecium M48	*aac (6')-leaph (2")-la*	[26]

[a]American Type Culture Collection.

Table 2. Primer pairs used to detect resistance genes in PCR assays.

Target gene	Primer sequence (5'→3')	Product size (bp)	Reference
tet(M)	1-GTTAAATAGTGTTCTTGGAG 2-CTAAGATATGGCTCTAACAA	657	[46]
tet(L)	1-CATTTGGTCTTATTGGATCG 2-ATTACACTTCCGATTTCGG	475	[46]
tet(O)	1-AGGGGGTTCTTTATGGCTG 2-CGTGAGAGATATTCCTGCG	223	This study
erm(B)	1-CCGAACACTAGGGTTGCTC 2-ATCTGGAACATCTGTGGTATG	139	This study
erm(A)	1-TAACATCAGTACGGATATTG 2-AGTCTACACTTGGCTTAGG	200	This study
mef	1- AGTATCATTAATCACTAGTGC 2-TTCTTCTGGTACTAAAAGTGG	348	[47]
blaZ	1-ACTTCAACACCTGCTGCTTTC 2-TAGGTTCAGATTGGCCCTTAG	240	[26]
aac (6')-Ie aph (2")-Ia	1-GAGCAATAAGGGCATACCAAAAATC 2-CCGTGCATTTGTCTTAAAAAACTGG	505	[48]

0.3984×10^{-9} ng of the 23S rDNA target sequence. The enterococcal abundance in the amplified samples was then calculated by the following formula: amplicon weight (ng)/$(0.0996 \times 10^{-9} \times 4)$. Although it is well known that multiple copies of enterococcal 23S rDNA are found in the *Enterococcus* genome (*E. faecalis* and *Enterococcus faecium* contain 4 and 6 copies, respectively), the number of 23S rRNA gene copies per genome has not been determined in all species. The use of 4 copy numbers for qPCR analyses of enterococcal populations in marine samples may introduce a bias, potentially affecting assay accuracy; however, it is currently used worldwide for qPCR determinations involving enterococci [29], thus allowing comparisons with other studies.

Final counts were expressed as cells/g of wet sediment.

One-way analysis of variance (ANOVA) was used to test for differences in the abundance of benthic enterococci in sediments before and after antibiotic exposure. Differences were considered significant at P values <0.05.

Results

Sediment Analysis

Sediment samples were collected from three stations: an area used for feed administration (St. 1), an area in the same pond located 20 m downstream of the feeding area (St. 2), and an area upstream of the farm that was therefore not influenced by aquaculture activities (St. 3). Major differences in the main environmental characteristics were noted in sediments from the three stations, but especially between those from the farm (St. 1 and St. 2) and those from the control station. The former were dominated by the silt–clay fraction (<63 μm, 93% and 89%) and characterized by a very high organic matter content (28.3 mg/g and 24.3 mg/g), whereas the latter were characterized by a lower percentage of silt-clays (73%) and a lower organic matter concentration (13.4 mg/g).

Optimization of qPCR for the Enumeration of Enterococci

The qPCR assay developed in this study showed very high reproducibility and repeatability. Intra- and interassay reproducibility were both very satisfactory; CVs were 1.8% (at the concentration of 10^{-6} ng of the target gene), 1.4% (10^{-7} ng), 0.5% (10^{-8} ng), and 1.2% (10^{-9} ng) for intra-assay and 1.0% (of 10^{-6} ng), 1.3% (10^{-7} ng), 3.4% (10^{-8} ng), and 1.5% (10^{-9} ng) for interassay comparisons. The LOD of the qPCR assay was 5.225×10^{-9} ng, corresponding to 52 copies of the 23S rRNA gene of *E. faecalis* ATCC 29212 (reference strain),

corresponding to 13 cells. The qPCR showed a linear dynamic range over the DNA concentrations tested for the target gene. The average efficiency of the qPCR reaction was 109.2%, while the average regression coefficient of R^2 was 0.97. Melt curve analysis showed a clear and reproducible melting peak between 80.5 and 81°C.

Enumeration of Enterococci

The abundance of culturable enterococci estimated with the MF technique was respectively 1.12×10^2, 1.00×10^2, and 0.12×10^2 CFU/g in sediments from St. 1, St. 2, and St. 3, whereas total enterococcal abundance estimated by qPCR was respectively 5.03×10^5, 1.69×10^5, and 5.70×10^5 cells/g.

The same sediments were analyzed by qPCR after *in vitro* incubation with antibiotic to assess the response of enterococci to selective pressure and the effect of this exposure on the AR gene profile of the sediment. The abundance of benthic enterococci in sediments collected inside the farm increased significantly in presence of AMP and CN, but not of TET and ERY (Figure 2). The increase was 4 (AMP) and 8 times (CN) ($p<0.01$) in samples from St. 1, and 11 times (AMP and CN; $p<0.01$) in those from St. 2. Enterococcal abundance did not increase in sediments from St. 3 (Figure 2).

Antibiotic Susceptibility Testing of the Isolated Enterococci

A total of 476 isolates (225 from St. 1, 228 from St. 2, and 23 from St. 3) were recovered with the MF technique, and 250 were streaked on SB medium; 150 cultures showing good growth were used in antibiotic susceptibility tests. Neither MICs above the resistance breakpoint nor high-level resistance to CN was detected.

Direct Detection of Resistance Genes, before and after Antibiotic Enrichment

We sought the TET, macrolide, AMP, and aminoglycosides resistance genes more frequently detected in enterococci. Two newly-developed multiplex PCRs were used to detect directly selected TET [tet(M), tet(L) and tet(O)] and macrolide [erm(B), erm(A) and mef] resistance genes in sediment samples; AMP (blaZ) and aminoglycoside [(aac (6')-Ie aph (2")-Ia] resistance genes were sought using individual assays. tet(M) and tet(L) were detected at all 3 stations, whereas genes coding for ERY, AMP and CN resistance were never found (Table 3). No difference in

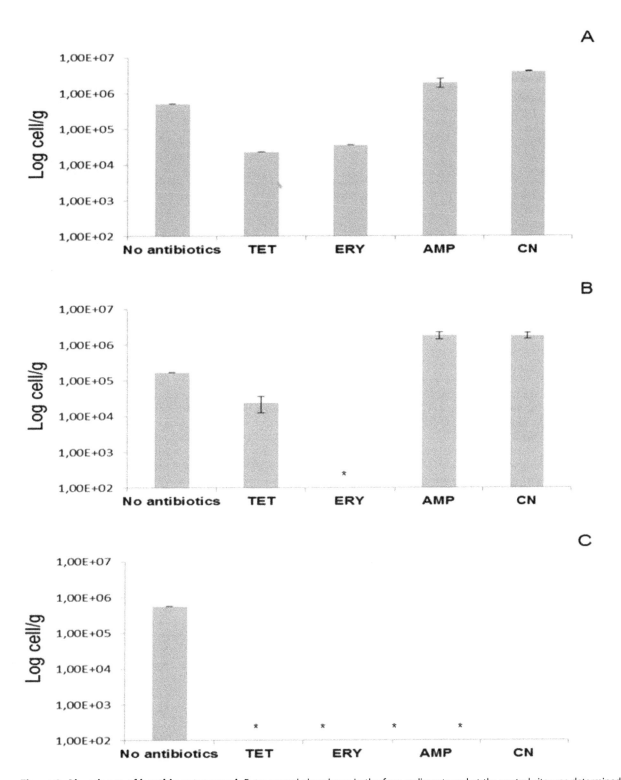

Figure 2. Abundance of benthic enterococci. Enterococcal abundance in the farm sediments and at the control site was determined by qPCR before and after incubation with antibiotic-supplemented BHI broth. A, St. 1; B, St. 2 and C, St. 3. *Not detectable i.e.<LOD of the qPCR assay.

the AR gene profiles was seen after incubation with AMP and CN, while considerable changes were observed after incubation with TET and ERY. After growth in presence of TET, *tet*(L) and *tet*(M) were no longer detectable in any sample nor in those from St. 3, respectively; *tet*(O) became detectable in the sample from St. 2. Incubation with ERY resulted in detection of *erm*(B)

in sediments from St. 1 and St. 2 and of *mef* in those from St. 1 and St. 3.

Discussion

Enterococci are among the major etiological agents of hospital-associated infections [30]. They are characterized by a proneness

Table 3. Resistance genes detected before and after sediment incubation in antibiotic-supplemented BHI broth.

Station	Resistance genes							
	Before antibiotic exposure				After antibiotic exposure*			
	TET	ERY	AMP	CN	TET	ERY	AMP	CN
St.1	tet(M),tet(L)	-	-	-	tet(M)	erm(B),mef	-	-
St.2	tet(M),tet(L)	-	-	-	tet(M),tet(O)	erm(B)	-	-
St.3	tet(M),tet(L)	-	-	-	-	mef	-	-

*detected both in sediment and in broth.

to acquire resistance determinants [31] and by rapid adaptation to environmental conditions [11,32,33]. Aquaculture is believed to contribute to the spread and persistence of AR in the environment and indeed antibiotic-resistant bacteria have frequently been detected at aquaculture sites [4,34]. The study of resistant enterococci in sediment under aquaculture farms has the potential to disclose important information about the ecology of these bacteria in habitats outside the normal host and about the environmental factors, that can contribute to their evolution in the marine environment.

Several studies have addressed the dynamics and abundance of enterococci in seawater and sediment using different approaches, including MF and molecular assays based on qPCR [29,35] and RT-qPCR [8,36]. However, little is known of the impact of fish farms on the origin and spread of antibiotic-resistant strains and related AR genes [4,23]. We found higher counts of benthic enterococci with qPCR than with culture methods in line with previous data showing that cultivation-based techniques underestimate bacterial abundance in marine samples, due to large amounts of nonculturable cells [8]. Furthermore, the finding of a greater amount of culturable enterococci within the farm than in control sediments indicates that benthic enterococci under aquaculture sites may be more metabolically active. This could depend on large inputs of labile organic nutrients connected with farming activities, described by other researchers [37] and found in the present work, where the concentration of sedimentary organic matter and the silt–clay fraction found in the breeding pond were greater than the one determined in the control station. Since enterococci are part of the fish gut microbiota [38], the accumulation of fecal matter in the sediment beneath the fish farm could also directly contribute to the amount and diversity of the enterococci recovered from the farm.

Different environmental factors may have influenced the discrepancies in enterococcal counts found between samples collected inside and outside the farm. Local bird populations feeding on aquaculture might be involved in delivery of fecal material, with its burden of intestinal, possibly antibiotic-resistant, enterococci [39]. Local waste impacting the various sites differently may also be implicated, although no landfills are found close to the farm area. A contribution from bird fecal material cannot of course be excluded, but it is probably an inherent risk factor in fish farms.

This study focused on assessing the presence of antibiotic-resistant enterococci, to gain insights into the impact of fish-farming activities on the presence and spread of resistant strains.

The lack of use of antibiotics, declared by the owner of the farm, does not contrast with our results. Indeed no antibiotic-resistant enterococcal strains were isolated before the antibiotic-enrichment step, while in a farm where antibiotics had not been used over only the previous two years we recovered 12% of resistant strains [4].

Antibiotic exposure induced considerable changes in the abundance of benthic enterococci in farm (St. 1 and St. 2) compared to control sediments (St. 3). AMP and CN clearly favored enterococcal growth in farm sediments, as demonstrated by the fact that the number of bacteria detected after antibiotic exposure exceeded the one obtained before exposure; in contrast, St. 3 samples were qPCR-negative, yielding enterococcal counts below the method's sensitivity threshold. Exposure to TET stimulated enterococcal growth in samples from both farm stations; ERY exerted a similar effect on St. 1 sediments, despite the fact that the number of enterococci was lower there than in untreated samples. Even though direct counts cannot of course be compared with counts performed after growth in rich medium, the comparison may nonetheless provide indirect evidence of the abundance of resistant bacteria in the original sediments. The low counts obtained after exposure to TET and ERY can be explained by the presence of a small fraction of enterococci resistant to these antibiotics or by the slow growth rate of resistant bacteria.

None of the enterococcal isolates was resistant to any of the antibiotics tested, including TET; however, tet(M) and tet(L) were detected by direct PCR in sediments from all sites before antibiotic enrichment. This is not surprising, because tet genes are widely disseminated in the environment even in the absence of antimicrobial use [4,40]; moreover, an environmental evolution of TET resistance has recently been suggested [41]. The lack of TET-resistant enterococci in any of our sediment samples may probably be explained by considering that tet genes are also carried by non-enterococcal strains, including autochthonous marine bacteria. The presence of dormant bacteria, including enterococci, is a further possibility that could moreover explain the *ex novo* detection of tet(O), erm(B), and mef after incubation in antibiotic-supplemented rich medium, a source of readily available nutrients which coupled to incubation at 37°C may have provided a suitable environment for bacterial reactivation and growth.

Differences in resistance gene profiles before and after exposure to antibiotics and rich medium were particularly evident in the sediments collected under the farm (St. 1 and St. 2) incubated with ERY. Erythromycin may have selected and revived macrolide-resistant bacteria, making macrolide resistance genes, i.e. erm(B) and mef, detectable by PCR. AR gene profiles were also altered after incubation with TET, with *ex novo* detection of tet(O) and loss of tet(L), whereas tet(M) was uniformly detected. These findings may be explained by the fact that whereas tet(O) and tet(M) are ribosomal protection genes conferring high-level resistance, tet(L) codes for antibiotic efflux, which is characterized by a low-level of resistance. The present data indicate a possible contribution of aquaculture practices to the selection of genetic determinants conferring high-level resistance. The failed detection of *erm*(B) in sediments from the control station seems to corroborate this hypothesis.

Despite being limited to a single farm, our data indicate that aquaculture environments may not only select for resistant strains when using antibiotics, as reported previously [2,42], but also influence the metabolic activity of benthic enterococci due to the abundance of organic carbon sources. Since the owner denied all antibiotic use and we found no resistant strains before the enrichment step, these data suggest that aquaculture may constitute a reservoir of resistance genes irrespective of antibiotic use. This view is supported by the data obtained from the control station, where the lack of enterococci after antibiotic exposure is consistent with the lower abundance of resistant strains outside than inside the farm.

In conclusion, our findings indicate that aquaculture has the potential to affect antibiotic-resistant benthic enterococcal populations and that fish-farm sediments can contain AR genes of putative enterococcal origin. Moreover, besides hosting active and culturable strains, fish-farm sediments could constitute reservoirs of dormant resistant enterococci capable of quick reactivation following new nutrient inputs into the system. The hypothesis

agrees with recent studies suggesting that dormancy generates a seed bank, i.e. a reservoir of dormant bacteria that can eventually revive under different environmental conditions [43]. The antibiotic-resistant enterococcal seed bank found in fish-farm sediments might constitute an underrated health risk and stresses the need for long-term monitoring of the effects of aquaculture operations on potentially pathogenic microbes, to assess their antimicrobial resistance properties and the potential for their spread and transmission to different bacterial species.

Acknowledgments

The authors are grateful to Dr. Andrea Brenciani for supplying the reference strains *Streptococcus pyogenes* 7008, to Dr. Claudio Palmieri and to Dr. Stefano Bompadre for technical assistance.

Author Contributions

Conceived and designed the experiments: ADC GML FB. Performed the experiments: ADC GML CV SP ST. Analyzed the data: ADC GML CV SP ST. Contributed reagents/materials/analysis tools: ST PP. Wrote the paper: ADC GML CV FB.

References

1. Luna GM, Vignaroli C, Rinaldi C, Pusceddu A, Nicoletti L, et al. (2010) Extraintestinal *Escherichia coli* carrying virulence genes in coastal marine sediments. Appl Environ Microbiol 76: 5659–5668.
2. Seyfried EE, Newton RJ, Rubert KF 4th, Pedersen JA, McMahon KD (2010) Occurrence of tetracycline resistance genes in aquaculture facilities with varying use of oxytetracycline. Microb Ecol 59: 799–807.
3. Vignaroli C, Luna GM, Rinaldi C, Di Cesare A, Danovaro R, et al. (2012) New sequence types and multidrug resistance among pathogenic *Escherichia coli* isolates from coastal marine sediments. Appl Environ Microbiol l78: 3916–3922.
4. Di Cesare A, Vignaroli C, Luna GM, Pasquaroli S, Biavasco F (2012) Antibiotic-resistant enterococci in seawater and sediments from a coastal fish farm. Microb Drug Resist 18: 502–509.
5. Cabello FC (2006) Heavy use of prophylactic antibiotics in aquaculture: a growing problem for human and animal health and for the environment. Environ Microbiol 8: 1137–1144.
6. Kümmerer K (2009) Antibiotics in the aquatic environment–a review–part II. Chemosphere 75: 435–441.
7. Stachowiak M, Clark SE, Templin RE, Baker KH (2010) Tetracycline-Resistant *Escherichia coli* in a Small Stream Receiving Fish Hatchery Effluent. Water Air Soil Pollut 211: 251–259.
8. Luna GM, Dell'Anno A, Pietrangeli B, Danovaro R (2012) A new molecular approach based on qPCR for the quantification of fecal bacteria in contaminated marine sediments. J Biotechnol 157: 446–453.
9. Jeng HC, England AJ, Bradford HB (2005) Indicator organisms associated with stormwater suspended particles and estuarine sediment. J Environ Sci Health A Tox Hazard Subst Environ Eng 40: 779–791.
10. Pianetti A, Bruscolini F, Sabatini L, Colantoni P (2004) Microbial characteristics of marine sediments in bathing area along Pesaro-Gabicce coast (Italy): a preliminary study. J Appl Microbiol 97: 682–689.
11. Badgley BD, Nayak BS, Harwood VJ (2010) The importance of sediment and submerged aquatic vegetation as potential habitats for persistent strains of enterococci in a subtropical watershed. Water Res 44: 5857–5866.
12. Heaney CD, Sams E, Wing S, Marshall S, Brenner K, et al. (2009) Contact with beach sand among beach goers and risk of illness. Am J Epidemiol 170: 164–172.
13. Cattaneo C, Casari S, Bracchi F, Signorini L, Ravizzola G, et al. (2010) Recent increase in enterococci, viridans streptococci, *Pseudomonas* spp. and multiresistant strains among haematological patients, with a negative impact on outcome. Results of a 3-year surveillance study at a single institution. Scand J Infect Dis 42: 324–332.
14. Sundsfjord A, Willems R (2010) *Enterococcus* research: recent development and clinical challenges. Clin Microbiol Infect 16: 525–526.
15. Manolopoulou E, Sarantinopoulos P, Zoidou E, Aktypis A, Moschopoulou E, et al. (2003) Evolution of microbial populations during traditional Feta cheese manufacture and ripening. Int J Food Microbiol 82: 153–161.
16. Champagne J, Diarra MS, Rempel H, Topp E, Greer CW, et al. (2011) Development of a DNA microarray for enterococcal species, virulence, and antibiotic resistance gene determinations among isolates from poultry. Appl Environ Microbiol 77: 2625–2633.
17. Paulsen IT, Banerjei L, Myers GS, Nelson KE, Seshadri R, et al. (2003) Role of mobile DNA in the evolution of vancomycin-resistant *Enterococcus faecalis*. Science 299: 2071–2074.
18. Fisher K, Phillips C (2009) The ecology, epidemiology and virulence of Enterococcus. Microbiology 155: 1749–1757.
19. Wright ME, Abdelzaher AM, Solo-Gabriele HM, Elmir S, Fleming LE (2011) The inter-tidal zone is the pathway of input of enterococci to a subtropical recreational marine beach. Water Sci Technol 63: 542–549.
20. Korajkic A, Brownell MJ, Harwood VJ (2011) Investigation of human sewage pollution and pathogen analysis at Florida Gulf coast beaches. J Appl Microbiol 110: 174–83.
21. Lata P, Ram S, Agrawal M, Shanker R (2009) Enterococci in river Ganga surface waters: propensity of species distribution, dissemination of antimicrobial-resistance and virulence-markers among species along landscape. BMC Microbiol 9: 140.
22. Moore DF, Guzman JA, McGee C (2008) Species distribution and antimicrobial resistance of enterococci isolated from surface and ocean water. J Appl Microbiol 4: 1017–1025.
23. Petersen A, Dalsgaard A (2003) Species composition and antimicrobial resistance genes of *Enterococcus* spp., isolated from integrated and traditional farms in Thailand. Environ Microbiol 5: 395–402.
24. Clinical and Laboratory Standards Institute (2009) Methods for Dilution Antimicrobial Susceptibility Tests for Bacteria That Grow Aerobically; Approved Standard–Eighth Edition Document M07-A8, Vol. 29, No 2. Clinical and Laboratory Standards Institute, Wayne, PA.
25. Hynes WL, Ferretti JJ, Gilmore MS, Segarra RA (1992) PCR amplification of streptococcal DNA using crude cell lysates. FEMS Microbiol Lett 73: 139–142.
26. Garofalo C, Vignaroli C, Zandri G, Aquilanti L, Bordoni D, et al. (2007) Direct detection of antibiotic resistance genes in specimens of chicken and pork meat. Int J Food Microbiol 113: 75–83.
27. Cao Y, Griffith JF, Dorevitch S, Weisberg SB (2012) Effectiveness of qPCR permutations, internal controls and dilution as means for minimizing the impact of inhibition while measuring *Enterococcus* in environmental waters. J Applied Microbiol 113: 66–75.
28. Bustin SA, Benes V, Garson JA, Hellemans J, Huggett J, et al. (2009) The MIQE guidelines: minimum information for publication of quantitative real-time PCR experiments. Clin Chem 55: 611–622.
29. Srinivasan S, Aslan A, Xagoraraki I, Alocilja E, Rose JB (2011) *Escherichia coli*, enterococci, and *Bacteroides thetaiotaomicron* qPCR signals through wastewater and septage treatment. Water Res 45: 2561–2572.
30. Hidron AI, Edwards JR, Patel J, Horan TC, Sievert DM, et al. (2008) NHSN annual update: antimicrobial-resistant pathogens associated with healthcare-associated infections: annual summary of data reported to the National Healthcare Safety Network at the Centers for Disease Control and Prevention, 2006–2007. Infect. Control Hosp Epidemiol 29: 996–1011.
31. Arias CA, Contreras GA, Murray BE (2010) Management of multidrug-resistant enterococcal infections. Clin Microbiol Infect 16: 555–562.
32. Leavis HL, Willems RJ, van Wamel WJ, Schuren FH, Caspers MP et al. (2007) Insertion sequence-driven diversification creates a globally dispersed emerging multiresistant subspecies of E. faecium. PLoS Pathog 3: e7.
33. Laverde Gomez JA, Hendrickx AP, Willems RJ, Top J, Sava I, et al. (2011) Intra- and interspecies genomic transfer of the *Enterococcus faecalis* pathogenicity island. PLoS One 6: e16720.

34. Tamminen M, Karkman A, Lõhmus A, Muziasari WI, Takasu H, et al. (2011) Tetracycline Resistance Genes Persist at Aquaculture Farms in the Absence of Selection Pressure. Environ Sci Technol 45: 386–391.

35. Ferretti JA, Tran HV, Cosgrove E, Protonentis J, Loftin V, et al. (2011) Comparison of *Enterococcus* density estimates in marine beach and bay samples by real-time polymerase chain reaction, membrane filtration and defined substrate testing. Mar Pollut Bull 62: 1066–1072.

36. Bergeron P, Oujati H, Catalán Cuenca V, Huguet Mestre JM, Courtois S (2011) Rapid monitoring of *Escherichia coli* and *Enterococcus* spp. in bathing water using reverse transcription-quantitative PCR. Int J Hyg Environ Health 214: 478–484.

37. Pusceddu A, Fraschetti S, Mirto S, Holmer M, Danovaro R (2007) Effects of intesive mariculture on sediment biochemistry. Ecological Applications 17: 1366–1378.

38. Barros J, Igrejas G, Andrade M, Radhouani H, López M, et al. (2011) Gilthead seabream (Sparus aurata) carrying antibiotic resistant enterococci. A potential bioindicator of marine contamination?. Mar Poll Bull 62: 1245–1248.

39. Radhouani H, Igrejas G, Pinto L, Gonçalves A, Coelho C, et al. (2011) Molecular characterization of antibiotic resistance in enterococci recovered from seagulls (Larus cachinnans) representing an environmental health problem. J Environ Monit 13: 2227–2233.

40. Pallecchi L, Bortoloni A, Paradisi F, Rossolini GM (2008) Antibiotic resistance in the absence of antimicrobial use: mechanisms and implications. Expert Rev Anti infect Ther 6: 725–732.

41. Martinez JL, Sánchez MB, Martínez-Solano L, Hernandez A, Garmendia L, et al. (2009) Functional role of bacterial multidrug efflux pumps in microbial natural ecosystems. FEMS Microbiol Rev 33: 430–449.

42. Yu DJ, Lai BS, Li J, Ma YF, Yang F, et al. (2012) Cornmeal-induced resistance to ciprofloxacin and erythromycin in enterococci. Chemosphere 89: 70–75.

43. Lennon JT, Jones SE (2011) Microbial seed banks: the ecological and evolutionary implications of dormancy. Nat Rev Microbiol 9: 119–130.

44. Vignaroli C, Zandri G, Aquilanti L, Pasquaroli S, Biavasco F (2011) Multidrug-resistant enterococci in animal meat and faeces and co-transfer of resistance from an *Enterococcus durans* to a human *Enterococcus faecium*. Curr Microbiol 62: 1438–1447.

45. Brenciani A, Ojo KK, Monachetti A, Menzo S, Roberts MC, et al. (2004) Distribution and molecular analysis of *mef*(A)-containing elements in tetracycline-susceptible and -resistant *Streptococcus pyogenes* clinical isolates with efflux-mediated erythromycin resistance. J Antimicrob Chemother 54: 991–998.

46. Aarestrup FM, Agerso Y, Gerner-Smidt P, Madsen M, Jensen LB (2000) Comparison of antimicrobial resistance phenotypes and resistance genes in *Enterococcus faecalis* and *Enterococcus faecium* from humans in the community, broilers, and pigs in Denmark. Diagn Microbiol Infect Dis 37: 127–137.

47. Sutcliffe J, Grebe T, Tait-Kamradt A, Wondrack L (1996) Detection of erythromycin-resistant determinants by PCR. Antimicrob Agents Chemother 40: 2562–2566.

48. Kao SJ, You I, Clewell DB, Donabedian SM, Zervos MJ, et al. (2000) Detection of the high-level aminoglycoside resistance gene *aph*(2″)-*Ib* in *Enterococcus faecium*. Antimicrob Agents Chemother 44: 2876–2879.

11

Nervous Necrosis Virus Replicates Following the Embryo Development and Dual Infection with Iridovirus at Juvenile Stage in Grouper

Hsiao-Che Kuo[1,2,3,◐]**, Ting-Yu Wang**[1,◐]**, Hao-Hsuan Hsu**[1,3]**, Peng-Peng Chen**[1]**, Szu-Hsien Lee**[4,5]**, Young-Mao Chen**[1,2,3]**, Tieh-Jung Tsai**[1]**, Chien-Kai Wang**[6]**, Hsiao-Tung Ku**[7,8]**, Gwo-Bin Lee**[4,5,9]*****, Tzong-Yueh Chen**[1,2,3]*****

1 Laboratory of Molecular Genetics, Institute of Biotechnology, National Cheng Kung University, Tainan, Taiwan, 2 Research Center of Ocean Environment and Technology, National Cheng Kung University, Tainan, Taiwan, 3 Agriculture Biotechnology Research Center, National Cheng Kung University, Tainan, Taiwan, 4 Institute of Nanotechnology and Microsystems Engineering, National Cheng Kung University, Tainan, Taiwan, 5 Department of Engineering Science, National Cheng Kung University, Tainan, Taiwan, 6 Division of Environmental Health and Occupational Medicine, National Health Research Institutes, Zhunan, Miaoli, Taiwan, 7 Research Division I, Taiwan Institute of Economic Research, Taipei, Taiwan, 8 Office for Energy Strategy Development, National Science Council, Taipei, Taiwan, 9 Department of Power Mechanical Engineering, National Tsing Hua University, Hsinchu, Taiwan

Abstract

Infection of virus (such as nodavirus and iridovirus) and bacteria (such as *Vibrio anguillarum*) in farmed grouper has been widely reported and caused large economic losses to Taiwanese fish aquaculture industry since 1979. The multiplex assay was used to detect dual viral infection and showed that only nervous necrosis virus (NNV) can be detected till the end of experiments (100% mortality) once it appeared. In addition, iridovirus can be detected in a certain period of rearing. The results of real-time PCR and *in situ* PCR indicated that NNV, in fact, was not on the surface of the eggs but present in the embryo, which can continue to replicate during the embryo development. The virus may be vertically transmitted by packing into eggs during egg development (formation) or delivering into eggs by sperm during fertilization. The ozone treatment of eggs may fail to remove the virus, so a new strategy to prevent NNV is needed.

Editor: Jean-Pierre Vartanian, Institut Pasteur, France

Funding: This research was funded and supported by Sanfong Technology Co. Ltd. and the Council of Agriculture, Taiwan (97AS-14.3.1-BQ-B6, 98AS-5.4.2-BQ-B1, 99AS-5.3.1-ST-aC to TYC). The funders had no role in study design, data collection and analysis, decision to publish, or preparation of the manuscript.

Competing Interests: Yes, the authors have the following competing interest. This research was partly funded and supported by Sanfong Technology Co. Ltd. There are no patents, products in development or marketed products to declare.

* E-mail: ibcty@mail.ncku.edu.tw (TYC); gwobin@pme.nthu.edu.tw (GBL)

◐ These authors contributed equally to this work.

Introduction

There are two common seen viruses in farmed groupers. Nervous necrosis virus (NNV), a two-single-stranded RNA piscine nodavirus, can cause damage to the central nervous system [1,2] and results in high mortality rates (80–100%) of hatchery-reared larvae and juveniles [3–6]. Iridoviruses, a virus with double-stranded DNA [7,8], can cause serious diseases in poikilothermic vertebrates, fish, amphibians, or reptiles [9] and have had a significant negative impact on modern aquaculture and wildlife conservation [10]. In the laboratory, intraperitoneal challenge of healthy juvenile grouper with grouper iridovirus resulted in cumulative mortality of 100% within 11 days [11]. Since 1995, this virus has almost caused epizootics in grouper fish farms of southern Taiwan, where it has caused up to 60% mortality [11]. In addition, farmed groupers are susceptible to infection by *Vibrio anguillarum*, a bacterial pathogen found in marine and freshwater fish species which can cause a terminal hemorrhagic septicemia known as vibriosis leading to high mortality rates [12,13].

It has been known that piscine nodavirus can be transmitted vertically from the broodfish via the eggs [1,14,15]. The ozone and other chemicals have been applied to treat the eggs in order to break virus particles on the surface of the eggs and produce virus-free eggs [16]. However, juveniles from ozone-treated eggs still have high mortality due to virus infection. In the case of Atlantic halibut, *Hippoglossus hippoglossus*, disinfection by ozonation of sea water still has the cumulative mortality rate of 100% to larvae in 44 days [17]. The high mortality rate makes the NNV transmission pathway remain a mystery.

To reveal this mystery, our efforts have focused on the development of microfluidics-based multiplex RT-PCR for detection of pathogens in farmed fish [18], specifically nodavirus, iridovirus and *V. anguillarum* which mainly cause diseases to grouper fish. Initially, to understand the infection status of different pathogens and possible interactions in grouper fish farm we had applied the multiplex RT-PCR system. The present study also incorporates detection of expression of the Mx gene. This sequence encodes a cytoplasmic protein with activity against a number of viruses [19,20]. Interferon can induce the Mx gene expression which has also been used as a molecular marker for type I interferon production [21,22]. Similarly, grouper Mx gene expression is induced by piscine viruses (e.g., nodavirus and iridovirus) infection but not by bacterial infection; thus, Mx gene expression can be used to monitor viral infection in grouper [23].

From the information regarding these infectious pathogens within fish farm, we were aiming to identify the vertical transmission pathway of NNV.

Results

Evaluation of multiplexing assay

The microfluidic chip device [18] can distinguish the amplified pathogen signals from the DNA marker (Figure S2). An ethidium bromide-stained gel of the four fragments, amplified by the RT-PCR reactions are shown in Fig. 1A. 300-bp, 238-bp, 202-bp and 171-bp fragments were observed as expected, corresponding to products amplified from sequences encoding major capsid protein of iridovirus, flagellin A of *V. anguillarum*, grouper Mx, and RdRp and protein B2 of nodavirus. When all four templates and primer pairs were pooled and subjected to RT-PCR, four bands of the expected sizes were present and clearly distinguished (Fig. 1A, lane 1). The same results were obtained by CE on separation of four signals from pathogens and Mx gene (Fig. 1B).

Evaluation of assay on *ex vivo* samples

The results showed that the amplified PCR products (Figure S3) were expected regarding to challenging experiments. NNV was detected in group 1 fish (lane 1, Fig. S3), *V. anguillarum* was detected in group 2 fish (lane 2, Fig. S3), iridovirus was detected in group 3 fish (lane 3, Fig. S3), NNV and *V. anguillarum* were detected in group 4 fish (lane 4, Fig. S3), NNV and iridovirus were detected in group 5 fish (lane 5, Fig. S3), iridovirus and *V. anguillarum* were detected in group 6 (lane 6, Fig. S3), NNV, iridovirus and *V. anguillarum* were detected in group 7 fish (lane 7, Fig. S3).

Monitoring of pathogens contamination in fish farm

The multiplex RT-PCR method was applied to investigate the infection status of individual fish rearing tank for three times, and each time for sampling continuously in 27 days (three repeats in Table 1). The RT-PCR assay did not detect *V. anguillarum* in any of these grouper fish samples but Mx gene expression was detected in most of the cases (except Exp. II-23 and Exp. III-19 were shown as negative). In Exp. I, iridovirus was first detected on the 6th day while NNV was detected on the 9th day. The NNV signal was detected continuously for the rest of the experiment but the signal of iridovirus only appeared for one week (day 6 to day 12) (Figure 2). Different story was observed in Exp. II, no iridovirus was detected but NNV signal came out from the second day of the experiment (Figure 2). In Exp. III, NNV signal appeared from the first day of the experiment. In this experiment, iridovirus was detected from day 5 to day 22 of the experiment. However, in Exp. III, both iridovirus and NNV signals were not detected on the 19th and the 20th day of the experiment. This may be due to sampling and false negative results.

NNV continued to replicate and localized in the embryo

The amount of virus was increased during the development of the embryo (Figure 3) and even after they hatch (data not shown).

The results also showed that, NNV is localized inside the embryo but not on the surface of the egg (Figure 4D–4H). Unfertilized (Figure 4B) and NNV-free eggs (Figure 4C) showed no signal of NNV (Figure 4B–4H).

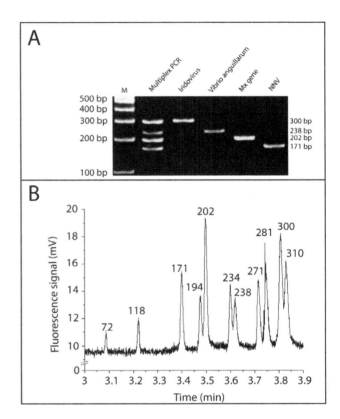

Figure 1. Optimization of the multiplex RT-PCR system. The microfluidic RT-PCR system is comprised of a microfluidic control module and a micro temperature control module [18]. The current design of microfluidic system is capable for rapid detection of four target genes (two viruses, one bacteria and one host immune response gene). The cDNA of samples can be uniformly distributed into the subsequent four PCR reaction chambers to detect four genes simultaneously. **A.** 100 ng of cDNA was amplified with each set of primers separately and with the mixture of the primers. Lane 2, Iridovirus; lane 3, *Vibrio anguillarum*; lane 4, Mx; lane 5, NNV (nodavirus); lane 1, multiprimer reactions simultaneously in a single tube. **B.** The electropherograms of the RT-PCR products from the single-tube reaction. The sizes of the DNA marker are 310 bp, 281 bp, 271 bp, 234 bp, 194 bp, 118 bp and 72 bp.

Discussion

The major recently reported pathogens, e.g., iridovirus, NNV and *V. anguillarum*, to grouper fish were chosen to be investigated by microfluidic chip in this study. The assay also incorporates a host immune response marker, Mx which encodes an interferon-induced member of the dynamin superfamily of large GTPases [23,24]. The gene can be used as an indicator to monitor fish to general viral exposure, such as latent or unknown viral pathogens. Vibriosis, caused by *V. anguillarum*, was an important infectious disease in cultured marine fish in Taiwan in the late 1970s and early 1980s [25]. In the present survey however, no *V. anguillarum* infection was detected, indicated the improvements of grouper aquaculture over the past 30 years. Circulating and regular exchange of fresh sea water can maintain the bacteria below disease-causing concentration in culturing tank. In addition, the improvement of fish farm management (record and regular check and identify for pathogens) is also a key factor to prevent the disease caused by bacteria.

Although vibriosis may have almost been terminated in farmed groupers, other diseases caused by virus are still causing trouble to

Table 1. Results of mutliplex RT-PCR assay for individual aquaculture farms.

Sample[‡]	Presence of PCR Product				Date (day/mon/yr)
	Nodavirus	Iridovirus	V. anguillarum	Mx gene	
Exp. I					
Exp.I-1	−	−	−	+	16/12/2008
Exp.I-2	−	−	−	+	17/12/2008
Exp.I-3	−	−	−	+	18/12/2008
Exp.I-4	−	−	−	+	19/12/2008
Exp.I-5	−	−	−	+	20/12/2008
Exp.I-6	−	+	−	+	21/12/2008
Exp.I-7	−	+	−	+	22/12/2008
Exp.I-8	−	+	−	+	23/12/2008
Exp.I-9	+	+	−	+	24/12/2008
Exp.I-10	+	+	−	+	25/12/2008
Exp.I-11	+	+	−	+	26/12/2008
Exp.I-12	+	+	−	+	27/12/2008
Exp.I-13	+	−	−	+	28/12/2008
Exp.I-14	+	−	−	+	29/12/2008
Exp.I-15	+	−	−	+	30/12/2008
Exp.I-16	+	−	−	+	31/12/2008
Exp.I-17	+	−	−	+	01/01/2009
Exp.I-18	+	−	−	+	02/01/2009
Exp.I-19	+	−	−	+	03/01/2009
Exp.I-20	+	−	−	+	04/01/2009
Exp.I-21	+	−	−	+	05/01/2009
Exp.I-22	+	−	−	+	06/01/2009
Exp.I-23	+	−	−	+	07/01/2009
Exp.I-24	+	−	−	+	08/01/2009
Exp.I-25	+	−	−	+	09/01/2009
Exp.I-26	+	−	−	+	10/01/2009
Exp.I-27	+	−	−	+	11/01/2009
Exp. II					
Exp.II-1	−	−	−	+	16/02/2009
Exp.II-2	+	−	−	+	17/02/2009
Exp.II-3	+	−	−	+	18/02/2009
Exp.II-4	+	−	−	+	19/02/2009
Exp.II-5	+	−	−	+	20/02/2009
Exp.II-6	+	−	−	+	21/02/2009
Exp.II-7	+	−	−	+	22/02/2009
Exp.II-8	+	−	−	+	23/02/2009
Exp.II-9	+	−	−	+	24/02/2009
Exp.II-10	+	−	−	+	25/02/2009
Exp.II-11	+	−	−	+	26/02/2009
Exp.II-12	+	−	−	+	27/02/2009
Exp.II-13	+	−	−	+	28/02/2009
Exp.II-14	+	−	−	+	01/03/2009
Exp.II-15	+	−	−	+	02/03/2009
Exp.II-16	+	−	−	+	03/03/2009
Exp.II-17	+	−	−	+	04/03/2009
Exp.II-18	+	−	−	+	05/03/2009

Table 1. Cont.

Sample[‡]	Presence of PCR Product				Date (day/mon/yr)
	Nodavirus	Iridovirus	V. anguillarum	Mx gene	
Exp.II-19	+	−	−	+	06/03/2009
Exp.II-20	+	−	−	+	07/03/2009
Exp.II-21	+	−	−	+	08/03/2009
Exp.II-22	+	−	−	+	09/03/2009
Exp.II-23	+	−	−	−	10/03/2009
Exp.II-24	+	−	−	+	11/03/2009
Exp.II-25	+	−	−	+	12/03/2009
Exp.II-26	+	−	−	+	13/03/2009
Exp.II-27	+	−	−	+	14/03/2009
Exp. III					
Exp.III-1	+	−	−	+	18/04/2009
Exp.III-2	+	−	−	+	19/04/2009
Exp.III-3	+	−	+	+	20/04/2009
Exp.III-4	+	−	−	+	21/04/2009
Exp.III-5	+	+	−	+	22/04/2009
Exp.III-6	+	+	−	+	23/04/2009
Exp.III-7	+	+	−	+	24/04/2009
Exp.III-8	+	+	−	+	25/04/2009
Exp.III-9	+	+	−	+	26/04/2009
Exp.III-10	+	+	−	+	27/04/2009
Exp.III-11	+	+	−	+	28/04/2009
Exp.III-12	+	+	−	+	29/04/2009
Exp.III-13	+	+	−	+	30/04/2009
Exp.III-14	+	+	−	+	01/05/2009
Exp.III-15	+	+	−	+	02/05/2009
Exp.III-16	+	+	−	+	03/05/2009
Exp.III-17	+	+	−	+	04/05/2009
Exp.III-18	+	+	−	+	05/05/2009
Exp.III-19	−	−	−	−	06/05/2009
Exp.III-20	−	−	−	+	07/05/2009
Exp.III-21	+	+	−	+	08/05/2009
Exp.III-22	+	+	−	+	09/05/2009
Exp.III-23	+	−	−	+	10/05/2009
Exp.III-24	+	−	−	+	11/05/2009
Exp.III-25	+	−	−	+	12/05/2009
Exp.III-26	+	−	−	+	13/05/2009
Exp.III-27	+	−	−	+	14/05/2009

[‡]All of the surveyed sites were indoor grouper aquaculture farms with maintained seawater at controlled temperature (29°C) and pH (7.37 to 7.49). The location and species were Linyuan(LY) and *E. coioides*, respectively. Total number of fish sampled: Six fish were collected and pooled for each site.

the grouper aquaculture. To solve this problem, advanced techniques have been introduced to modern aquaculture for detection and monitoring the potential viral pathogens. Here, we introduced the microfluidic chip system to grouper aquaculture. The microfluidic chip system is able to do multiplex assay. Instead of previous target RNA2 of nodavirus [3], the overlapping sequence between RNA1 and RNA3 was chosen for this study.

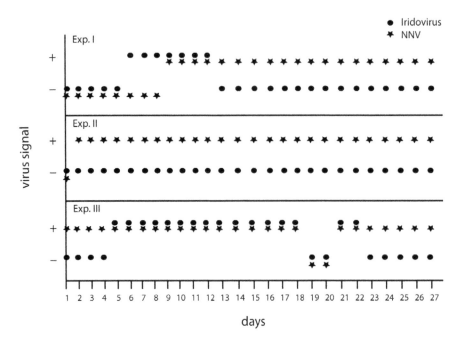

Figure 2. The RT-PCR on microfluidic chip to detect irodovirus and nordavirus (NNV) during juvenile rearing. Three experiments (Exp. I, Exp. II and Exp. III) have been proceed, sampled and monitored continuously for 27 days for each experiments.

The advantages of using this target in the current work were, (1) the size of the intended PCR product had to be distinct from each other, (2) the sequence of the primer pairs had to be compatible with other primer pairs in the "one-pot" PCR reaction,(3) the PCR conditions had to be the same for all the primer sets and the most important of all was that RNA1 and RNA3 are both expressed in the early stage of infection [26].

The multiplex assay also detected dual viral infection (Figure 2, Exp. I and Exp. III). In the case of mosquito cells, it can accommodate at least 3 viruses simultaneously, which provides an opportunity for genetic exchange between diverse viruses [27]. Different strains of virus from different hosts may also be due to genetic exchange. Interestingly, in our observation, only NNV can be detected till the end (100% mortality) of experiments once it appeared. For the other detection, iridovirus was not detected in

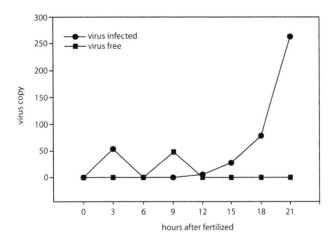

Figure 3. Real-time PCR of NNV-free and NNV-infected grouper eggs. The amount of virus was increased following the development of embryo.

Exp. II but could be detected in certain period of time (6th–12th day in Exp. I and 5th–22nd in Exp. III). This phenomenon needs to be investigated in more detail to clarify whether NNV can influence the replication of iridovirus or not.

We have demonstrated that the problem caused by NNV in grouper aquaculture is much more serious than by iridovirus. It was reported that vertical transmission is the only infection pathway for salmon anemia virus (ISAV) [28,29]. In the case of NNV in grouper fish farming, the difficulty to remove the virus from fish farm may be due to the vertical transmission pathway involved in spread of the virus. It has been suggested that the treatment of grouper eggs with ozone or other chemicals can remove the virus from the eggs [16]. However, juveniles from ozone-treated eggs can still be infected by the virus, and the cumulative mortality rate is even close to 100% [17](our own observation, data not shown). In fact, NNV is not localized on the surface of the eggs (Figure 4) but it is present inside the embryo and may continue to replicate during the embryo development (Figures 3 and 4).

The virus may have been packed into the eggs since egg development (formation) or it has been delivered into the eggs by the sperms during fertilization. The research on hepatitis B virus has shown that the viral DNA can pass through the zona and oolemma and enter into oocytes by sperm [30,31]. Furthermore, it was also reported that the virus can be absorbed into the fish sperm and vertically transmitted [32]. Or alternatively, the NNV genome might integrate into host genome. There were evidences indicating that some virus species, such as adenoviral vectors, could integrate into host genome and pass to embryo through vertical transmission pathway [33]. Likewise, HBV and HIV-1 also can integrate into host embryonic genome [30,34]. Grouper genome containing NNV genome also can explain why most of the broodstocks and wild marine fish are positive for betanodavirus [35]. The results from Mx gene expression (Exp. I, 1st–5th day) showed in some cases that the gene was highly expressed without viral (neither NNV nor iridovirus) infection. These results

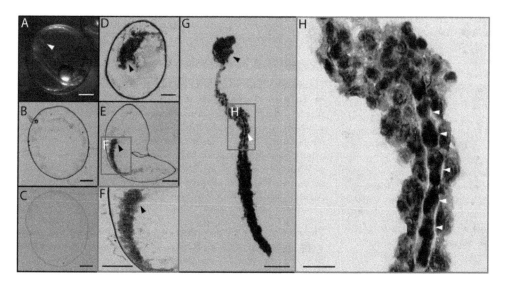

Figure 4. *In situ* **RT-PCR of grouper eggs (15 h after spawning). A,** grouper egg. The embryo has been developed. White arrow indicated the head of the embryo. **B,** section of non-fertilized grouper egg (no embryo). **C,** section of virus-free grouper egg (negative control). **D–F,** section of virus-containing grouper egg (positive control). The dark precipitation indicated the signal of NNV. Black arrow heads indicated the heads of the embryos. Figure 4F was the magnification image from the red box in the figure 4E. **G–H,** section of virus-containing grouper embryo. The dark precipitation indicated the signal of NNV (Black arrow head; also indicated the head of the embryo) Figure 4H was the magnification image from the red box in the figure 4G. The spinal cord can be visualized in **H** (white arrow heads) the middle line of the embryo) and showed the signal of NNV. Bar = 0.2 μm (**A–G**); bar = 0.1 μm (**H**).

may due to the physical status of when the viral infection gets started or close to the end of sampling individuals. In mammalian, when treated with interferon (mimic the virus infection), Mx was detectable in 3 h, and its expression had continuously increased for another 5 h and decreased thereafter [36]. Moreover, Mx could not be detected before 3 h and after 20 h [36].

In conclusion, NNV can vertically transmit from parental grouper fish to the progenies at the embryo stage. The virus could be brought into the egg by sperm during fertilization. Or one possibility was that the virus genome may have already integrated into some of the host fish genome. The ozone treatment of eggs may fail to remove the virus as there is no sign of virus on the surface of the egg. The grouper fish genome (approximately 1 Giga-bases) sequencing has been initiated in our lab which will be completed recently. The genome information should lead to a better understanding of the interactions between the pathogens and host.

Materials and Methods

Grouper fish sampling and challenge

Epinephelus coioides (orange-spotted grouper) were collected from a grouper fish farm in Linyuan, Taiwan from 2008 to 2009 (Table 1). Three individual experiments were performed in 3–4 weeks and continuously sampled for 27 days. The juveniles used were 25–30 days post hatching. For each sample, six live juveniles were randomly selected (whole fish which were then killed and pooled before submitted to experiment) and immersed in TRIzol Reagent™ (Molecular Research Center, Inc. Ohio, USA) solution directly. Twelve asymptomatic groupers ranging in size from 0.5–3 inches were obtained from grouper aquaculture farm and maintained in a 1×0.5×0.5 m³ fish tank with constant aeration and a water temperature of 29°C. A commercial diet was fed daily to satiation. These fish were used as negative controls (after sacrificed, these asymptomatic fish showed no signs of viral/bacterial infection). The study proposal was reviewed by the grant application committee and granted by the Bureau of Animal and Plant Health Inspection and Quarantine, Council of Agriculture, Taiwan, who specifically approved the experiments.

In order to test the multiplex RT-PCR assay on *ex vivo* specimens, healthy grouper fish were challenged intraperitoneal injection with different pathogens; five-days post-infection RNA was extracted from the challenged fish and subjected to microfluidic chip-based multiplex RT-PCR.

Egg samples

The grouper eggs were collected from the egg collection net within 1 hour after spawning. Three different broodstock grouper aquaculture farms in southern Taiwan (Fangliao, Shueidiliao, Jiadong; for the location was indicated in Figure S1) were sampled, and the eggs were then transported alive to a laboratory at National Cheng Kung University. The eggs were maintained in a 1×0.5×0.5 m³ fish tank with constant aeration and a water temperature of 28±2°C. The experiments were repeated thrice, and the NNV-free eggs were collected as a control group. Samples were collected every three hours until they hatch. The samples were then subjected to real-time PCR for NNV detection.

Nucleic acid extraction

Each pool of 6 fish (whole fish) was homogenized in liquid nitrogen and subjected to RNA extraction using TRIzol Reagent™ (Molecular Research Center, Inc. Ohio, USA) according to the manufacturer's instructions. RNA samples were used immediately or stored at −80°C.

cDNA synthesis, PCR and RT-PCR on microfluidic chip

The multiplex RT-PCR assay was performed using microfluidic chips [18]. The microfluidic chip device, which can be easily transported to the fish farms, allowed on-site analysis of the fish samples, and individual PCR signals could be distinguished from the DNA marker (Figure S2). The use of microfluidic chips, including cDNA synthesis, PCR and RT-PCR techniques were described previously [18]. RNA samples were treated with DNase I (New England Biolabs) prior to cDNA synthesis. RNA and cDNA were quantified using an Ultrospec 3300 Pro spectrophotometer (Amersham Biosciences, Piscataway, NJ, USA); nucleic acids were diluted using sheared salmon sperm DNA (5 ng mL^{-1}) as a carrier. PCR consisted of denaturation (94°C for 7 min), amplification(40 cycles of 94°C for 40 sec, 60°C for 40 sec and 72°C for 40 sec), and elongation (72°C for 7 min). PCR products were transported automatically to the CE sample reservoir by the last set of micropumps [18].

The primer sets used for amplification were the same as those described previously [18]. IridoMCP-F and IridoMCP-R were designed for iridovirus major capsid protein, VAFA-F and VAFA-R for *V. anguillarum* flagellin A, gMx-F and gMx-R for grouper Mx and NodaRNA1-F and NodaRNA1-R for both nodavirus RdRp (RNA1) and protein B2 (RNA 3, a subgenomic RNA transcribed from the 3′ end of RNA1).

In situ RT-PCR

To confirm our real-time PCR results, the grouper eggs were then subjected to *in situ* RT-PCR. Fifteen-hour-after-spawning-eggs were collected for *in situ* RT-PCR. The eggs were fixed overnight in freshly prepared 10% formaldehyde solution and embedded in paraffin. Four-micrometer-thick sections were cut using a Leica CM 1900 microtome (Leica Microsystems, Nussloch, Germany), mounted on a polylysin-coated slide, deparaffinized in xylene for 5 min, and dehydrated with a graded series of ethanol solutions (30, 50, 70, 85, 95, and 3×100%; each containing 0.85% NaCl). Samples were then stained with hematoxylin-eosin (H&E). *In situ* RT-PCR was modified from a previously described regimen [3,37] and was done with a DIG probe synthesis kit and DIG nucleic acid detection kit (Roche Applied Science, Mannheim, Germany) according to the manufacturer's instructions. The sections were visualized using an Axiovert 40 microscope (Carl Zeiss, Gottingen, Germany), and images were captured using a SPOT RT3 camera (Spot Imaging Solutions, Sterling Heights, MI).

In vivo infection

Nervous necrosis viruses were isolated from the diseased fish as described previously [3], and *V. anguillarum* was obtained from BCRC (Bioresource Collection and Research Center, Taiwan) 12908. *V. anguillarum* was originally isolated from marine fish. For *in vivo* infection, virus was prepared in 20 μL (1×10^5 TCID$_{50}$mL^{-1}), and challenge was performed by intraperitoneal (i.p.) injection. *V. anguillarum* infection was done by immersion of sea water which contained 10^8 cells mL^{-1}. The bacteria were freshly prepared by growing in Tryptic Soya Broth (TSB)+1.5% NaCl at 28°C for 6 h. Fish were then sampled once the clinical signs were observed and the organs were isolated for detection of virus/bacteria. There were eight experimental groups each containing 6 grouper fish. Group 1 fish were challenged with NNV, group 2 fish were challenged with *V. anguillarum*, group 3 fish were challenged with iridovirus, group 4 fish were challenged with NNV and *V. anguillarum* at the same time, group 5 fish were challenged with NNV and iridovirus at the same time, group 6 fish were challenged with iridovirus and *V. anguillarum* at the same time, group 7 fish were challenged with NNV, iridovirus and *V.*

anguillarum at the same time, group 8 fish were used as uninfected controls.

Supporting Information

Figure S1 The locations of sampling grouper fish farms in Taiwan. Taiwan is located between tropical and subtropical regions. The grouper aquacultures are mainly gathered in southern Taiwan due to the grouper fish preferred warm water temperature. Black circles indicate five major regions of sampling grouper aquaculture farms. The open circle indicates the location of National Cheng Kung University (NCKU). (Modified from a map available at http://mapsof.net under a Creative Commons Attribution-ShareAlike 1.0 License.)

Figure S2 The electropherograms of the RT-PCR products from purified RNA. The minimum concentration detected on the CE module was 3 copies/μL (Kuo *et al.* unpublished). Each sample includes a mixture of DNA markers and RT-PCR products obtained from the infected grouper. Eleven peaks (corresponding to the DNA markers, added to samples post-amplification) and a single peak of the RT-PCR product from grouper were resolved successfully within 4 min. Panels A–D show the 300-bp, 238-bp, 202-bp, and 171-bp PCR products generated from samples containing iridovirus, *V. anguillarum*, the Mx gene and nodavirus, respectively. **A.** Detection of iridovirus (I) in infected fish sample by microfluidic chip. The primers, derived from the gene encoding the iridovirus major capsid protein, amplify a 300-bp fragment. **B.** Detection of bacteria (V) from infected fish sample by microfluidic chip. The primers, derived from a *V. anguillarum* flagellin A gene sequence, amplify a 238-bp fragment. **C.** Detection of grouper Mx gene (MX) expression in infected fish sample by microfluidic chip. The set of primers amplifies a 202-bp fragment from the grouper Mx. **D.** Detection of nodavirus (NNV) in infected fish sample by microfluidic chip. The primers, derived from the gene encoding RNA-dependent RNA polymerase and protein B2 of nodavirus, amplify a 171-bp fragment.

Figure S3 Evaluation of multiplex RT-PCR with *ex vivo* samples (infected fish). M. DNA marker; lane 1, NNV (nodavirus); lane 2, *V. anguillarum*; lane 3, iridovirus; lane 4, NNV (nodavirus)+*V. anguillarum*; lane 5, NNV (nodavirus)+iridovirus; lane 6, *V. anguillarum*+iridovirus; lane 7, NNV (nodavirus)+*V. anguillarum*+iridovirus.

Acknowledgments

We would like to thank Mr. Nai-Heng Yu, Mr. Fu-Ping Chang, Tekho Co. Ltd. (An-Pin Live Fish Center), and Fish Breeding Association Taiwan for kindly providing fish samples for use in this study.

Author Contributions

Conceived and designed the experiments: TYC GBL. Performed the experiments: HCK TYW HHH PPC SHL YMC TJT. Analyzed the data: HCK TYW. Wrote the paper: TYC HCK TYW. Assisted in drafting of text and figures of manuscript: TYC HCK TYW. Revised manuscript critically for important intellectual content: TYC HCK TYW CKW HTK.

References

1. Breuil G, Pépin JFP, Boscher S, Thiéry R (2002) Experimental vertical transmission of nodavirus from broodfish to eggs and larvae of the sea bass, Dicentrarchus labrax (L.). J Fish Dis 25: 697–702.

2. Mori K, Nakai T, Muroga K, Arimoto M, Mushiake K, et al. (1992) Properties of a new virus belonging to nodaviridae found in larval striped jack (Pseudocaranx dentex) with nervous necrosis. Virology 187: 368–371.

3. Kuo HC, Wang TY, Chen PP, Chen YM, Chuang HC, et al. (2011) Real-time quantitative PCR assay for monitoring of nervous necrosis virus infection in grouper aquaculture. J Clin Microbiol 49: 1090–1096.

4. Skliris GP, Krondiris JV, Sideris DC, Shinn AP, Starkey WG, et al. (2001) Phylogenetic and antigenic characterization of new fish nodavirus isolates from Europe and Asia. Virus Res 75: 59–67.

5. Munday BL, Nakai T (1997) Nodaviruses as pathogens in larval and juvenile marine finfish. World J Microbiol Biotechnol 13: 375–381.

6. Munday BL, Nakai T, Nguyen HD (1994) Antigenic relationship of the picorna-like virus of larval barramundi, Lates calcarifer Bloch to the nodavirus of larval striped jack, Pseudocaranx dentex (Bloch & Schneider). Aust Vet J 71: 384–385.

7. Jakob NJ, Muller K, Bahr U, Darai G (2001) Analysis of the first complete DNA sequence of an invertebrate iridovirus: coding strategy of the genome of Chilo iridescent virus. Virology 286: 182–196.

8. Tsai CT, Ting JW, Wu MH, Wu MF, Guo IC, et al. (2005) Complete genome sequence of the grouper iridovirus and comparison of genomic organization with those of other iridoviruses. J Virol 79: 2010–2023.

9. Williams T, Barbosa-Solomieu V, Chinchar VG (2005) A decade of advances in iridovirus research. Adv Virus Res 65: 173–248.

10. Huang Y, Huang X, Liu H, Gong J, Ouyang Z, et al. (2009) Complete sequence determination of a novel reptile iridovirus isolated from soft-shelled turtle and evolutionary analysis of Iridoviridae. BMC Genomics 10: 224.

11. Chou HY, Hsu CC, Peng TY (1998) Isolation and characterization of a pathogenic iridovirus from cultured grouper (Epinephelus sp.) in Taiwan. Fish Pathol 33: 201–206.

12. Emmy E (1987) Vibriosis: Pathogenicity and pathology. A review. Aquaculture 67: 15–28.

13. Myhr E, Larsen JL, Lillehaug A, Gudding R, Heum M, et al. (1991) Characterization of Vibrio anguillarum and closely related species isolated from farmed fish in Norway. Appl Environ Microbiol 57: 2750–2757.

14. Samuelsen OB, Nerland AH, Jorgensen T, Schroder MB, Svasand T, et al. (2006) Viral and bacterial diseases of Atlantic cod Gadus morhua, their prophylaxis and treatment: a review. Dis Aquat Organ 71: 239–254.

15. Kai YH, Su HM, Tai KT, Chi SC (2010) Vaccination of grouper broodfish (Epinephelus tukula) reduces the risk of vertical transmission by nervous necrosis virus. Vaccine 28: 996–1001.

16. Arimoto M, Sato J, Maruyama K, Mimura G, Furusawa I (1996) Effect of chemical and physical treatments on the inactivation of striped jack nervous necrosis virus (SJNNV). Aquaculture 143: 15–22.

17. Grotmol S, Totland GK (2000) Surface disinfection of Atlantic halibut Hippoglossus hippoglossus eggs with ozonated sea-water inactivates nodavirus and increases survival of the larvae. Dis Aquat Organ 39: 89–96.

18. Lien KY, Lee SH, Tsai TJ, Chen TY, Lee GB (2009) A microfluidic-based system using reverse transcription polymerase chain reactions for rapid detection of aquaculture diseases. Microfluid Nanofluid 7: 795–806.

19. Verrier ER, Langevin C, Benmansour A, Boudinot P (2011) Early antiviral response and virus-induced genes in fish. Dev Comp Immunol 35: 1204–1214.

20. Ohta T, Ueda Y, Ito K, Miura C, Yamashita H, et al. (2011) Anti-viral effects of interferon administration on sevenband grouper, Epinephelus septemfasciatus. Fish Shellfish Immunol 30: 1064–1071.

21. Pakingking RJ, Mori KI, Sugaya T, Oka M, Okinaka Y, et al. (2005) Aquabirnavirus-induced protection of marine fish against piscine nodavirus infection. Fish Pathol 40: 125–131.

22. Leong JC, Trobridge GD, Kim CH, Johnson M, Simon B (1998) Interferon-inducible Mx proteins in fish. Immunol Rev 166: 349–363.

23. Chen YM, Su YL, Lin JH, Yang HL, Chen TY (2006) Cloning of an orange-spotted grouper (Epinephelus coioides) Mx cDNA and characterisation of its expression in response to nodavirus. Fish Shellfish Immunol 20: 58–71.

24. Haller O, Stertz S, Kochs G (2007) The Mx GTPase family of interferon-induced antiviral proteins. Microbes Infect 9: 1636–1643.

25. Song YL, Chen SN, Kou GH (1988) Serotyping of Vibrio anguillarum strains isolated from fish in Taiwan. Fish Phathol 23: 185–189.

26. Fenner BJ, Thiagarajan R, Chua HK, Kwang J (2006) Betanodavirus B2 is an RNA interference antagonist that facilitates intracellular viral RNA accumulation. J Virol 80: 85–94.

27. Kanthong N, Khemnu N, Pattanakitsakul SN, Malasit P, Flegel TW (2010) Persistent, triple-virus co-infections in mosquito cells. BMC Microbiol 10: 14.

28. Melville KJ, Griffiths SG (1999) Absence of vertical transmission of infectious salmon anemia virus (ISAV) from individually infected Atlantic salmon Salmo salar. Dis Aquat Organ 38: 231–234.

29. Nylund A, Plarre H, Karlsen M, Fridell F, Ottem KF, et al. (2007) Transmission of infectious salmon anaemia virus (ISAV) in farmed populations of Atlantic salmon (Salmo salar). Arch Virol 152: 151–179.

30. Ali BA, Huang TH, Xie QD (2005) Detection and expression of hepatitis B virus X gene in one and two-cell embryos from golden hamster oocytes in vitro fertilized with human spermatozoa carrying HBV DNA. Mol Reprod Dev 70: 30–36.

31. Huang TH, Zhang QJ, Xie QD, Zeng LP, Zeng XF (2005) Presence and integration of HBV DNA in mouse oocytes. World J Gastroenterol 11: 2869–2873.

32. Mulcahy D, Pascho RJ (1984) Adsorption to fish sperm of vertically transmitted fish viruses. Science 225: 333–335.

33. Larochelle N, Stucka R, Rieger N, Schermelleh L, Schiedner G, et al. (2011) Genomic integration of adenoviral gene transfer vectors following transduction of fertilized mouse oocytes. Transgenic Res 20: 123–135.

34. Wang D, Li LB, Hou ZW, Kang XJ, Xie QD, et al. (2011) The integrated HIV-1 provirus in patient sperm chromosome and its transfer into the early embryo by fertilization. PLoS One 6: e28586.

35. Gomez DK, Sato J, Mushiake K, Isshiki T, Okinaka Y, et al. (2004) PCR-based detection of betanodaviruses from cultured and wild marine fish with no clinical signs. J Fish Dis 27: 603–608.

36. Meier E, Fah J, Grob MS, End R, Staeheli P, et al. (1988) A family of interferon-induced Mx-related mRNAs encodes cytoplasmic and nuclear proteins in rat cells. J Virol 62: 2386–2393.

37. Nuovo GJ (1995) In situ PCR: protocols and applications. PCR Methods Appl 4: S151–167.

12

Discrimination between Weaned and Unweaned Atlantic Cod (*Gadus morhua*) in Capture-Based Aquaculture (CBA) by X-Ray Imaging and Radio-Frequency Metal Detector

Ekrem Misimi*, Svein Martinsen¤, John Reidar Mathiassen, Ulf Erikson

SINTEF Fisheries and Aquaculture, Trondheim, Norway

Abstract

The aim of this study was to investigate the feasibility of two detection methods for use in discrimination and sorting of adult Atlantic cod (about 2 kg) in the small scale capture-based aquaculture (CBA). Presently, there is no established method for discrimination of weaned and unweaned cod in CBA. Generally, 60–70% of the wild-caught cod in the CBA are weaned into commercial dry feed. To increase profitability for the fish farmers, unweaned cod must be separated from the stock, meaning the fish must be sorted into two groups – unweaned and weaned from moist feed. The challenges with handling of large numbers of fish in cages, defined the limits of the applied technology. As a result, a working model was established, focusing on implementing different marking materials added to the fish feed, and different technology for detecting the feed presence in the fish gut. X-ray imaging in two modes (planar and dual energy band) and sensitive radio-frequency metal detection were the detection methods that were chosen for the investigations. Both methods were tested in laboratory conditions using dead fish with marked feed inserted into the gut cavity. In particular, the sensitive radio-frequency metal detection method with carbonyl powder showed very promising results in detection of marked feed. Results show also that Dual energy band X-ray imaging may have potential for prediction of fat content in the feed. Based on the investigations it can be concluded that both X-ray imaging and sensitive radio-frequency metal detector technology have the potential for detecting cod having consumed marked feed. These are all technologies that may be adapted to large scale handling of fish from fish cages. Thus, it may be possible to discriminate between unweaned and weaned cod in a large scale grading situation. Based on the results of this study, a suggestion for evaluation of concept for in-situ sorting system is presented.

Editor: Ronald Thune, Louisiana State University School of Veterinary, United States of America

Funding: This work was funded by Fisheries and Aquaculture Research Fund – FHF through the project "Capture-based Aquaculture". The funders had no role in study design, data collection and analysis, decision to publish, or preparation of the manuscript.

Competing Interests: The authors have declared that no competing interests exist.

* E-mail: ekrem.misimi@sintef.no

¤ Current address: Nekton AS, Smøla, Norway

Introduction

The peak season of the commercial cod fisheries in Norway is from January until April, following the migration of mature cod into the Norwegian fjords [5]. During this period, the market supplies of fresh cod increase up to ten times the average yearly supply, whilst in December the market demand for fresh cod exceeds the supply, due to the use of cod as a raw material in Norwegian and European traditional cuisine. The seasonal dependence and variation of the cod fisheries has triggered strategies such as farming of cod, and conservation of the wild caught cod through capture-based aquaculture (CBA). Products deriving from these activities are more suited in terms of volume and quality for the high-demand market in December. Since the dynamics between fisheries, aquaculture and markets is constantly fluctuating, strong year-groups of wild cod have resulted in increased total catch, from 215.000 tons in 2008 to 340.000 tons in 2011. This increase has resulted in diminishing price levels, from

16 NOK (appr. 2 €) per kg round weight (RW) to 11 NOK (appr. 1.50 €) per kg RW in average. This has made commercial cod farmers unable to cope with the lower prices, as the production costs of farmed cod in Norway, normally exceed 18 NOK (2.50 €) per kg RW. The expected production cost from capture based aquaculture might be lower, given efficient conversion of wild caught cod to formulated feed.

The present CBA of cod, with feeding of wild-caught cod, derives from extended trials in the 1980's [2]. The basic principles of CBA are gentle capture of fish in seine nets and transfer to holding tanks inside fishing vessels. Fish are transferred from the vessels to special flat-bottom cages, specifically developed for storing and resting of cod. Although the operations are performed with caution, the cod can experience physiological changes like gas bladder expansion and severe stress, but the physiological normality is usually established 12 to 24 hours after capture [2].

The captured fish are presented different diets: a) moist feed (frozen herring, mackerel, and capelin); b) semi moist feed

Table 1. Dilution ranges for the metal and glass bead markers used in the investigations/experiments.

Marker weight(%)	Marker weight:20 gr feed sample	Marker weight: 30 gr feed sample
0.1	0.02	0.03
0.5	0.1	0.15
1	0.2	0.3
5	1	1.5
10	2	3

(vacuum soaked commercial fish feed, premix of fishmeal and fish oil); and c) dry feed (commercial fish feed). Captured wild cod traditionally prefer moist feed, and complete weaning (adaption to moist feed) will normally occur shortly after introduction [15]. Moist feed consists of frozen blocks of herring and capelin that slowly thaw and disintegrate inside the cage, with the fish foraging upon the thawed feed. Seen from the cod farmer's point of view, the optimal feed source is commercial dry feed, in terms of nutritionally adequate content, supporting rapid growth, extended storage stability, availability, non-disease carrier and logistics [16]. Fish are reluctant when presented dry feed and the maximum proportion of fully-weaned cod individuals, weaned from moist feed, and varies between 40–70% [16]. Weaned fish achieve satisfying growth, and reportedly increase their weight from 1.5–2 up to 5–6 kg after 5 months. Unweaned fish will not forage on dry feed, resulting in loss of weight and generally poor condition. To achieve viable economy in the cod CBA industry, an efficient method for discriminating unweaned and weaned fish is necessary. Presently, there is no established method for discrimination between weaned and unweaned cod in CBA.

X-ray radiography has been reported as a suitable technique for studying fish feeding, digestion, measuring feed intake and fish growth [14,8,9,10]. In [18], they report use of X-radiography and X-ray dense markers for quantitative determination of gastrointestinal content. [12] used X-ray dense lead glass beads for estimation of individual feed intake of Chinook salmon by X-radiography, while [19] used a similar technique for estimation of feed intake and growth of Atlantic salmon.

The aim of this work was to investigate the required technology platform needed for sorting of wild cod, enabling efficient discrimination between weaned and unweaned cod in CBA. The technological approach involved testing potential methods for in-situ inspection of the gut content after feed intake consisting of commercial dry feed containing markers. Two aspects of the sorting procedure were studied: a) the marking of fish and b) evaluation of detection technology for efficient discrimination. Although the presence of fish feed in fish gut can be possible to detect by use of X-ray images [11], other fish organs such as liver, swim bladder, and gut cavity liquids can disturb the imaging of fish feed. Therefore, marking of fish feed is seen as a potential solution to enhance the detection capability of both X-ray and other detection methods for detection of fish feed.

Two detection technologies that may be suitable for the discrimination between weaned and unweaned cod were evaluated: 1) X-ray imaging in two modes (planar X-ray and dual energy band), with an integrated image analysis, and 2) sensitive radio-frequency metal detection. Five different types of markers were used for discrimination purposes. In addition, the potential of dual energy band X-ray imaging for prediction of fat content in the feed was investigated.

Materials and Methods

Three experiments were conducted, each investigating a separate detection method for the inspection of gut content to facilitate sorting of weaned and unweaned cod. No human participants were used in the study. The fish was delivered by commercial fishery boats that caught wild cod at the coast of Central Norway. From before, they have all necessary permissions to fish, so no specific permission was needed for this study. Fish was delivered to SINTEF SeaLab packed on ice in styrofoam boxes. The wild cod is not an endangered fish species either in Norway or globally.

X-ray Imaging of Fish Fed with Commercial Feed with High Contrast Feed Markers

Fish. Wild Atlantic cod *(Gadus morhua)* $(2.1 \pm 0.5$ kg, n = 20) were caught at the coast of Central Norway. The fish were killed and packed in ice before being shipped to SINTEF Sealab in Trondheim. Their intestinal content was maintained in the body cavity before proceeding with the investigations.

Feed and markers. Fish dry feed [21] was mixed with crude herring oil added as coating for evenly distribution of commercially available markers and to provide a sticky surface of the pellet. The amount of marked feed was 1.0–1.5% of the fish body weight. The markers were chosen so that they were able to provide a good contrast for X-ray image analysis. In addition to metal-based markers shown in Table 1, Ballotini glass beads of 0.6 mm [24] were used at the same dilution ranges (Table 1).

The chosen commercially available markers were added to the fish feed at a dilution range from 0.1% to 10% of feed portion weight (Table 1). The size of each feed portion inserted into the fish was between from 20–30 g of fish feed, corresponding to 1–1.5% of fish round weight. This amount corresponds to a typical meal portion size for a large cod [1,13].

Image acquisition and processing. A line scan X-ray setup [7] was used for acquisition of X-ray images. Prior to the experiment, an image acquisition PC was connected to the Ishida apparatus so that the generated images could be saved directly and online into the PC disk. Line scans were stitched in the X-ray machine, resulting in a sequence of 5–6 images of 90×332 pixels per fish while fish was moving on the conveyer belt. These images were then stitched to a full image (Figure 1a) by implementing a stitching function in Matlab R2011b Image Processing Toolbox [23]. The energy level used for acquisition of X-ray images was 60 kV.

Dual-energy X-ray for Detection of Different Feed Density (Fat Content)

Fish. Wild-caught cod $(2.5 \pm 0.6$ kg, n = 10) were killed, bled and chilled in ice. The fish were transported to the Curato X-Ray

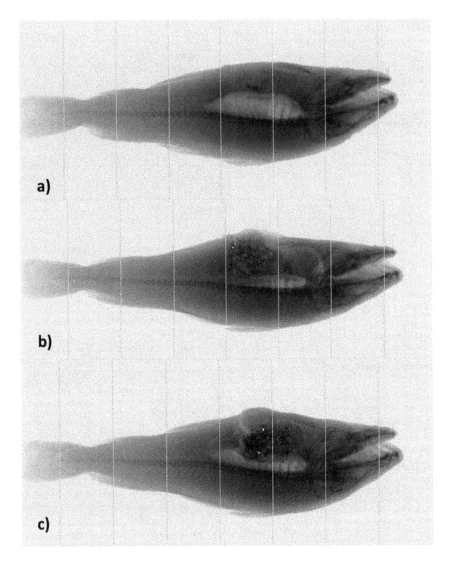

Figure 1. X-ray images of Atlantic cod generated from the Ishida X-ray machine: a) image of a fish without feed, images of a fish with glass beads at b) 0.1% (0.03 g) and c) 10% (3 g) of weight percent of total feed content in gut.

Clinic in Trondheim. Their intestinal content was maintained in the body cavity while the experiments were carried out. To determine the feasibility, of using dual-energy X-ray imaging for discriminating weaned and unweaned cod, a single cod was used for some preliminary experiments.

Feed and markers. Feed samples were weighed and prepared in small plastic bags. Each sample weighed approximately 25 g. The samples were taken from three commercial fish feed diets; Skretting Optiline 2500 (30% fat), Skretting Nutra Parr (22% fat), and BioMar Classic Marine 800 (18% fat). Prior to X-ray imaging, the feed samples were inserted into the fish abdominal cavity through an incision, and the incision was then carefully collated.

Image acquisition and processing. The X-ray images for dual energy X-ray analysis were taken using the commercial medical planar X-ray apparatus [22]. Configuration, setup of parameters, placement and introduction of fish for the imaging system was similar to that described in [20]. To simulate the dual energy X-ray imaging system, fish were imaged with two different X-ray energy levels. A total of 21 planar X-ray images were acquired at both 40 kV (low energy image) and 70 kV (high

energy image) (Figure 2). Sets of images of fish without feed, and with NiCr, Steel grit and Cr-Ni alloy markers were generated and saved. The images were transferred then offline into a PC and processed. The Region of Interest (ROI) for dual energy band X-ray image analysis was chosen manually.

These particular voltages were chosen based on the results reported in [20]. Image processing and analysis was done off-line after all images had been acquired. The regions of the images, containing the feed, were segmented from the rest of the fish. Subsequent to this, the regional statistics (consisting on comparing the mean and standard deviation of the image intensity differences between 40 kV and 70 kV images) of the fish regions with feed were compared to those of the same anatomical regions in fish without feed. Features used during the dual energy X-ray image analysis are defined in Table 2, while their mean values are given in Table 3. To test prediction of fat for the fish diets with 18%, 22% and 30% of fat, fat prediction model was developed by using the set of features defined in Table 2 and linear regression. The set of features was extracted from ROI containing fish diets of the abovementioned fat percentages.

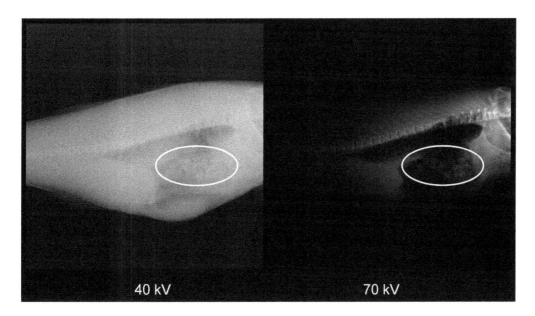

Figure 2. Dual energy X-ray images of Atlantic cod: a) X-ray image at 40 kV, b) X-ray image at 70 kV.

Radio-frequency Metal Detection of Marked Feed in Cod

Fish. Wild-caught cod (2.0 ± 0.8 kg, n = 15) caught at the coast of Central Norway, were killed, packed in ice, and shipped to the ACT Group Lab (Jessheim, Norway). As before, the intestines were kept intact during the experiment.

Feed and markers. Fish feed from [21] was mixed with four different markers (Table 4), and crude herring oil to get a sticky pellet surface and even distributions of marker particles. A dilution series was created to establish the lower detection limit of each marker. The trial feed with markers were packed into small plastic bags in 20 g portions (approximately 1% of fish body weight), equal to a potential daily feed portion, and inserted within the abdominal cavity of the fish. The abdominal cavity was collated by applying plastic strips around the fish body.

Radio-frequency metal detector. The radio-frequency metal detector used for this trial was a CEIA MS21 industrial metal detector [25] supplied by the ACT Group. The detector had a high precision and high sensitivity and was equipped with compensation technology for cancelling external environmental disturbances. The metal detector was arranged over a conveyer belt, approximately 2 m long. The signal strength was given in decibels, and lower detection limit was set to 10 dB (red line in Figure 3). The instrument was calibrated prior to the test by analyzing fish without marked feed and by adjusting the instrument for external disturbance. The metal detector was arranged over a conveyer belt to enable relatively fast and continuous insertion of fish into the detection unit. The sampling/acquisition period per fish was approximately 2 sec, but it can potentially be decreased by optimizing the system setup.

Statistical Analysis

To develop the linear regression model for fat prediction of fish diets inserted in the fish gut cavity, the Minitab [26] statistical software was used. The image features, mean and standard deviation of the X-ray image intensity (Table 2, 3), extracted from dual energy X-ray images, were fed into the Minitab and fat prediction models were generated. The combinations used, in form of multiplication and ratio of features, are motivated from the fundamental principle in dual energy X-ray imaging where the high energy and low energy images are combined in multiplication and division image processing operations [4].

Results and Discussion

X-ray

In Figure 1 and 4 are shown the X-ray images of fish without feed versus fish with feed mixed with glass beads (Figure 1b, 1c) and metal-based markers (Figure 4b, 4c) at the different dilution ranges (Table 1). Figure 1a shows a good visibility of the swim bladder. In Figure 1b and 1c there are also a good visibility of fish feed although its segmentation was not straightforward. The white dots were a result of automatic detection and segmentation of glass

Table 2. Extracted and generated features used for dual energy X-ray image analysis.

$M_{40\ kV}$, $S_{D40\ kV}$	Mean Intensity and standard deviation for ROI image at 40 kV
$M_{70\ kV}$, $S_{D70\ kV}$	Mean Intensity and standard deviation for ROI image at 70 kV
$M_{40/70} = M_{40\ kV}/M_{70\ kV}$	Ratio between mean intensities of the ROI images at 40 kV and 70 kV
$M_{40*70} = M_{40\ kV} * M_{70\ kV}$	Multiplication of mean intensities of the ROI images at 40 kV and 70 kV
$S_{40/70} = S_{40\ kV}/S_{70\ kV}$	Ratio between standard deviation for ROI images at 40 and 70 kV
$S_{40*70} = S_{40\ kV}*S_{70\ kV}$	Multiplication of standard deviation for ROI images at 40 and 70 kV

Table 3. Mean values of the extracted and generated features from the dual energy X-ay 40 kV and 70 kV ROI images of fish diets of the specific fat percentages (18%, 22%, 30%).

Fat(%)	40 kV mean	40 kV std	70 kV mean	70 kV std	40/70 mean	40/70 std	40*70 mean	40*70 std
18(n=6)	2587,65	85,70	37,44	7,11	81,71	25,39	91755,42	35862,59
22(n=6)	2594,17	105,27	40,71	11,06	70,19	17,34	106254,71	32482,88
30(n=6)	2627,30	134,30	45,67	16,72	65,40	21,80	122088,94	49793,78

beads by the Ishida X-ray machine at a certain threshold. It is seen that increase in concentration of glass markers did not result in a better discrimination (Figure 1c).

In Figure 4b and 4c are shown the X-ray images of cod with feed mixed with different concentrations of metal markers. Figure 4b shows a fish with feed containing 0.1% marker (0.03 g), while Figure 4c shows a fish with feed containing 10% marker (3 g). As it is seen in the Figure 4c, the image segmentation of particles was effective at the 10% marker concentration. This is in line with the conclusions of the preliminary study [11], who suggest that the feed could be added particles that absorb X-ray to increase contrast and thereby enable the discrimination of cod weaned to feed.

The advantage of X-ray as a non-invasive method for sorting and discrimination ability at the 10% particle metal concentration (3 g) is shown. Some potential challenges for the methods could be: 1) the need for metal particles to be mixed into the fish feed; 2) to achieve the necessary accuracy, each fish must be singulated prior to imaging. The singulation of fish can be technically solved by use of commercially available singulators. Although prospects of sorting large amounts of fish from a cage containing typically 50.000–100.000 individuals may seem challenging for a concept based on X-ray imaging, there already are commercially available X-ray imaging systems that can cope with harsh environment such as X-ray imaging systems for reservoir studies or pipes provided by InnospeXion [6].

Dual energy X-ray

An example of the visibility of undigested feed [21], with 30% fat, is seen in Figure 2, showing that the feed is visible in both the 40 kV and 70 kV images. Despite the visibility of the feed structure, in the form of cuboid chunks, we conclude that the individual variations in cod thickness, and variations in feed spatial distribution within the gut, makes planar dual-energy X-ray imaging unsuited for the purpose of measuring feed content in the gut. However, the method is promising for measurement of fat percentage in the fish feed diets, assuming the precise extraction of the ROI to be analyzed. Depending on the set of features, the prediction of fat from both linear and multiple linear regression varies from poor $R^2 = 0,4476$ (use of only one feature $M_{40*70} = M_{40\,kV} * M_{70\,kV}$) to $R^2 = 0,9029$ (use of all generated features) (Table 5). To obtain fat percentage measurements with higher accuracy and throughout the fish gut volume, a dual-energy computed tomography (CT) imaging approach may be convenient, as indicated in previous research [17].

Radio-frequency Metal Detector

The radio-frequency metal detector gave consistent measurements (Table 6). All markers gave detection responses above the detection limit for metal concentrations between 0.1 and 5%. Steel grit marker at 0.1% (0.03 g) concentration provided the strongest signals, while carbonyl iron powder did so at approximately 0.7% (2.1 g) (Figure 3). The steel grit particles were relatively large, with sharp edges and not of uniform shape. The carbonyl-iron powder could be an excellent marker due to its low particle size (a few μm), round particle shape, adaptability to processing in commercial feed mill technology, and benign interaction with the cod digestive system.

Due to efficiency and detection accuracy, radio-frequency metal detectors combined with feed-added carbonyl iron powder at low concentrations seemed like a robust and efficient system for detection of feed in an *in vivo* context. The radio-frequency metal detector could easily be coupled with a conveyer belt or a pipe

Table 4. Particle size of markers used for metal detection.

Marker	Particle size (mm)	Magnetism
Aluminium oxide (AlO2)	0.4–0.6	No
Carbonyl iron (Fe)	0.006–0.009 (powder)	Yes
Steel grit	0.2–0.4	No
Chrome – Nickel alloy (Cr-Ni)	0.6	No

system where fish can be inspected and sorted regardless of positioning and speed.

Use of Fish Feed and Markers

The use of fish feed with added markers (NiCr, AlO$_2$, Carbonyl Iron, Steel Grit, Ballotini glass beads), as opposed to other marking methods, provided an effective and low-cost method for marking of fish. The general advantage of using this method was that the marked feed could be manufactured in a commercial feed extruder, and the feed could be distributed to all fish in the cohort at once. One potential disadvantage be the variable marking efficiency when feed is unevenly distributed or not accepted by the fish.

Biological Hazards and Fish Welfare

Since we worked with dead fish under laboratory conditions, possible biological hazards and animal welfare issues related to the application of the different markers were not tested in this study.

In practice with live fish, however, it is hardly conceivable that markers such as steel grit, aluminum oxide and Cr-Ni alloys are suitable in fish feed since these compounds are either abrasive, they can have sharp edges, or they can be carcinogen to humans. As such, these compounds might harm the digestive system in fish and their use can therefore result in poor fish welfare. If absorbed in the flesh, they might constitute a hazard to the consumers. Although the abovementioned markers don't have a foreseeable practical use in live fish, we used them for method development as they were commercially available. One of the main hypothesis whether the use of markers would increase the contrast in detection of marked feed to be able to discriminate cod weaned to feed.

On the other hand, results imply that carbonyl iron has an excellent foreseeable practical use in live fish due to the fact that it does not pose biological hazard to live fish. In human beings, single doses of 1–10 g carbonyl iron were tolerated with no evidence of toxicity and only minor gastrointestinal side effects [3].

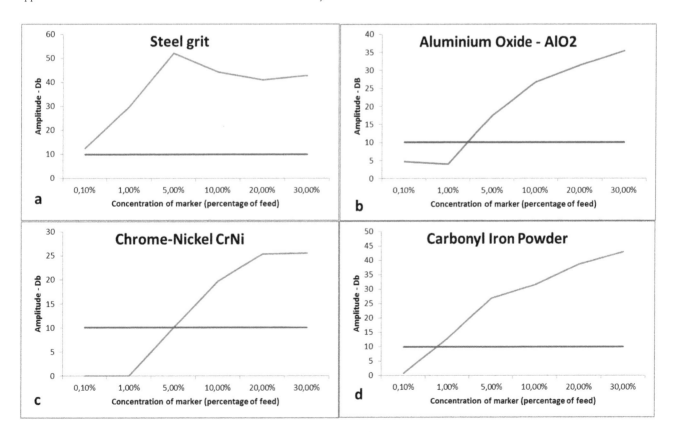

Figure 3. Amplitude of radio frequency metal detector response vs concentration of metal-based markers a) Steel grit, b) Aluminium oxide (AlO$_2$), c) Chrome-nickel (CrNi), d) Carbonyl iron powder. The red curve is the detection limit at 10 dB.

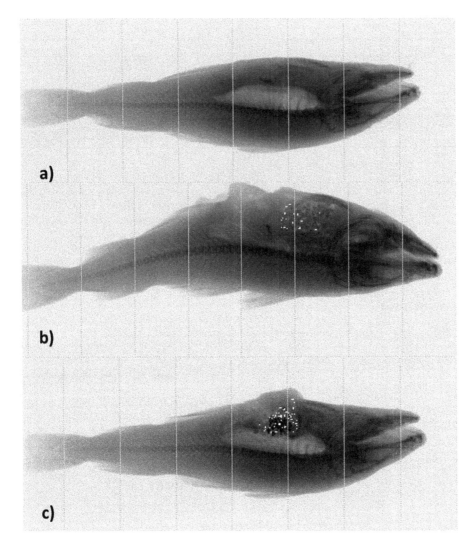

Figure 4. X-ray images of Atlantic cod generated from the Ishida X-ray machine: a) image of fish without feed, b) image of fish with Chrome-Nickel (CrNi) markers at 0.1% (0.03 g) weight percent of total feed content in gut; c) image of fish with Chrome-Nickel (CrNi) markers at 10% (3 g) weight percent of total feed content in gut.

Future work will be concentrated on developing trials with live fish and using carbonyl iron powder as feed marker.

Concept for Implementation of Sorting System

With a detection/discrimination subsystem based on either X-ray imaging or radio frequency metal detection, a suggested concept for implementation of a sorting system for discriminating between weaned and unweaned cod is shown (Figure 5). The discrimination process begins by feeding all the cod with commercial dry feed mixed with a small concentration of carbonyl iron powder. Once the feeding ceases, the cod are pumped from the cage into the sorting system. By mounting a fish pump, detection and sorting unit on a raft, the fish farmer could cut the costs by integrating the discrimination subsystem into the existing pump and sorting system. After drain-off of water, a conveyer belt can be used to transport the fish to the X-ray imaging system or the radio-frequency metal detection unit. Based on the result from the detection unit, with weaned cod being detected by the presence of the metallic carbonyl iron powder in the ingested dry feed, the weaned and unweaned cod can be transported to

Table 5. Prediction of fat of fish diets inserted into fish gut cavity with dual energy band X-ray image features.

Feature set	Prediction model	R^2
$M_{40*70} = M_{40\ kV} * M_{70\ kV}$	Fat(%) = 2,42+0,000196*M_{40*70}	0,4476
All features	Fat(%) = 517+0,00005* 40*70 M −0,203*40 kVmean −0,383 * 40 kV std +0,97* 70 kVmean −15,1 *70 kV std −0,12* 40/70 mean +1,29* 40/70 std +0,00566* 40*70 std	0,9029

Table 6. Response values from the metal detector for different marker types and concentrations (detection limit at 10 dB).

Concentration (%)	AlO₂ (dB)	CrNi (dB)	Steel grit (dB)	Carbonyl Iron powder (dB)
0.10	4.62	0	12.42	0.9
1.00	3.9	0	29.68	**12.84**
5.00	17.52	9.96	52.02	**26.9**
10.00	26.74	19.7	44.22	**31.36**
20.00	31.48	25.34	41	**38.64**
30.00	35.46	25.54	42.78	**42.86**

It is seen that the response value for Carbonyl Iron powder is above the limit for concentrations higher than 1%.

separate cages. From this point on, weaned cod can thus be fed with commercial dry feed, whereas the unweaned cod are fed with the more costly moist feed.

An evaluation of investment costs for the concept solutions based on X-ray, dual energy X-ray and metal detector discrimination methods is reported in [16]. This analysis takes into account the investment need for pumping system, singulation unit, raft, flexible pipes, detection unit, conveyer belts, control software, housing, and costs related to feed production. The cheapest concept solution is the one based on metal detector concept with an estimated cost of 43806 € (361000 NOK), while the concepts based on X-ray are estimated at an approximate cost of 62918 € (518500 NOK).

Conclusions and Future Work

The aim of this work was to investigate the feasibility of X-ray imaging and radio frequency metal detection methods for use in discrimination and sorting of wild cod into two groups – those that are unweaned and those that are weaned from moist feed. Based on the current research, we conclude that both X-ray imaging and radio frequency metal detection were promising methods for discrimination between unweaned and weaned wild cod. In the present investigation, the best results were obtained with the method based on the use of carbonyl iron powder in the commercial dry fish feed combined with the high sensitive radio-frequency metal detector technology. Metal detector based concept was also better when it comes to investment cost,

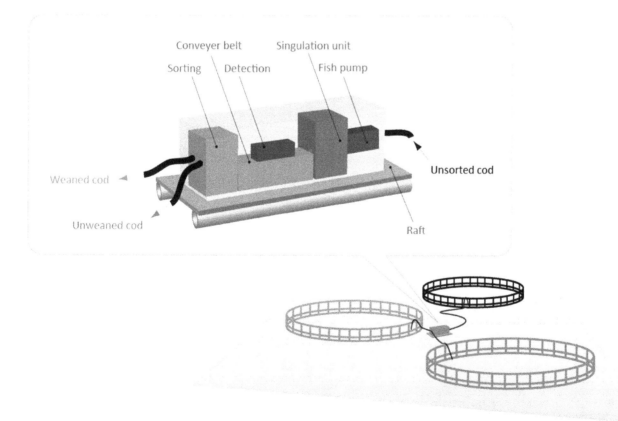

Figure 5. Principle diagram of the concept of sorting system for in situ discrimination of weaned and unweaned Atlantic cod in capture-based aquaculture (CBA) systems.

capacity, accuracy, and use in water. Therefore, as future work we suggest building of a prototype based on metal detector concept, and evaluation of the concept for sorting of live fish in the controlled laboratory conditions. Another important aspect of these trials would be to document the potential biological hazards of carbonyl iron powder to live fish.

Acknowledgments

Dr. Emil Veliyulin is greatly acknowledged for his contribution in preliminary studies that led to the present work. So are Dr. Egis Dauksas for his help in the X-ray investigation experiment, research scientist Andreas Myskja Lien for his work on Figure 5, research scientist Morten Bondø for validation of statistics in Minitab, and Philip Laurence from Ishida Europe for his assistance with the Ishida IX-GA X-ray machine.

Author Contributions

Conceived and designed the experiments: SM EM JRM. Performed the experiments: SM EM. Wrote the paper: EM SM UE JRM. Analyzed the data and images: EM SM JRM. Contributed with fish welfare and biological hazards section: UE.

References

1. Braaten B (1984) Growth of cod in relation to fish size and ration. In: Dahl E, Danielssen D, Moksness E, Solemdal P, editors. The propagation of cod. Institute of Marine Research, Arendal, Norway, 677–710.

2. Dreyer BM, Nøstvold BH, Midling KØ, Hermansen Ø (2008). Capture-based aquaculture of cod. In: Lovatelli A, Holthus PF, editors. *Capture-based aquaculture*. Global overview. FAO Fisheries Technical Paper 508, Rome, 183–198.

3. Gordeuk VR, Brittenham GM, McLaren CE, Hughes MA, Keating LJ (1986) Carbonyl iron therapy for iron deficiency anemia. Blood 67: 745–752.

4. Hamamtsu website: Available http://www.hamamatsu.com/resources/pdf/sys/SFAS0023E06_C10800s.pdf.Accessed 2014 February 20.

5. Hermansen Ø, Isaksen JR, Dreyer B (2012) Challenging spatial and seasonal distribution of fish landings – Experiences from vertically integrated trawlers and delivery obligations in Norway. Marine Policy 36: 206–213.

6. InnospeXion website. Available: http://www.innospexion.dk/index.php/en/pipe-inspection-systems. Accessed 2014 February 20.

7. Ishida Ltd website. Available: http://www.ishida.com/products/ix_g/ix_g_features.html. Accessed 2014 February 20.

8. Jobling M, Christiansen JS, Jørgensen EH, Arnesen AM (1993) The application of X-radiography in feeding and growth studies with fish: a summary of experiments conducted on Arctic charr. Reviews in Fisheries Science 1: 223–237.

9. Jobling M, Arnesen AM, Baardvik BM, Christiansen JS, Jørgensen EH (1995) Monitoring feeding behavior and food intake: methods and applications. Aquaculture Nutrition 1: 131–143.

10. Jobling M, Coves D, Damsgård B, Kristiansen HR, Koskela J, et al. (2001) Techniques for measuring feed intake. In: Houlihan D, Boujard T, Jobling M, editors. Food Intake in Fish. Oxford: Blackwell Science.

11. Martinsen S, Kirkhus T (2009) Codgrade. SINTEF Report Norway : SFH80F094027.

12. Pirhonen J, Schreck CB, Reno PW, Ögüt H (2003) Effect of fasting on feed intake, growth and mortality of chinook salmon, Oncorhynchus tshawytscha, during and induced Aeromonas salmonicida epizootic. Aquaculture 216: 31–38.

13. Refstie S, Førde-Skjærvik O, Rosenlund G, Rørvik KA (2006) Feed intake, growth, and utilization of macronutrients and amino acids by 1- and 2-year old Atlantic cod (*Gadus morhua*) fed standard or bioprocessed soybean meal. Aquaculture 255: 279–291.

14. Santos JD, Jobling M (1991) Factors affecting gastric evacuation in cod, Gadus morhua L., fed single-meals of natural prey. J Fish Biol 38(5): 697–713.

15. Sæther BS, Bjørn PA, Midling KØ, Nilsen R, Jacobsen R, et al. (2009) Fangstbasert Akvakultur, Tilvenning av villtorsk til tørrfôr (Capture-based aquaculture, weaning of wild cod to dry feed). NOFIMA report in Norwegian 4/2009. 25p. http://www.nofima.no/filearchive/Rapport%2004-2009.pdf.

16. Sæther BS, Noble C, Humborstad OB, Martinsen S, Veliyulin E, et al. (2012) Fangstbasert akvakultur: Mellomlagring, oppfôring og foredling av villfanget fisk. NOFIMA report 14/2012.

17. Segtnan VH, Høy M, Sørheim Ø, Kohler A, Lundby F, et al. (2009) Noncontact salt and fat distributional analysis in salted and smoked salmon fillets using X-ray computed tomography and NIR interactance imaging, J Agriculture and Food Chemistry 57: 1705–1710.

18. Talbot C, Higgins PJ (1983) A radiographic method for feeding studies on fish using metallic iron powder as a marker. J Fish Biol 23: 211–220.

19. Toften H, Arnesen AM, Jobling M (2003) Feed intake, growth and ionoregulation in Atlantic salmon (Salmo salar L.) smolts in relation to dietary addition of a feeding stimulant and time of seawater transfer. Aquaculture 217: 647–662.

20. Veliyulin E, Misimi E, Bondø M, Vebenstad PA, Østvik SO (2011) A simple method for weight estimation of whole herring *(Clupea harengus)* using planar X-ray imaging. Journal of Food Science 76: E328–E331.

21. Skretting Optiline website. Available: http://www.skretting.no/internet/SkrettingNorway/webInternet.nsf/wprId/BC7D3F9750A5066DC1257433003E6D02!OpenDocument. Accessed 2013 November 24.

22. SIEMENS Axiom Aristos FDX website. Available: http://www.medwrench.com/?equipment.view/equipmentNo/4418/Siemens/Axiom-Aristos-FX-Plus/. Accessed 2014 February 10.

23. Mathworks website. Available: http://www.mathworks.se/. Accessed 2013 November 24.

24. Potters Industries website. Available: http://www.pottersbeads.com/. Accessed 2014 March 30.

25. Heat and Control website. Available: http://www.heatandcontrol.com/. Accessed 2014 February 20.

26. Minitab website. Available: http://www.minitab.com/en-us/. Accessed 2014 March 30.

13

Modelling the Impact of Temperature-Induced Life History Plasticity and Mate Limitation on the Epidemic Potential of a Marine Ectoparasite

Maya L. Groner[1]*, **George Gettinby**[2], **Marit Stormoen**[3], **Crawford W. Revie**[1], **Ruth Cox**[1]

1 Centre for Veterinary Epidemiological Research, Department of Health Management, Atlantic Veterinary College, University of Prince Edward Island, Charlottetown, Prince Edward Island, Canada, **2** Department of Mathematics and Statistics, University of Strathclyde, Glasgow, Scotland, United Kingdom, **3** Centre of Epidemiology and Biostatistics, Norwegian School of Veterinary Science, Oslo, Norway

Abstract

Temperature is hypothesized to contribute to increased pathogenicity and virulence of many marine diseases. The sea louse (*Lepeophtheirus salmonis*) is an ectoparasite of salmonids that exhibits strong life-history plasticity in response to temperature; however, the effect of temperature on the epidemiology of this parasite has not been rigorously examined. We used matrix population modelling to examine the influence of temperature on demographic parameters of sea lice parasitizing farmed salmon. Demographically-stochastic population projection matrices were created using parameters from the existing literature on vital rates of sea lice at different fixed temperatures and yearly temperature profiles. In addition, we quantified the effectiveness of a single stage-specific control applied at different times during a year with seasonal temperature changes. We found that the epidemic potential of sea lice increased with temperature due to a decrease in generation time and an increase in the net reproductive rate. In addition, mate limitation constrained population growth more at low temperatures than at high temperatures. Our model predicts that control measures targeting preadults and chalimus are most effective regardless of the temperature. The predictions from this model suggest that temperature can dramatically change vital rates of sea lice and can increase population growth. The results of this study suggest that sea surface temperatures should be considered when choosing salmon farm sites and designing management plans to control sea louse infestations. More broadly, this study demonstrates the utility of matrix population modelling for epidemiological studies.

Editor: Erik Sotka, College of Charleston, United States of America

Funding: This research was undertaken thanks to funding from the Canada Excellence Research Chairs Program (http://www.cerc.gc.ca/chairholders-titulaires/gardner-eng.aspx), the SALMODIS project (http://www.sintef.no/salmodis), and the Atlantic Innovation Fund (http://www.acoa-apeca.gc.ca/eng/ImLookingFor/ProgramInformation/AtlanticInnovationFund/Pages/AtlanticInnovationFund.aspx). The funders had no role in study design, data collection and analysis, decision to publish, or preparation of the manuscript.

Competing Interests: The authors have declared that no competing interests exist.

* E-mail: mgroner@upei.ca

Introduction

Many marine pathogens are capable of causing dramatic population-, community- and ecosystem-level shifts and the patterns of infection are frequently associated with temperature [1], [2], [3], [4], [5]. In particular, high temperatures are often associated with increased frequency or severity of infection, as a result of altered development and survival of the pathogen, physiological changes in the host and range expansions [1], [2], [6], [7], [8]. Understanding the role that temperature plays in the epidemiology of marine diseases is important for predicting and potentially mitigating infestations and may be important for forecasting disease risk in a climate change context.

Quantifying the influence of temperature on infections in marine environments is challenging. For many marine pathosystems, there is a lack of baseline data on how temperature influences epidemiological patterns and those that exist are often confounded with other influential water quality information (e.g., salinity, circulation) [1], [4]. In addition, temperature can influence the host and the pathogen separately, and these effects

may differ among life history stages. In many cases only some of these interactions are understood or the etiologic agent of disease is unknown [1], [2], [6], [9]. Despite these challenges, water temperature often follows well-defined seasonal patterns and its effects should be predictable.

Open-pen aquaculture may offer a unique opportunity to understand the role of temperature on marine diseases. In particular, because these systems often control spatial and temporal variation in host densities, they can be used to examine the role of temperature in influencing pathogen life history and virulence. One case where temperature may be especially influential is that of sea louse (*Lepeophtheirus salmonis*) infestations on salmonids. Sea lice are an ectoparasite of farmed and wild salmonids (Atlantic salmon (*Salmo salar*), steelhead (*Oncorhynchus mykiss*), and Pacific salmon (*Oncorhynchus* spp.)) and infestations have been associated with declines in returns of adult wild salmonids [10], [11], [12]. Sea lice have very plastic life history responses to temperature. For example, the generation time of sea lice has been estimated to range between 50 days at 12°C and 114 days at 7°C [13], suggesting that infestations may increase in response to warmer temperatures. Nonetheless, the role that temperature plays in sea louse infestations is not clear. While controlled laboratory manipulations consistently find strong effects of temperature on sea louse development [14], effects of temperature on the population dynamics of sea lice are only detectable in some field data [15], [16], [17].

Many fish farms experience substantial economic losses due to morbidity of infested stock as well as the use of expensive chemotherapeutants to control sea lice [18]. A number of methods have been pursued within the salmon industry to control sea lice infestations on farms. These include adoption of integrated pest management approaches in which management areas, defined by hydrological boundaries, are fallowed periodically to break the sea lice reinfection cycle and all salmon in the management area are restricted to a single age cohort to avoid infection between age-classes. In addition to these practices, chemotherapeutant treatments are often necessary to control sea lice [19], [20]. While they have the potential to be very effective at reducing densities of attached sea lice (chalimus and mobiles) [21], [22], the success rates of chemical treatments often vary and in some cases numerous treatments are required to control sea louse populations [20]. Additional concerns with chemical treatments arise because they can be expensive [18], are stressful to salmon [23], have potentially detrimental environmental impacts [24], can be hazardous to the workers that dispense them and are proving to be less effective over time because sea lice have evolved resistance [20] [25]. Different treatments target different stages of sea lice and while both temperature and the stage targeted may influence the efficacy of a treatment, the role of these factors has not been throroughly investigated.

A range of modelling techniques have been used to evaluate sea louse population growth over time, including delay differential equation models [26], [27], individual-based models [28], advection-diffusion models [29], system dynamic models [30], and stochastic Monte Carlo simulation models [31]. While many of these models include temperature, we are only aware of one that has explicitly examined the effect of temperature variation on population demography and vital rates [14].

Matrix population models provide a useful tool for exploring the interactions between life history, temperature and population demography. Matrix population models can be manipulated to incorporate life history variation, stochasticity, environmental-dependencies and population feedbacks (e.g. density dependence) [32]. Moreover, analytical tools are well-developed for understanding the contribution of all of these factors to population demographics [32]. For example, elasticity analysis can be used to examine the effect of proportional changes to contributions of life stages (defined as matrix elements) on population growth, while sensitivity analysis can be used to examine the effect of absolute changes in life stage properties on population growth. The elasticities of population growth to changes in matrix elements can be used to predict the effectiveness of stage-targeted control methods, while the sensitivities of population growth to changes in matrix elements can provide insight in predicting how a population will evolve in response to selection at a specific life stage [32]. Comparison of elasticities and sensitivities of matrices constructed for the same organism at different temperatures can be used to understand how temperature-induced life history plasticity may alter population demographics. While population matrix models have a long history of use in conservation biology and pest management [33], [34], [35], they have rarely been used to understand the epidemiology of marine pathogens or parasites [36].

In this study we use stochastic matrix population models to understand the influence of temperature on the population growth, reproduction and demography of sea lice (L. salmonis) on farmed Atlantic salmon. We use sensitivity and elasticity analyses to understand the contribution of each life stage to population growth. We also examine how density-dependent mating and the rate that larval sea louse attach to hosts influence these patterns. Finally we evaluate these results in terms of the effect of temperature on population growth and effective control of sea lice.

Materials and Methods

Matrix Construction

To evaluate the effects of temperature, seasonality and the host attachment rate on sea louse demography, we created stage-structured population projection matrices (PPM) for female sea lice based on parameters from the literature. The model does not explicitly include Atlantic salmon hosts because they are not expected to influence the epidemiology of sea lice. This is because they exhibit little immune response to sea lice [37] and are maintained at constant densities throughout the salt water production phase.

L. salmonis transition through nine recognised life stages [38]. After hatching from the egg, the sea louse goes through three unattached stages during which it does not feed: nauplii (2 stages) and copepodid. Once the copepodid finds a host, it develops through two chalimus stages, two preadult and one adult stage. In our model, we reduced the life cycle to seven stages that reflect biologically important transitions: egg, larvae (consisting of nauplii I and II and copepodid), chalimus (stages I and II), preadult (I and II) and three adult phases which will be referred to as gravid I, between-clutch and gravid II (Figure 1). The three adult phases are separated here because they differ in terms of fecundity. Transitions from stage to stage occur in one direction, with the exception that females can transition from gravid II to between-clutch, and then back to gravid II, reflecting observations that females can produce up to 11 successive pairs of egg strings [39]. The population projection matrix represents daily transitions and operates on the life stage state vector with elements [egg, larvae, chalimus, preadult, gravid I, between-clutch and gravid II] (Figure 1).

Entries on the diagonal (P_i) indicate the proportion of individuals remaining in a stage, entries on the sub- and super-diagonal (G_i) indicate the proportion of individuals developing into a new stage and F_5 and F_7 indicate the fecundity of gravid I and gravid II adult females, respectively. P_2 and G_2 are a function of the rate that sea lice attach to the host (γ, described below), and G_5 and G_6 and G_7 are a function of egg hatching and development (described below). The remaining P_i and G_i elements, together with the F_i elements, are defined as shown:

$$P_i = 1 - \frac{1}{\delta_{ij}} * (1 - \mu_i), \begin{cases} \text{for } i = \{1,3\text{-}6\}, j = i+1 \\ \text{for } i = 7, j = 6 \end{cases} \quad (1)$$

$$G_i = \left(\frac{1}{\delta_{ij}}\right) * (1 - \mu_i), \{\text{for } i = \{1,3,4\}, j = i+1 \quad (2)$$

$$F_i = \frac{\omega_i}{2} * \phi, \, i = \{5,7\} \quad (3)$$

where δ_{ij} = time to develop from stage i to stage j, μ_i = mortality rate at stage i, and ω_i = number of viable eggs in the clutch

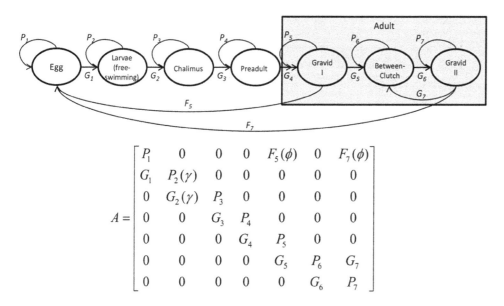

$$A = \begin{bmatrix} P_1 & 0 & 0 & 0 & F_5(\phi) & 0 & F_7(\phi) \\ G_1 & P_2(\gamma) & 0 & 0 & 0 & 0 & 0 \\ 0 & G_2(\gamma) & P_3 & 0 & 0 & 0 & 0 \\ 0 & 0 & G_3 & P_4 & 0 & 0 & 0 \\ 0 & 0 & 0 & G_4 & P_5 & 0 & 0 \\ 0 & 0 & 0 & 0 & G_5 & P_6 & G_7 \\ 0 & 0 & 0 & 0 & 0 & G_6 & P_7 \end{bmatrix}$$

Figure 1. Diagram of stage-structured population projection matrix that is used in simulations. P_i indicates the probability of staying in a stage, G_i indicates the probability of transitioning to another stage and F_i indicates fecundity. Survival and attachment of free-swimming larvae are a function of the rate that they attach to hosts (γ) and fecundity is a function of the probability of mating (φ).

produced at stage i. This number is multiplied by the probability of mating (φ) and divided by two because the matrix only considers females and assumes a 1:1 sex ratio.

Parameters

Developmental transitions. With the exception of copepodids, developmental transitions of sea lice were temperature-dependent. We parameterized developmental times in our model based on a review by Stien et al. 2005 [14] of existing data on temperature-dependent development in *L. salmonis* (Table 1). This review uses developmental rates across a temperature range to parameterize a modified Belahrádek equation [40]:

$$\tau_{ij}(T) = \left(\frac{\beta_{1ij}}{(T - 10 + \beta_{1ij}\beta_{2ij})} \right)^2 \tag{4}$$

where τ_{ij} is the minimum required developmental time for individual i in stage j at temperature T. β_{1ij} is a shape parameter and β_{2ij}^{-2} is the average for τ at $10°C$. Variation in developmental rates was incorporated by randomly selecting values for β_{1ij} and β_{2ij} from a normal distribution with mean and standard deviation taken from [14]. A stage-specific constant, v_j, was added to this

temperature-dependent estimate, to represent additional time beyond the minimum developmental time needed to make developmental transitions. Total developmental time (δ_{ij}) of individual i at stage j is:

$$\delta_{ij} = \tau_{ij} + v_j \tag{5}$$

Values of v were from [14] and the developmental rates were the inverse of the developmental times. There is no evidence that copepodid developmental rates are temperature-dependent, so all copepodids in this model developed in 4.6 days, the mean value as estimated by [14]. The developmental rate of larvae was calculated as the inverse of the sum of nauplii and copepodid developmental times.

Survival. Survival estimates ($1-\mu_i$) for all stages except larvae were stochastically drawn from a triangular distribution defined by the minimum, maximum and most probable survival times based on data from [41] (Table 2). As with many invertebrates, there is little evidence that survival of sea lice is directly dependent upon temperature [14], [26].

Table 1. Parameters used for matrix model calculations (equations 1–8).

Life Stage	β_1 (Standard Error)	β_2 (Standard Error)	v
Eggs	41.98 (2.85)	0.338 (0.012)	2
Nauplii	24.79 (1.43)	0.525 (0.017)	0
Chalimus	74.7 (33.64)	0.236 (0.007)	0.85
Pre-adult Female	67.47 (20.36)	0.177 (0.006)	0.34

Parameters used in equations 4 and 5 to estimate developmental rates of each stage (from [14]).

Table 2. Survival rates for each life stage of the sea louse.

Parameters	Mean	Lower	Upper	Source
Egg Viability	0.90	0.75	0.96	[39]
Nauplii Survival (daily)	0.83			[14]
Chalimus Survival (daily)	0.992	0.98	0.997	[41]
Preadult Survival (daily)	0.965	0.953	0.98	[41]
Adult Survival (daily)	0.965	0.904	0.997	[41]

With the exception of nauplii, survival estimates for individuals were randomly drawn from a triangular distribution with lower and upper values shown.

Survival of larvae was the product of nauplii survival (calculated as above) and copepodid survival. Survival of copepodids depends upon the attachment of copepodids to a host, a process which is sensitive to local host densities and abiotic factors including currents and salinities [15], [17], [42]. Daily survival rates of copepodids (S_c) depend on the attachment rate (γ) as well as the development time (δ_c) such that:

$$S_c = \gamma^{\frac{1}{\delta_c}} \tag{6}$$

Since the attachment rate of copepodids varies considerably in nature, we simulated scenarios with several different values for γ (described in *Analyses*).

Fecundity. Sea lice reproduce sexually and females have two external egg strings in each clutch that are attached to them until hatching. Egg string production and hatching are synchronized on an individual louse. Estimates of egg viability, clutch size and time to hatching were based on data from [39]. Sea lice are estimated to have 152 ± 31 (mean $\pm$ SD) eggs per egg string in the first clutch and 296 ± 100 eggs per egg string in subsequent clutches [39]. Estimates of clutch size were chosen from these normal distributions and multiplied by two to account for both egg strings. The number of eggs produced was then multiplied by the estimated viability (Tables 1 and 2) and divided by two because the model only tracks female members of the population. There is little evidence for an effect of temperature or clutch order on egg viability [39].

Because the first clutch of eggs is substantially smaller than subsequent clutches we divided adult female stages into three parts, gravid I to represent the first extrusion of eggs (represented by F_5), gravid II to represent subsequent extrusions of eggs (represented by F_7) and between-clutch to represent the time between the extrusion of eggs. After completing the gravid I stage (P_5), individuals will alternate between the between-clutch and gravid II stages represented by P_6, P_7, G_6 and G_7 in the PPM.

The time spent between clutches was the sum of the time needed for eggs to develop after egg string extrusion (δ_{egg}-1) and the time between the hatching of one clutch and the release of the next egg string (ζ). Both of these are temperature-dependent. The latter parameter was estimated with the following relationship:

$$\varsigma = \max(1, -1.2T + 19.64) \tag{7}$$

This equation is based on hatching times at 7.2°C and 12.2°C [39] and the observation that the shortest time between clutches, which is observed at temperature > 15°C, is ~24 hours [43]. Egg extrusion (η)(e.g., the time in the gravid I or gravid II stages) was set at one day [43]. G_5, G_6 and G_7 are therefore defined as:

$$G_5 = G_7 = \frac{1}{\eta} * (1 - \mu_{adult}) \tag{8}$$

$$G_6 = \frac{1}{\zeta + \delta_{eggs} - \eta} * (1 - \mu_{adult}) \tag{9}$$

Density-dependence. Recent models of sea louse mating suggest that reproduction in sea lice is limited by mate availability when the abundance of sea lice is low [31], [44]. We incorporated this density-dependent effect, also called depensation or an Allee effect [32], in some iterations of the model by reducing the fecundity by the probability of mating (ϕ), which was calculated based on the ratio of adult sea lice to hosts.

To include density-dependent mating in our model, we used a variation of the model presented by [45]. The model assumes parasites are distributed on hosts according to the negative binomial distribution. This distribution is suited for dioecious parasites that aggregate together. It simulates the probability that a female will mate (ϕ) as a function of the mean number of adult lice on a host (m, calculated here as twice the number of adult females, therefore assuming and equal sex ratio) and a parameter describing overdispersion of sea lice among hosts (k) such that:

$$k = \frac{m}{VMR - 1} \tag{10}$$

where VMR is the variance to mean ratio of adult sea lice on hosts. Because sea lice are polygamous [46], we used a variation of this model that assumes that parasites coaggregate and that mating occurs for all females when there is at least one male on a host (i.e. complete promiscuity):

$$\phi(m,k) = 1 - (1-\alpha)^{1+k}(1 - \frac{\alpha}{2})^{-1-k} \tag{11}$$

where $\alpha = m/(m+k)$. The model assumes an equal sex ratio [47]. In the special case where $k \to \infty$, VMR = 1 and the lice assume a Poisson distribution among hosts with probability of mating simplifying to:

$$\phi(m) = 1 - e^{-\frac{m}{2}} \tag{12}$$

For models where this density-dependent effect was included, fecundity estimates F_5 and F_7, were multiplied by ?? .

Analyses

We calculated PPMs for a number of relevant scenarios, including a range of fixed temperatures and larval attachment rates, density-dependent mating, and yearly temperature profiles (described below). Depending upon the scenario, we examined some or all of the following demographic parameters: population growth (λ), reproductive rate (R_0), generation time, and the sensitivity and elasticity of population growth to changes in matrix elements.

In our equations, λ was equal to the rate of population change over a day (i.e. the dominant eigenvalue of the matrix **A**). R_0 was equal to the average number of offspring by which an egg will be replaced within its lifetime (i.e. the rate by which the population increases from one generation to the next) and generation time was equal to the time necessary to produce the number of offspring predicted by R_0 [32]. Population growth is stable when $\lambda = 1$, decreases for $\lambda < 1$ and increases for $\lambda > 1$. Sensitivity was calculated as the effect of absolute changes to matrix elements on the population growth rate,

$$s_{ij} = \left(\frac{\partial \lambda}{\partial a_{ij}}\right) \tag{13}$$

while elasticities were calculated as the effect of proportional changes in matrix elements on the population growth rate,

$$e_{ij} = \left(\frac{\partial \log \lambda}{\partial \log a_{ij}} \right) \qquad (14)$$

where a_{ij} indicates the matrix element [32].

In addition to the above analyses we performed a number of simulations to examine how various starting conditions affected population dynamics. All analyses were implemented in R (v. 2.15.0) using the 'popbio' package [48]. Details about how stochastic effects were included into each analysis are included below. Matrices showing means and standard deviations for all fixed temperature scenarios are in supplemental appendix S1. Annotated R-code is available in supplemental code S1, S2, S3, S4, S5, S6, S7 and S8.

Effects of fixed temperatures and larval attachment rates. In order to understand the effect of temperature on population growth rate, we calculated the matrix **A** for the following water temperatures, 4°C, 8°C, 12°C, 16°C and 20°C. These temperatures are within the range typically experienced by sea lice [49]. While temperatures colder than 4°C are also likely to occur in some locations (e.g. [17]), there are no data available to parameterize life history traits at these values.

The proportion of copepodids that attach to a host varies considerably in nature as a function of host behaviour and water salinity, hydrodynamics and light availability [50], [51], [52]. While there is some evidence that temperature may influence attachment rates of copepodids, it was not conclusive enough to include in the model [50], [51]. In order to measure how this variation influences sea lice populations, for each specific temperature matrix, we calculated the matrix **A** with different values for the attachment rate ($\gamma = 0.001$, 0.01, 0.1, 0.5, 0.9), at each of the fixed temperature profiles.

Fixed temperature scenarios included demographic stochasticity where variation can be estimated from the literature. Individual louse developmental, survival and fecundity estimates in the PPM were calculated independently in each run (see descriptions above) and clutch sizes were drawn from normal distributions (described in *developmental transitions*). Survival estimates were drawn from triangular distributions (described in *survival*). For all fixed scenarios, we calculated the matrix **A** 1000 times to create a distribution of matrices which we used to calculate the mean and 95% confidence intervals for the intrinsic population growth rate, the net reproductive rate and the generation time.

We calculated the sensitivity and elasticity of λ to matrix elements. Graphical displays of these results show the sums of elasticities or sensitivities associated with a life stage. For example, the sensitivity of λ to larval sea lice is the sum of the sensitivity of G_2 and P_2 and the elasticity of λ to fecundity is the sum of the elasticities of λ to F_5 and F_7. Matrices of means and standard deviations for elasticities and sensitivities in all fixed temperature scenarios and larval attachment rates are shown in supplemental appendix S2 and S3.

See supplemental appendix S4 for calculations of the proportion of individuals at each life stage at equilibrium and simulations to determine the time until equilibrium was reached (Figures S1 and S2 in supplemental appendix S4).

Effects of temperature on density-dependent mating. Density-dependent effects on population growth result in nonlinear models and the analyses described above are therefore not applicable. In order to understand how temperature and the larval attachment rate described above influence density-dependent mating, we ran simulations using density-dependent terms for fecundity elements F_5 and F_7 of the projection matrix. In each simulation we began with a population of 0.1 adult gravid I

females per host. The model was simulated at the constant temperatures described above until the threshold of 3 adult female lice per host was crossed. This threshold was chosen because at this abundance mating success is nearly 95% for all models (4A) and mate limitation begins to have negligible effect on population growth. This was simulated by calculating matrix values associated with each life stage 100 times and using the means of these values for $VMR = 1$, 1.3 and 2 as well as in cases where density-dependent mating was not modelled.

Effects of seasonality. Understanding the demographic properties of sea louse populations at a single temperature is useful for developing a conceptual understanding of the effect of temperature; however, sea lice live in environments that typically experience substantial seasonal temperature variation. It is unclear how variation from 'typical' seasonal patterns might influence population growth rates. Therefore, we evaluated the intrinsic population growth rate across seasons for a variety of temperature scenarios. The baseline temperature profile was a sine curve fitted to a mean of temperature data collected at 33 fish farm sites in Scotland over a 5 year period [53]. We varied this temperature profile to create the following scenarios: cold and warm years (all temperatures 2°C below or above baseline), a year with cold winters (winter minimum is 2°C below baseline), a year with warm summer (summer maximum is 2°C above baseline), a year of 'more seasonal' temperatures (same mean temperature, but minimum and maximum temperatures are 2 degrees more extreme than in the baseline scenario) and a year of 'most seasonal' temperatures (same mean temperature, but the minimum winter and maximum summer temperatures are 4.5°C more extreme than in the baseline scenario). In all seasonal scenarios, we started the simulation in spring (i.e. day 1 began 120 days into the calendar year).

We calculated yearly PPMs for each temperature profile and each of five larval attachment rates ($\gamma = 0.001$, 0.01, 0.1, 0.5, 0.9). To do this we constructed a PPM for each day ($\boldsymbol{B}_{day}$) based on γ and the predicted temperature for that day. An overall PPM for the year was calculated by multiplying PPMs such that:

$$A = B_{365} * B_{364} * \ldots \ldots B_2 * B_1 \qquad (15)$$

This process was repeated 1000 times and the intrinsic population growth ($\lambda_{daily} \pm$ 95% confidence intervals) was calculated from the resulting **A** matrices.

Effects of treating different life stages across seasons. Chemical and biological treatments to control sea lice target different life stages are used in response to elevated sea louse abundances. Instead of simulating specific types of treatments as has been attempted in previous models (e.g. [27], [28]) we model a generic treatment that will provide insight as to how temperature and the stage being targeted influence the effect of a treatment on λ_{daily}. To do this, we create a treatment matrix, **H**, which is the identity matrix, with the exception that targeted stages are $1-e$, where e is the efficacy of the treatment. For example, this matrix targets all adults:

$$H = \begin{bmatrix} 1 & 0 & 0 & 0 & 0 & 0 & 0 \\ 0 & 1 & 0 & 0 & 0 & 0 & 0 \\ 0 & 0 & 1 & 0 & 0 & 0 & 0 \\ 0 & 0 & 0 & 1 & 0 & 0 & 0 \\ 0 & 0 & 0 & 0 & 1-e & 0 & 0 \\ 0 & 0 & 0 & 0 & 0 & 1-e & 0 \\ 0 & 0 & 0 & 0 & 0 & 0 & 1-e \end{bmatrix}.$$

This matrix is then incorporated into the most extreme seasonal temperature matrices (described above) such that:

$$A = B_{365} * B_{364} * \ldots\ldots B_2 * H * B_1 \qquad (16)$$

where **H** is inserted when the treatment should occur. Scenarios were run in which a single treatment targeting one or combinations of several life-stages was delivered between day 1 and day 365.

The efficacy of treatments varies considerably from farm to farm. For example, the field efficacy of emamectin benzoate has been estimated to be anywhere between 60 and 99% in naive populations [54], but may be considerably less in resistant populations [55]. For the purposes of this study we set $e = 0.95$. Because we expect that analyses performed with lower efficacy will have qualitatively similar results, we only explore one value for e.

For these analyses, we quantified mean effects on population growth. To construct matrices, we quantified 1000 development times, survival estimates and (where relevant) fecundity and viability estimates and used the mean values in these distributions to calculate B_{day} matrices.

Results

Effects of Fixed Temperatures and Larval Attachment Rates

Increasing temperature caused λ and R_0 to increase and generation time of sea lice to decrease (Figure 2), though the extent of these changes depended upon the larval attachment rate. Both temperature and the larval attachment rate caused λ to increase. When $\gamma = 0.001$, λ_{daily} was 0.99 at 4°C and 1.14 at 20°C. When $\gamma = 0.5$, λ_{daily} was 1.013 at 4°C and 1.28 at 20°C. The generation time of sea lice was between three and four times longer at 4°C compared to 20°C and increased with lower values of γ. For example, when $\gamma = 0.001$, the generation time was 107 days at 4°C and 28 days at 20°C. When $\gamma = 0.5$, the generation time was 73 days at 4°C and 23 days at 20°C. The generation time was longer with lower values of γ. R_0 was greatest when the larval attachment rate and temperature were high. For example, when $\gamma = 0.001$, R_0 increased from 0.25 at 4°C to 38 at 20°C. When $\gamma = 0.5$, R_0 increased from 2.6 at 4°C to 321 at 20°C.

Sensitivity analysis showed that, in general, λ is most sensitive to the survival and development of preadults (Figure 3).Sensitivity to this life stage is greatest when the attachment rate is low and the temperature is high. At high attachment rates and low temperatures, λ is most sensitive to the survival and development of larval sea lice.

Elasticity analysis shows that λ is most sensitive to proportional changes in matrix elements associated preadults and chalimus (Figure 3). The elasticity values associated with these terms decrease slightly with an increasing larval attachment rate and are relatively insensitive to temperature.

Density-dependent Results

The probability of mating increases with the abundance of adult female lice (Figure 4A). The rate of this change is much faster when parasites are aggregated. When adult female lice are at low abundances the probability of mating increases when females are aggregated; however, above an abundance of one, aggregation has little effect on the probability of mating. Above abundances of three adult female lice, the probability of mating approaches 1.

The time necessary to reach the depensation threshold of three adult female lice per fish is shortest when lice are aggregated and the temperature is high (Figure 4B). At 20°C, a population of lice with an initial abundance of 0.1 adult females per louse will reach the threshold in 30 days for $VMR = 2$ and 44 days for $VMR = 1$. At colder temperatures the population growth is so slow that sea lice will take over a year to reach the depensation threshold. This occurs at 5°C when $VMR = 2$ and at 10°C when $VMR = 1$.

Effects of Seasonality

Both increases in yearly mean temperature and yearly temperature variation caused an increase in λ_{daily} (Figure 5A and 5B); however increases in yearly mean temperatures had a greater effect. For example, when $\gamma = 0.1$ per day, a 2°C increase in the yearly mean temperature caused estimates of λ_{daily} to increase from 1.085 to 1.119, while a 2°C decrease in the yearly mean temperature caused λ_{daily} to decrease to 1.055. Increasing the extreme values for one season caused a similar change. A 2°C decrease in the winter minimum caused λ_{daily} to decrease to 1.072, while a 2°C increase in the summer maximum temperature by 2°C caused λ_{daily} to change to 1.103 . Increasing the variation in temperature across the year also caused λ to increase (Figure 5C and 5D). When $\gamma = 0.1$, λ_{daily} increased to 1.089 in the more seasonal temperature profile and 1.096 in the most seasonal temperature profile.

For all temperature profiles, increasing the larval attachment rate caused λ_{daily} to increase. For example in the average seasonal temperature profile, λ_{daily} is 1.037, 1.057, 1.085, 1.114 and 1.130 for attachment rates of 0.001, 0.01, 0.1, 0.5 and 0.9 respectively.

Effects of Treating Different Life Stages Across Seasons

Analysis of the effect of single stage-targeted treatments across time show that both the life stage targeted and the time of treatment application can influence the effect of a treatment on λ (Figure 6). Treatments targeting eggs, chalimus, preadults and adults are similarly effective in the summer; however, in the winter treatments targeting chalimus and preadults are most effective. Treatment efficacy is far greater when several stages are targeted simultaneously.

Discussion

Sea lice are a prominent marine ectoparasite that threaten the productivity of salmon farming and are associated with increased mortality in wild salmon [12], [13]. The matrix models presented here suggest that temperature can increase the rate at which infestations establish and develop on farmed salmonids as a result of increased reproductive success and development at high temperatures and because the dampening effect of mate limitation on population growth is more quickly overcome at higher temperatures. The life history of *L. salmonis* has been studied in much more detail than many other sea lice in the Caligidae family and construction of similar matrix models for other species of sea lice may not be possible due to the lack of data needed for parameterization [56]; however, many aspects of sea louse life history are similar across species including temperature-dependent

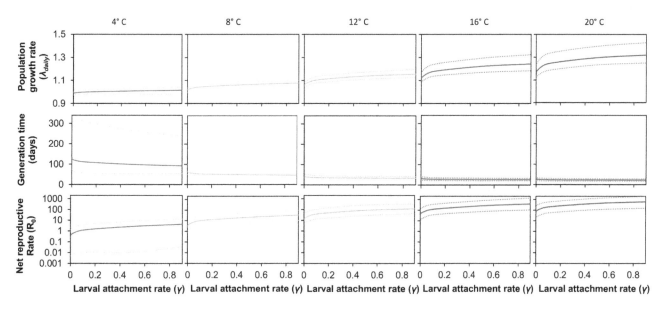

Figure 2. Effects of temperature and larval attachment rate on λ, R₀ and generation time. Dotted lines indicate 95% confidence intervals. Population growth is positive when λ is greater than 1.

development, high fecundity and the existence of a free-swimming larval stage that has an endogenous energy supply [39], [56]. Therefore it is likely that the temperature-dependent trends found in this study extend to other species in the Caligidae family.

Increased temperature causes more rapid development in sea lice for every life stage except copepodids, in which endogenous energy supplies provided to the egg and the probability of finding a host constrain survival and development [39]. One of the more dramatic results of increased development is the larger increase in net reproductive rate of sea lice. At high temperatures, they produce more surviving offspring in a shorter time than at low temperatures. Collectively these increases in development and fitness drive the rapid increase in population growth that occurs with increasing sea surface temperature.

The temperature ranges addressed in this study are within the ranges experienced on farms. Typical sea surface temperatures on

salmon farms range from 1–14°C in Atlantic Canada, 6–18°C in Ireland and 1–20°C in some Norwegian fjords [49]; however, the majority of research on sea lice is focused on relatively moderate temperatures (e.g. between 6°C and 14°C). This study suggests that extreme temperatures are critical for determining the growth rate of a population. In addition, the wide confidence intervals for generation time and R₀ at 4°C and in population growth at 20°C suggests that stochastic effects are more dominant at extreme temperatures and population trajectories at these temperatures may be more challenging to predict. The effect of extreme temperatures depends not only on their specific value, but on the entire profile of the seasonal variation. Increased overall variation does not have as large an effect on population growth as does an increase in mean temperature.

Despite the strong effect of temperature seen in this model and in laboratory studies [14], effects of temperature on sea louse

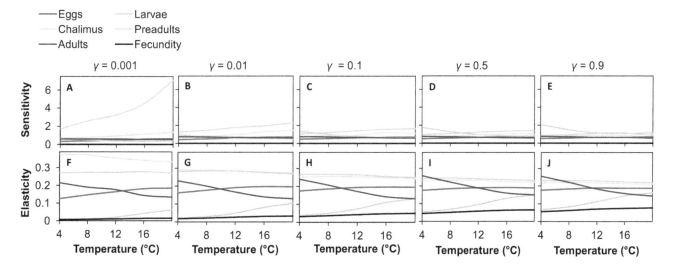

Figure 3. Sensitivities and elasticities of population growth rate (λ) to matrix elements. The sum of the elasticities of matrix values for surviving in (Pᵢ) and transitioning out of (Gᵢ) the same stages are presented.

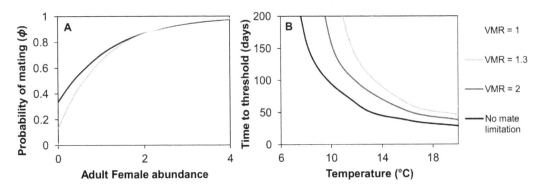

Figure 4. Effect of temperature on population growth when mate limitation is included. Figure A shows the probability of mating as a function of the abundance of adult females per host. Figure B shows the time to reach the threshold abundance of three female lice per fish (at which point mate-limitation is negligible) when sea lice abundances on hosts have a variance to mean ratio (*VMR*) of 1, 1.3 and 2. For all scenarios, the initial fish lice density was 0.1 fish per host and the attachment rate of larvae to the host was 0.1.

infestations are not always detected in analyses of sea louse infestations in the field. Two studies conducted on field data of sea louse infestations on farms in Norway [17], [57] and one study conducted in British Columbia [49] found that temperature was positively correlated with sea louse abundance on farmed and wild sea trout. In contrast, other studies, conducted in British Columbia [58], Scotland [16] and Norway [15] found no detectable effect of temperature on sea louse abundances. One explanation for this apparent discrepancy may be that seasonal temperature decreases in the winter masked the effects of warmer summer temperatures. In addition, other factors, such as the attachment rate of

copepodids may be influencing infestations more than temperature. This study suggests that, in order to detect a temperature effect on sea louse infestations, both the mean and the range of temperatures across a year must be considered.

Mate limitation is also influential in determining the rate at which a new population of sea lice increases. Mate limitation is especially pronounced at cold temperatures, during which more than a year may be required for new populations to reach abundances where mate limitation does not occur, if this threshold is reached at all. We made many assumptions about sea louse mating that require further research to quantify empirically. In

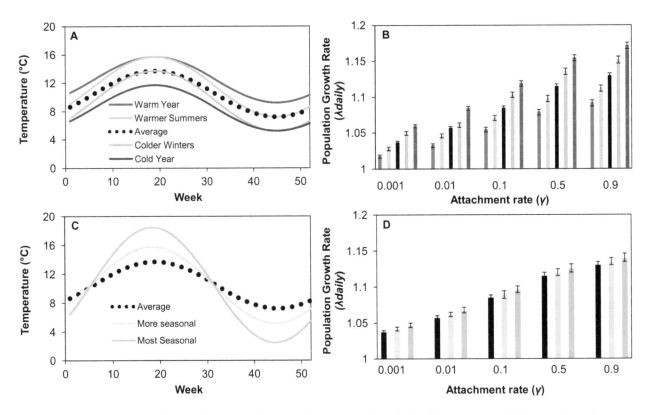

Figure 5. Temperature profiles used for population projection matrices their effect on λ_{daily}. Minimum and maximum seasonal temperatures were altered relative to a baseline temperature (A) and temperature variance was increased relative to a baseline temperature (C). The daily population growth rate for each of these scenarios is shown (B and D). Baseline temperatures are averages of five years of temperature data from 33 farm sites in Scotland [53]. For all parameters means ± 95% confidence intervals are shown.

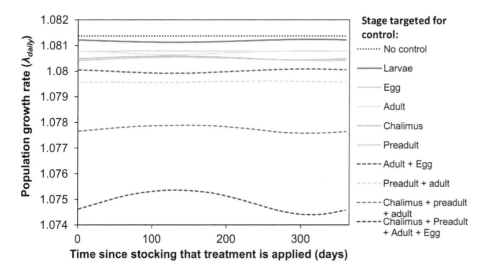

Figure 6. Effect of delivering a single control treatment at different times of the year on the population growth rate (λ_{daily}) of sea lice. The y-axis indicates the time of year since stocking that the single control treatment was delivered. All simulations assume a spring stocking of salmon smolts, so day 1 is May 1. A treatment efficacy of 95%, a larval attachment rate of 0.1 and the most seasonal temperature scenario were used for all calculations. Results for treatments applied to different life stages are shown.

particular, the assumptions of high promiscuity among males and females, equal sex ratios and co-aggregation of males and females all can dramatically influence mating success of parasites [45] and are not well quantified for sea lice [47]. In addition, host switching among preadult males is predicted to be density-dependent and may influence sex ratios and aggregation [59]. Skewed sex ratios have been found in some sea louse populations [60], but not others [47] and more work is needed to quantify this variation, identify underlying environmental and biological factors, and predict the consequences of this variation for population demographics.

While high louse abundances may cause negative density-dependent effects (i.e., over-compensation), we did not include these effects in our model. Over-compensation could occur if population growth is reduced when sea louse abundances on the host exceed thresholds above which immune responses in the host may limit attachment by the parasite (for example in *O. gorbuscha* [61]), or as a result of mortality of the infested host; however, there is limited evidence that Atlantic salmon exhibit effective immune responses against sea lice [37]. Moreover, it is unclear whether morbidity and mortality of salmon with high levels of infestations (e.g., mean annual levels greater than 20–30 sea lice per salmon) will feed back to influence the epidemiology of sea lice [59], [60].

The numerous methods used to control sea louse infestations on salmon farms must balance the negative effects of control strategies including costs [18], stress to salmon [23], environmental impacts of treatments [24], and evolution of drug resistance [20] against the positive result of sea louse control. The analyses performed in this study may inform specific recommendations for effective sea louse control. Elasticity analyses and analyses of single treatments in this study suggest that to be most effective, treatments for sea lice must target more than one stage. While all of the simulated control treatments reduced the population growth rate, none of them reduced lambda to levels lower than 1, at which point the population would start declining. This suggests that several treatments throughout the year is ideal for effective control in most situations. Treatments targeting chalimus and preadults are most effective at low temperatures and when the attachment rate of larvae is low. All treatments are less effective at controlling sea lice in the summer. Because we did not incorporate immigration of

free-swimming copepodids into the model, and this can be an important source of sea lice in some locations [42], it is also possible that we are underestimating the role that this stage may play in areas with high connectivity between farms.

Matrix elements with the highest sensitivities are most likely to influence fitness (λ) and evolutionary responses (i.e. drug resistance) may emerge faster when these stages are targeted for control. If heritable variation for resistance exists in selected sea louse populations, resistance to treatments may evolve more quickly when treatments are targeting the preadults. This is particularly relevant as there is interest in developing immuno-stimulants that may increase the ability of the host to reject attaching copepodids [62]. Such a control strategy has the potential to be successful because sea lice may be less likely to evolve resistance.

In addition to specific suggestions for more strategic use of treatments in response to sea louse monitoring, the temperature-dependent PPMs in this study give some insight into the potential for climate change to exacerbate sea louse infestations. While the level of uncertainty in projected oceanic conditions is high, many climate change models predict increases in temperate sea surface temperatures as a result of increased stratification, decreased upwelling and altered circulation [63]. However, increased temperature is only one component of projected changes to marine environments; oceans are predicted to experience sea-level rise, altered circulation, decreased salinity and decreased pH [63]. These other factors have the potential to influence sea louse epidemiology; increased circulation may decrease the attachment rate of infectious copepodids by transporting them away from susceptible hosts [64] and decreased salinity decreases attachment of copepodids [50], [65]. Further studies are needed to understand the net effects of the projected oceanic conditions on geographical ranges and potential of sea louse epizootics on both farmed and wild salmon. In addition, more work is needed to parameterize sea louse life history at high and low temperature extremes. For example, a maximum temperature threshold for *L. salmonis* has not been determined and it is possible that physiological costs of thermotolerance may decrease survival or fitness of sea lice at high temperatures [56].

More work is necessary to understand the effects of temperature variation on wild salmon. Increased temperature means and variation are predicted to increase sea louse infestations on farmed salmon, suggesting that careful consideration of temperature must be taken into account when sites for salmon farming are chosen, especially in areas that may be predisposed to high levels of exposure to sea lice due to hydrodynamic conditions and proximity to existing salmon farms or wild salmon migration routes. In contrast to salmon farms, the densities and locations of wild salmon fluctuate temporally and the costs of infestations may be mediated by other stressors that are less likely to occur on salmon farms (for example, food limitation and predation) [66]. Quantifying spatial movement and physiological states of wild salmon in relation to their sea lice exposure is an enormous challenge, but may be a major benefit to understanding and reducing conflicts between wild salmon conservation and salmon farming.

Conclusions

The model built in this study differs from many previously constructed agent-based and system-dynamic models of sea louse population dynamics; its' greatest utility is that it can be used to highlight specific characteristics of the life history of this ectoparasite that greatly influence its population growth. As such the results of this study may be informative for designing management programs to reduce the potential for sea louse epidemics especially when taking seasonal and inter-annual temperature variation into account. The broad recommendations that are generated by this study support the continued use of matrix population models for understanding the epidemiology of ectoparasites from a life history perspective.

Supporting Information

Appendix S1 Population projection matrices (means and standard deviations) of sea lice at 4°C, 8°C, 12°C, 16°C and 20°C.

Appendix S2 Matrices of sea lice sensitivities (means and standard deviations) at 4°C, 8°C, 12°C, 16°C and 20°C and for attachment rates of 0.001, 0.01, 0.1, 0.5 and 0.9.

Appendix S3 Matrices of sea lice elasticities (means and standard deviations) at 4°C, 8°C, 12°C, 16°C and 20°C and for attachment rates of 0.001, 0.01, 0.1, 0.5 and 0.9.

Appendix S4 Stable stage equilibrium and time to equilibrium.

Code S1 Instructions on the use of the annotated R-code.

Code S2 R-code for matrix model with fixed temperatures.

Code S3 R-code for matrix population models with seasonal variation.

Code S4 R-code for matrix population models with seasonal variation and treatments.

Code S5 R-code for matrix population project with density-dependent mating.

Code S6 R-code for parameters in the matrix model.

Code S7 R-code for parameters in the matrix model when there is density-dependent mating.

Code S8 R-code for the matrix A.

Acknowledgments

Susan Kalisz and the UPEI health management modelling group provided thoughtful comments about this study.

Author Contributions

Conceived and designed the experiments: MLG CWR GG RC MS. Performed the experiments: MLG. Analyzed the data: MLG CWR GG. Wrote the paper: MLG. Editing: MS RC.

References

1. Harvell CD, Mitchell CE, Ward JR, Altizer S, Dobson AP, et al. (2002) Climate warming and disease risks for terrestrial and marine biota. Science 296: 2158–2162.

2. Altizer S, Ostfeld RS, Johnson PT, Kutz S, Harvell CD (2013) Climate change and infectious diseases: from evidence to a predictive framework. Science, 341: 514–519.

3. Elston RA, Hasegawa H, Humphrey KL, Polyak IK, Hase CC (2008) Re-emergence of *Vibrio tubiashii* in bivalve shellfish aquaculture: severity, environmental drivers, geographic extent and management. Dis Aquat Org 82: 119–134.

4. Powell EN, Gauthier JD, Wilson EA, Nelson A, Fay RR, et al. (1992) Oyster disease and climate change. Are yearly changes in Perkinsus marinus parasitism in oysters (*Crassostrea virginica*) controlled by climatic cycles in the Gulf of Mexico?. Mar Ecol 13: 243–270.

5. Ruiz-Morenol D, Willis BL, Page AC, Weil E, Cróquer A, et al. (2012). Global coral disease prevalence associated with sea temperature anomalies and local factors. Dis Aquat Org 100: 249–261.

6. Roessig JM, Woodley CM, Cech JJ Jr, Hansen LJ (2004) Effects of global climate change on marine and estuarine fishes and fisheries. Rev Fish Biol Fish 14: 251–275.

7. Travers MA, Basuyaux O, Le Goïc N, Huchette S, Nicolas JL, et al. (2009) Influence of temperature and spawning effort on Haliotis tuberculata mortalities caused by *Vibrio harveyi*: an example of emerging vibriosis linked to global warming. Glob Change Biol 15: 1365–1376.

8. Vergeer LHT, Aarts TL, De Groot JD (1995) The 'wasting disease' and the effect of abiotic factors (light intensity, temperature, salinity) and infection with *Labyrinthula zosterae* on the phenolic content of *Zostera marina* shoots. Aquat Bot 52: 35–44.

9. Burge CA, Eakin CM, Friedman CS, Froelich B, Hershberger PK, et al. (2014) Climate change influences on marine infectious diseases: implications for management and society. Annu Rev Mar Sci 6: 1.1–1.29.

10. Connors BM, Krkošek M, Ford J, Dill LM (2010) Coho salmon productivity in relation to salmon lice from infected prey and salmon farms. J Appl Ecol 47: 1372–1377.

11. Krkošek M, Connors BM, Morton A, Lewis MA, Dill LM, et al. (2011) Effects of parasites from salmon farms on productivity of wild salmon. Proc Natl Acad Sci 108: 14700–14704.

12. Krkošek M, Revie CW, Gargan PG, Skilbrei OT, Finstad B., et al. (2013) Impact of parasites on salmon recruitment in the Northeast Atlantic Ocean. Proc R Soc Lond B 280: 20122359.

13. Tully O (1992) Predicting infestation parameters and impacts of caligid copepods in wild and cultured fish populations. Invertebr Repr Dev 22: 91–102.

14. Stien A, Bjørn PA, Heuch PA, Elston DA (2005) Population dynamics of salmon lice *Lepeophtheirus salmonis* on Atlantic salmon and sea trout. Mar Ecol Prog Ser 290: 263–275.

15. Heuch PA, Knutsen JA, Knutsen H, Schram TA (2002) Salinity and temperature effects on sea lice over-wintering on sea trout (*Salmo trutta*) in coastal areas of the Skagerrak. J Mar Biol Ass U K 82: 887–892.

16. Revie CW, Gettinby G, Treasurer JW, Grant AN, Reid SWJ (2002) Temporal, environmental and management factors influencing the epidemiological patterns of sea lice (*Lepeophtheirus salmonis*) infestations on farmed Atlantic salmon (*Salmo salar*) in Scotland. Pest Manag Sci 58: 576–584.

17. Jansen PA, Kristoffersen AB, Viljugrein H, Jimenez D, Aldrin M, et al. (2012) Sea lice as a density-dependent constraint to salmonid farming. - Proc R Soc Lond B 279: 2330–2338.

18. Costello MJ (2009) The global economic cost of sea lice to the salmonid farming industry. J Fish Dis 32: 115–118.

19. Rae GH (2002) Sea louse control in Scotland, past and present. Pest Manag Sci 58: 515–520.

20. Denholm I, Devine GJ, Horsberg TE, Sevatdal S, Fallang A, et al. (2002) Analysis and management of resistance to chemotherapeutants in salmon lice, *Lepeophtheirus salmonis* (Copepoda: Caligidae). Pest Manag Sci 58: 528–36.

21. Roth M, Richards RH, Dobson DP, Rae GH (1996) Field trials on the efficacy of the organophosphorus compound azamethiphos for the control of sea lice (Copepoda: Caligidae) infestations of farmed Atlantic salmon (*Salmo salar*). Aquaculture 140: 217–239.

22. Stone J, Sutherland IH, Sommerville CS, Richards RH, Varma KJ (2002) The efficacy of emamectin benzoate as an oral treatment of sea lice, *Lepeophtheirus salmonis* (Krøyer), infestations in Atlantic salmon, *Salmo salar* L. J Fish Dis 22: 261–270.

23. Burka JF, Hammell KL, Horsberg TE, Johnson GR, Rainnie DJ, et al. (1997) Drugs in salmonid aquaculture - A review. J Vet Pharmacol Ther 20: 333–349.

24. Burridge L, Weis JS, Cabello F, Pizarro J, Bostick K (2010) Chemical use in salmon aquaculture: A review of current practices and possible environmental effects. Aquaculture 306: 7–23.

25. Lees F, Baillie M, Gettinby G, Revie CW (2008) The Efficacy of Emamectin Benzoate against Infestations of *Lepeophtheirus salmonis* on Farmed Atlantic Salmon (*Salmo salar* L) in Scotland between 2002 and 2006. PLoS One 3: e1549.F.

26. Revie CW, Robbins C, Gettinby G, Kelly L, Treasurer JW (2005) A mathematical model of the growth of the sea lice, *Lepeophtheirus salmonis*, populations on farmed Atlantic salmon, *Salmo salar* L., in Scotland and its use in the assessment of treatment strategies. J Fish Dis 28: 603–613.

27. Gettinby G, Robbins C, Lees F, Heuch PA, Finstad B, et al. (2011) Use of a mathematical model to describe the epidemiology of *Lepeophtheirus salmonis* on farmed Atlantic salmon *Salmo salar* in the Hardangerfjord, Norway. Aquaculture 320: 164–170.

28. Groner ML, Cox R, Gettinby G, Revie CW (2013) Use of agent-based modelling to predict benefits of cleaner fish in controlling sea lice, *Lepeophtheirus salmonis*, infestations on farmed Atlantic salmon, *Salmo salar* L. J Fish Dis 36: 195–208.

29. Krkošek M, Lewis MA, Volpe JP (2005) Transmission dynamics of parasitic sea lice from farm to wild salmon. Proc R Soc Lond B 272: 689–696.

30. Frazer LN, Morton A, Krkošek M (2012) Critical thresholds in sea lice epidemics: evidence, sensitivity and subcritical estimation. Proc R Soc Lond B 279: 1950–1958.

31. Stormoen M, Skjerve E, Aunsmo A (2013) Modelling salmon lice, *Lepeophtheirus salmonis*, reproduction on farmed Atlantic salmon, *Salmo salar* L. J Fish Dis 36: 25–33.

32. Caswell H (2001) Matrix Population Models: Construction, analysis and interpretation. 2nd edn. Sunderland: Sinauer Associates, Inc.

33. Bommarco R (2001) Using matrix models to explore the influence of temperature on population growth of arthropod pests. Agric For Entomol 3: 275–283.

34. Govindarajulu P, Altwegg R, Anholt BR (2005) Matrix model investigation of invasive species control: bullfrogs on Vancouver Island. Ecol Appl 15: 2161–2170.

35. Dobson AD. Finnie TJ, Randolph SE (2011) A modified matrix model to describe the seasonal population ecology of the European tick *Ixodes ricinus*. J Appl Ecol 48: 1017–1028.

36. Gettinby G, Maclean S (1979) A matrix formulation of the life-cycle of liver fluke. Proc R Ir Acad B 79: 155–1167.

37. Fast MD, Muise DM, Easy RE, Ross NW, Johnson SC (2006) The effects of *Lepeophtheirus salmonis* infections on the stress response and immunological status of Atlantic salmon (*Salmo salar*). Fish Shellfish Immunol 21: 228–41.

38. Maran BAV (2013) The caligid life cycle: new evidence from *Lepeophtheirus elegans* reconciles the life cycles of *Caligus* and *Lepeophtheirus* (Copepoda: caligidae). Parasite 20: 15–36.

39. Heuch PA, Nordhagen JR, Schram TA (2000) Egg production in the salmon louse (*Lepeophtheirus salmonis* (Krøyer)) in relation to origin and water temperature. Aquacult Res 31: 805–814.

40. Belahrádek J (1935) Temperature and living matter. Protoplasma monograph 8.

41. Bjørn PA, Finstad B (1998) The development of salmon lice (*Lepeophtheirus salmonis*) on artificially infected post smolts of sea trout (*Salmo trutta*) Can J Zool 76: 970–977.

42. Rogers LA, Peacock SJ, McKenzie P, DeDominicis S, Jones, et al. (2013) Modeling Parasite Dynamics on Farmed Salmon for Precautionary Conservation Management of Wild Salmon. PLoS one 8: e60096.

43. Johannessen A (1978) Early stages of *Lepeophtheirus salmonis* (Copepoda Caligidae). Sarsia 63: 169–176.

44. Krkošek M, Connors BM, Lewis MA, Poulin R (2012) Allee effects may slow the spread of parasites in a coastal marine ecosystem. Am Nat 179: 401–12.

45. May RM (1977) Togetherness among Schistosomes: its effects on the dynamics of the infection. Math Biosci 35: 301–343.

46. Todd CD, Stevenson RJ, Reinardy H, Ritchie MG (2005) Polyandry in the ectoparasitic copepod *Lepeophtheirus salmonis* despite complex precopulatory and postcopulatory mate-guarding. Mar Ecol Prog Ser 303: 225–234.

47. Revie CW (2006) The development of an epi-informatics approach to increase understanding of the relationships between farmed fish and their parasites, PhD Thesis, Strathclyde: University of Strathclyde. 288 p.

48. Stubben CJ, Milligan BG (2007) Estimating and analyzing demographic models using the popbio package in R. J Stat Softw 22: 11.

49. Costello MJ (2006) Ecology of sea lice parasitic on farmed and wild fish. Trends Parasitol 22: 475–83.

50. Tucker CS, Sommerville C, Wootten R (2000) The effect of temperature and salinity on the settlement and survival of copepodids of *Lepeophtheirus salmonis* (Krøyer, 1837) on Atlantic salmon, *Salmo salar* L. J Fish Dis 23: 309–320.

51. Brooks KM (2005) The effects of water temperature, salinity, and currents on the survival and distribution of the infective copepodid stage of sea lice (*Lepeophtheirus salmonis*) originating on Atlantic salmon farms in the Broughton Archipelago of British Columbia, Canada. Rev Fish Sci 13: 177–204.

52. Genna RL, Mordue W, Pike AW, Mordue AJ (2005) Light intensity, salinity, and host velocity influence presettlement intensity and distribution on hosts by copepodids of sea lice, *Lepeophtheirus salmonis*. Can J Fish Aquat Sci 62: 2675–2682.

53. Revie CW, Gettinby G, Treasurer JW, Wallace C (2003) Identifying epidemiological factors affecting sea lice *Lepeophtheirus salmonis* abundance on Scottish salmon farms using general linear models. Dis Aquat Org 57: 85–95.

54. Armstrong R, MacPhee D, Katz T, Endris R (2000) A field efficacy evaluation of emamectin benzoate for the control of sea lice on Atlantic salmon. Can Vet J 41: 607–12.

55. Jones PG, Hammell KL, Dohoo IR, Revie CW (2012) Effectiveness of emamectin benzoate for treatment of *Lepeophtheirus salmonis* on farmed Atlantic salmon *Salmo salar* in the Bay of Fundy, Canada. Dis Aquat Org 102: 53–64.

56. Boxaspen K (2006) A review of the biology and genetics of sea lice. ICES J Mar Sci 63: 1304–1316.

57. Boxaspen K (1997) Geographical and temporal variation in abundance of salmon lice (*Lepeophtheirus salmonis*) on salmon (*Salmo salar* L.). ICES J Marine Sci 54: 1144–1147.

58. Saksida S, Constantine J, Karreman GA, Donald A (2007) Evaluation of sea lice abundance levels on farmed Atlantic salmon (*Salmo salar* L.) located in the Broughton Archipelago of British Columbia from 2003 to 2005. Aquacult Res 38: 219–231.

59. Connors BM, Lagasse C, Dill LM (2011) What's love got to do with it? Ontogenetic changes in drivers of dispersal in a marine ectoparasite. Behav Ecol 22: 588–593.

60. Grimnes A, Jakobsen PJ (1996) The physiological effects of salmon lice infection on post-smolt of Atlantic salmon. J Fish Biol 48: 1179–1194.

61. Jones SRM, Fast MD, Johnson SC, Groman DB (2007) Differential rejection of salmon lice by pink and chum salmon: disease consequences and expression of proinflammatory genes. Dis Aquat Org 75: 229–238.

62. Covello JM, Friend SE, Purcell SL, Burka JF, Markham RJF, et al. (2012) Effects of Orally Administered Immunostimulants on Inflammatory Gene Expression and Sea Lice (*Lepeophtheirus salmonis*) burdens on Atlantic salmon (*Salmo salar*). Aquaculture 366: 9–16.

63. Intergovernmental Panel on Climate Change (2007) Climate Change 2007: Impacts, Adaptation and Vulnerability (eds M. Parry, O. Canziani, J. Palutikof, P. van der Linden, and C. Hanson), Cambridge: Cambridge University Press.

64. Brooks KM (2005) The Effects of Water Temperature, Salinity, and Currents on the Survival and Distribution of the Infective Copepodid Stage of Sea Lice (*Lepeophtheirus Salmonis*) Originating on Atlantic Salmon Farms in the Broughton Archipelago of British Columbia, Canada. Rev Fish Sci 13: 177–204.

65. Bricknell IR, Dalesman SJ, O'Shea B, Pert CC, Luntz AJM (2006) Effect of environmental salinity on sea lice *Lepeophtheirus salmonis* settlement success. Dis Aquat Org 71: 201–12.

66. Groot C, Margolis L, Clark WC (1995) Physiological ecology of Pacific salmon. Vancouver: UBC Press.

14

The Positive Impact of the Early-Feeding of a Plant-Based Diet on Its Future Acceptance and Utilisation in Rainbow Trout

Inge Geurden[1]*, Peter Borchert[1,2], Mukundh N. Balasubramanian[1], Johan W. Schrama[1], Mathilde Dupont-Nivet[3], Edwige Quillet[3], Sadasivam J. Kaushik[1], Stéphane Panserat[1], Françoise Médale[1]

1 INRA, UR1067 NUMEA Nutrition, Métabolisme et Aquaculture, Aquapôle INRA, Saint Pée-sur-Nivelle, France, 2 Aquaculture and Fisheries Group, Wageningen Institute of Animal Sciences (WIAS), Wageningen University, Wageningen, The Netherlands, 3 INRA, UMR1313 GABI Génétique Animale et Biologie Intégrative, Jouy-en-Josas, France

Abstract

Sustainable aquaculture, which entails proportional replacement of fish-based feed sources by plant-based ingredients, is impeded by the poor growth response frequently seen in fish fed high levels of plant ingredients. This study explores the potential to improve, by means of early nutritional exposure, the growth of fish fed plant-based feed. Rainbow trout swim-up fry were fed for 3 weeks either a plant-based diet (diet V, V-fish) or a diet containing fishmeal and fish oil as protein and fat source (diet M, M-fish). After this 3-wk nutritional history period, all V- or M-fish received diet M for a 7-month intermediate growth phase. Both groups were then challenged by feeding diet V for 25 days during which voluntary feed intake, growth, and nutrient utilisation were monitored (V-challenge). Three isogenic rainbow trout lines were used for evaluating possible family effects. The results of the V-challenge showed a 42% higher growth rate (P = 0.002) and 30% higher feed intake (P = 0.005) in fish of nutritional history V compared to M (averaged over the three families). Besides the effects on feed intake, V-fish utilized diet V more efficiently than M-fish, as reflected by the on average 18% higher feed efficiency (P = 0.003). We noted a significant family effect for the above parameters (P<0.001), but the nutritional history effect was consistent for all three families (no interaction effect, P>0.05). In summary, our study shows that an early short-term exposure of rainbow trout fry to a plant-based diet improves acceptance and utilization of the same diet when given at later life stages. This positive response is encouraging as a potential strategy to improve the use of plant-based feed in fish, of interest in the field of fish farming and animal nutrition in general. Future work needs to determine the persistency of this positive early feeding effect and the underlying mechanisms.

Editor: Daniel Merrifield, The University of Plymouth, United Kingdom

Funding: This study was financed by the VEGE-AQUA project (FUI funding of French government), by institutional INRA funds, and by INRA/MRI (INRA-WUR Aquaculture Platform, post-doc grant MNB). The funders had no role in study design, data collection and analysis, decision to publish, or preparation of the manuscript.

Competing Interests: The authors have declared that no competing interests exist.

* E-mail: geurden@st-pee.inra.fr

Introduction

Sustainable feeding practices in intensive fish farming require further reductions in the use of dietary inputs from fisheries [1]. Yet, studies in search for alternatives to fishery-derived fishmeal and fish oil demonstrate high reductions in growth due to high inclusion levels of plant-protein in fish feed [2,3] as clearly shown in rainbow trout [4]. Low palatability of terrestrial plant-protein sources is considered a major constraint. For instance, alkaloids found in legumes such as peas and lupins reduce feed intake (FI) in rainbow trout without visible signs of adaptation [5,6]. Likewise, purified alcohol extracts (e.g. saponins) from soybean appear to be feeding deterrents [7]. Independent of their effect on FI, high levels of plant proteins have also been shown to depress the efficiency of feed utilization [4,8], pointing towards a digestive or metabolic problem. Other studies with salmonids showed that specific plant proteins such as soybean meal may provoke morphological changes and inflammation of the distal intestine [9,10]. In contrast, changes in the dietary lipid source only have a minor impact on feed utilization and FI in salmonid fish [11],

despite their capacity to discriminate and express specific feed preferences when given a choice among different feed oils [12,13].

Due to the overall poor understanding of the physiological reactions of fish to specific plant components, other strategies to expand the use of plant ingredients are being looked into. In this respect, selective breeding studies in rainbow trout demonstrate the large potential to exploit genetic variability for improving the growth of trout fed plant-based diets [14–17]. An alternative strategy to 'adapt the fish to the new feed', relatively under-explored in the field of fish nutrition, is by means of early nutritional intervention. In mammals, it is now well established that early nutrition may permanently alter the organism's physiology and metabolism. This phenomenon is believed to have evolved as a mechanism that allows the organism to fine-tune its physiology in an adaptive way to its early milieu [18–22]. The time frame in which the programming can occur is often confined to critical or sensitive periods early in life [20] such as during fetal [19] or early postnatal [23] nutrition. In fish, existing literature indicates that the early exposure to dietary factors such as high carbohydrate content [24] and changes in fatty acid profile [25,26] can induce persistent metabolic adaptations, at least at the molecular level.

Early nutritional events may not only influence an organism's metabolism or physiology, but also the development of sensory and cognitive systems [27]. Early exposure to quinine and citric acid, two substances innately aversive to rats, has been shown to reduce aversion to these tastes in rat later in life [28,29]. Similarly, early flavor experiences in humans have been found to program life-long flavor preferences [30]. In salmonids, the function of the chemosensory system involved in feeding arises early. After emergence from the substrate, young salmonids display a synchronized anatomical, physiological and behavioral development, vital for the transition from endogenous (yolk) to exogenous nutrition (usually 20–29 days post-hatch). Morphological evidence suggests that the olfactory system is functional as early as hatching [31]. Newly-hatched fry, which do not yet take food, already display nonspecific motor responses to olfactory stimuli [32]. The taste system arises later, but rapidly develops at the time of exogenous feeding with the spectrum of effective taste substances expanding with age [32]. To our knowledge, the possibility to orient later feed flavor acceptance by early life exposure to specific feeds remains unexplored in fish.

The present study explores the potential to improve the acceptance and/or the utilization of a feed rich in plant-ingredients in rainbow trout, by means of early exposure to the same plant-based feed during the first three weeks of exogenous feeding.

Materials and Methods

Ethics statement

The experiments were conducted following the Guidelines of the National Legislation on Animal Care of the French Ministry of Research (Décret 2001-464, May 29, 2001) and in accordance with the boundaries of EU legal frameworks, relating to the protection of animals used for scientific purposes (i.e. Directive 2010/63/EU). The author who performed animal experiments holds a personal license from the French Veterinary Services. The experiment was conducted at INRA NuMeA (UR1067) facilities, certified for animal services under the permit number A64.495.1 by the French veterinary services.

Experimental diets

Diets were manufactured at the INRA facility of Donzacq (France) using a twinscrew extruder (Clextral). The ingredient and analysed composition of both diets is given in Table 1. Diet M contained fishmeal and fish oil as protein and lipid source, respectively. Diet V contained a blend of palmseed, rapeseed and linseed oil, rich in saturated, mono-unsaturated and n-3 poly-unsaturated fatty acids, respectively, as lipid source. In order to avoid exceeding anti-nutrient threshold levels, we used a blend of wheat gluten, extruded peas, corn gluten meal, soybean meal and white lupin as protein sources. Synthetic L-lysine, L-arginine, dicalciumphosphate and soy-lecithin were added to diet V to correct the deficiency in essential amino acids, phosphorous and phospholipid supply. A mineral and a vitamin premix were added to both diets. Both diets fulfilled the known nutrient requirements of rainbow trout [33].

Biological material

Three isogenic heterozygous families (all individuals within a family share the same genotype) of rainbow trout (*Oncorhynchus mykiss*) were produced (C1-A22, C2-AB1 and C3-R23), expected

Table 1. Formulation, approximate crude protein (CP) levels of ingredients and analysed composition of the experimental diets M (fishmeal and fish oil-based) and V (all fishmeal and fish oil replaced by plant protein and plant oil sources).

Ingredients (g 100 g^{-1} diet)	Diet M	Diet V
Fish oil	8,5	-
Plant oil blend*	-	10,3
Fishmeal LT (CP 70%)	63	-
White lupinseed meal (CP 40%)	-	5,8
Corn gluten meal (CP 62%)	-	17,4
Soybean meal (CP 46%)	-	21,5
Wheat gluten (CP 80%)	-	25,6
Whole wheat (CP 10%)	25,4	5,1
Extruded dehulled peas (CP 24%)	-	3,1
Soy-lecithin	-	2,0
L-Arginine	-	1,0
L-Lysine	-	1,5
CaHPO4.2H20 (18%P)	-	3,6
Mineral and vitamin premix**	3,0	3,0
Analysed composition		
Dry matter (DM, % diet)	93,3	92,4
Crude protein (% DM)	52,1	50,5
Crude fat (% DM)	17,9	17,0
Gross energy (kJ g^{-1}DM)	22,3	22,3

*Consisting of (% blend): rapeseed oil (50), palm oil (30), linseed oil (20).
**INRA UPAE, 78352 Jouy en Josas, France.

to differ in their growth response to a plant-based feed (based on our own unpublished data). The three families were obtained by mating a single homozygous female line with males from three other homozygous lines [34]. The use of the same maternal line avoids effects associated with egg size and hatching time. Ova were collected from different females from the same line in order to produce a sufficient number of fish. The ova were carefully mixed and divided into three groups, each group being fertilized by gametes from one of the three male isogenic lines. Family differences are thus due to the genetic variability brought by the paternal lines.

Nutritional history and further pre-challenge phase

Hatching and first-feeding (23 days following hatching) took place at the INRA Lées-Athas fish farm, France (flow-through spring water, 7°C). For the first 21 days of exogenous feeding, the swim-up fry received either diet V or diet M, which was carefully distributed by hand on an hourly basis (8 to 10 meals per day) to duplicate groups (60 fry per tank). Each group was fed (7 min/meal) in slight excess. This early feeding period is referred to as 'nutritional history V or M' and fish from the respective nutritional histories are termed 'V- or M-fish' (Figure S1). The use of three genotypes (isogenic families) and two nutritional histories gave the following six treatment groups, C1V, C1M, C2V, C2M, C3V and C3M. During the period in between this early nutritional history period and the challenge test with diet V (V-challenge, see Figure S1), all groups were fed with diet M (hand feeding, 2 meals per day until visual satiation). Intermediate growth and feed intake was followed by weighing the fish groups and amount of feed distributed on a three-week basis. Survival was monitored daily. For accelerating the juvenile growth phase, fish were reared from ~2 g body weight at higher water temperature, 16.5°C (INRA Donzacq farm, flow-through spring water). Three weeks prior to the V-challenge, fish were transferred for acclimating to the INRA facilities of St. Pée-sur-Nivelle, consisting of a recirculating water unit of 24 tanks (70 L volume, 7 L/min water exchange rate, 16.5±1°C water temperature and artificial photoperiod set at 13 h light). The V-challenge was carried out with 4 replicate tanks (18 fish/tank) per treatment group and 3 replicate tanks (17 fish/tank) for treatments C1V and C3M for which a replicate tank was lost during acclimation (due to a blocked water inlet). Feed intake (FI) was recorded during the last 7 days of acclimation (diet M), once stabilized in all groups.

V-challenge

The V-challenge took place 7 months after the early nutritional history period (Figure S1). Here, all six treatment groups received diet V for 25 days. Rearing conditions were the same as during acclimation. For monitoring voluntary FI, two meals per day were carefully distributed by hand (diet V). Morning feeding started at 7:40 am and afternoon feeding at 2:00 pm. Specific care was taken to feed the groups to 'visual satiation'. Each tank was fed in three feeding rounds, the last until complete arrest of feeding activity. Fish were given approximately 15–20 min to recover appetite between each feeding round. The few pellets which remained unconsumed were counted and subtracted from the amount distributed, by multiplying their number with the mean pellet weight. We thus ensured that FI was recorded as precisely as possible.

For measuring initial (BWi) and final (BWf) body weight (BW), fish were counted and group-weighed at the start and end of the trial. Specific growth rate (SGR) was calculated as $100*(\ln(BWi)-\ln(BWf))/25$ days. Daily FI parameters, based on the total amount of food consumed divided by the number of days, were expressed on an individual basis (g/ind.day) or corrected for differences in growth, i.e. per 100 g average body weight (% BW.day) or per kg average metabolic body weight (g/kg met BW per day). Average BW was calculated as (BWi+BWf)/2 and average metabolic BW as $((BWf/1000*BWi/1000)^0.5)^0.8$. Feed efficiency (FE) was calculated as BW gain/total dry matter intake. For analysis of whole body composition, 6–8 fish (36-h unfed) per tank were sampled the day of initial and final weighing, killed by an overdose of anaesthesia (phenoxyethanol, 0.5 ml/l), frozen and kept at −20°C prior to biochemical analyses. Ground feed and whole fish samples (freeze-dried) were analysed for dry matter (105°C for 24 h), ash (combustion in a muffle furnace, 550°C for 12 h), protein (acid digestion, N×6.25, Kjeldahl Nitrogen analyser 2000, Fison Instruments, Milano, Italy), lipid content (petroleum ether extraction, Soxtherm, Gerhardt, Germany) and gross energy (adiabatic bomb calorimetry, IKA, Heitersheim, Germany). The retention efficiency of protein, lipid and energy was calculated as $100*(BWf*X-BWi*X)/FI*Y$, with X being the percentage protein or lipid or the amount of energy (kJ/g) in the fish and Y that in the feed.

Restricted V-challenge with focus on feed utilization efficiency (FE)

A restricted V-challenge was performed in order to investigate the effect of nutritional history on FE, independent of possible confounding effects related to differences in FI. For this, the fish received during four weeks an identical amount of diet V. We applied a restricted daily feed ration which was set at 0.75 g of feed per 100 g BW. Observations during the V-challenge suggested this

amount to be readily consumed by all groups. The amount of feed distributed was adjusted to the tank's biomass after two weeks, following an intermediate group-weighing. The ration was distributed by hand (2 meals/day) and special care was taken to ensure that all food distributed was consumed. The restricted V-challenge took place 13 months after the early diet V/M exposure. We used the remaining nutritional history M and V fish (only families C1 and C2 were available) from the same batch as in the V-challenge, kept at 7°C. As for the V-challenge, these had been fed with diet M from the end of early exposure until the first day of the restricted V-challenge. The four groups (C1M, C1V, C2M, C2V) were tested in duplicate tanks with 11 and 25 individuals per tank for the C1 and C2 treatments, respectively. The average BW of the fish at the start of the restricted V-challenge was not affected by nutritional history (P = 0.61) but was higher (P<0.05) in fish of family C2 than of family C1 (59.0 and 55.4 g, respectively). The restricted V-challenge was performed at the same temperature as the V-challenge (INRA Donzacq fish farm, flow-through spring water, 16.5°C).

Statistical analyses

Statistical analyses were performed using STATISTICA 7.0 (StatSoft Inc., Tulsa, USA). Data were tested for normality and homogeneity of variances by Kolmogorov-Smirnov and Bartlett tests, and then submitted to a two-way ANOVA to test the significance of the effects of nutritional history (N Hist), family (Fam) and their interaction (FxNH). In case of a significant effect (P<0.05), means were compared by Newman-Keuls post-hoc test.

Results

Performances of the fish during the pre-challenge phase

The percentage survival during the almost 8 month pre-challenge phase (including the nutritional history phase) was 89±6%. Survival was not significantly affected by nutritional history or family (P>0.05). The body weight of the fry at the end of the 3 weeks of first-feeding (nutritional history), was significantly (P<0.001) affected by nutritional history and by family: all M-fish were significantly bigger than V-fish. The early growth was highly dependent on the family in fish fed the V-, but not the M-diet, as shown by the statistical interaction between both factors (P<0.001), giving the following body weight ranking C3M, C2M and C1M (0.17 g)>C3V (0.12 g)>C2V (0.10 g)>C1V (0.08 g). The growth trajectory of the fish during the rest of the pre-challenge phase showed a similar pattern among all groups (Figure S2). At the end of the pre-challenge phase (start of the V-challenge), no effect of nutritional history was noted on the body composition (Table 2) nor on the body weight of the fish which ranged between 33.5 g and 42.1 g according to the family (Table 3). Fish dry matter content and protein level was family-dependent (C1 = C2>C3, Table 2). The average daily FI on diet M, measured at the end of the pre-challenge phase (last 7 days of acclimation), was unaffected by family or previous nutritional history (Figure 1).

Growth performance, feed intake and feed efficiency during V-challenge

The specific growth rate of the fish during the V-challenge (Figure 2) was on an average 42% higher in V- (1.9%/d) than in M-fish (1.3%/d). The whole body total lipid content was affected by nutritional history (V>M, Table 2). The major components of body weight gain, i.e. protein and lipid gain, were (averaged over the three families) 48 and 65% higher in V- compared to M-families, respectively (Table 3). There was a significant family

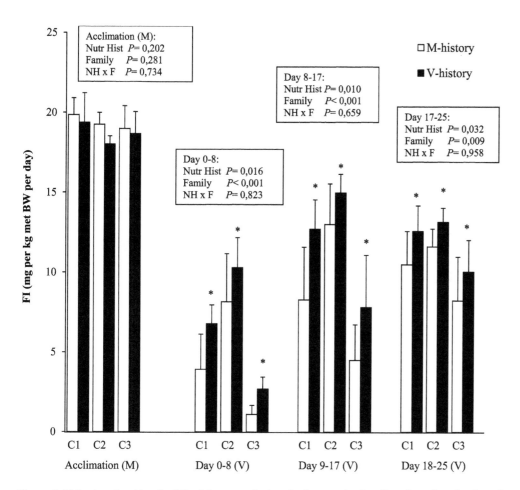

Figure 1. Voluntary feed intake (FI) of the trout during the last week of acclimation when feeding diet M and over three consecutive periods of the 25-day V-challenge with diet V. FI data represent means $\pm$ SEM (n = 4, except for C3M and C1V with n = 3) according nutritional history (M or V) and family (C1, C2, C3). For each period, the significance of the effects of nutritional history, family and their interaction (2-way ANOVA) is provided in the figure, * indicates a significant effect of nutritional history (V>M, p<0.05).

Table 2. The body weight (BW) of the trout at the start of the V-challenge and body composition of the fish at the start and at the end of the V-challenge.

Family	C1		C2		C3			P value (2-way ANOVA)		
Nutritional History	M	V	M	V	M	V	SEM	Fam	N Hist	FxNH
Body composition of the fish before the V-challenge										
Dry matter (% BW)	28,1	28,1	27,4	27,2	25,9	26,3	0,27	0,002	0,783	0,582
Protein (% BW)	14,0	13,5	13,8	14,3	13,1	13,4	0,14	0,008	0,591	0,013
Lipid (% BW)	11,5	11,2	10,5	10,5	10,2	10,3	0,15	0,067	0,496	0,627
Energy (kJ g^{-1} BW)	7,3	6,8	7,1	6,8	6,4	6,3	0,13	0,440	0,107	0,666
Body composition of the fish at the end of the V-challenge										
Dry matter (% BW)	29,3	28,9	30,1	30,5	27,4	28,8	0,28	0,002	0,257	0,245
Protein (% BW)	14,5	14,3	14,8	14,8	14,1	14,7	0,09	0,022	0,611	0,926
Lipid (% BW)	12,3	12,9	13,1	14,0	10,4	11,9	0,27	0,000	0,002	0,931
Energy (kJ g^{-1} BW)	7,8	7,9	8,5	8,6	7,0	7,9	0,13	0,000	0,003	0,011

Data represent treatment means according to their early nutritional history (M or V) and family (C1, C2, C3). P-values (2-way ANOVA) show the significance of the effects of nutritional history (N Hist), family (Fam) and their interaction (FxNH).

Table 3. Data on growth, feed intake and nutrient utilization of the trout during the 25-day V-challenge.

Family	C1		C2		C3			P-value (2-way ANOVA)		
Nutritional history	**M**	**V**	**M**	**V**	**M**	**V**	**SEM**	**Fam**	**N Hist**	**FxNH**
Growth parameters (g ind^{-1})										
Initial body weight	33,5	33,8	36,1	34,6	38,7	42,1	0,70	0,000	0,103	0,001
Final body weight	46,6	57,0	61,3	65,2	45,3	55,9	1,8	0,000	0,001	0,299
Protein gain	2,07	3,59	4,08	4,72	1,32	2,60	0,29	0,000	0,003	0,510
Lipid gain	1,92	3,56	4,20	5,48	0,79	2,34	0,37	0,000	0,001	0,910
Energy gain	122	218	264	323	69,2	176	20,7	0,000	0,001	0,643
Voluntary feed intake (FI)										
FI (g ind^{-1})	15,4	23,0	24,9	29,2	9,5	15,9	1,6	0,000	0,005	0,760
FI (% BW d^{-1})	1,52	2,02	2,03	2,33	0,89	1,29	0,12	0,000	0,011	0,834
FI (mg kg BW$^{-0.8}$ d^{-1})	8,1	11,2	11,5	13,3	4,7	7,1	0,7	0,000	0,005	0,769
Nutrient and energy utilization efficiency (% intake)										
Protein retention	28,4	33,5	35,1	34,6	29,3	35,1	0,8	0,069	0,017	0,117
Lipid retention	75,5	98,3	107,9	119,5	43,9	90,8	6,2	0,001	0,003	0,203
Energy retention	37,8	46,0	51,3	53,9	34,5	53,3	1,9	0,005	0,001	0,045

Data represent treatment means according to their early nutritional history (M or V) and family (C1, C2, C3). P-values (2-way ANOVA) show the significance of the effects of nutritional history (N Hist), family (Fam) and their interaction (FxNH).

effect on the latter parameters, C2>C1>C3, without family*nutritional history interaction.

Voluntary FI (g/fish), cumulated over the entire V-challenge (day 0–25), was 37% higher in V- compared to M-fish (Table 3). This difference was 27% and 30% when expressed per unit average body weight or per unit metabolic body weight,

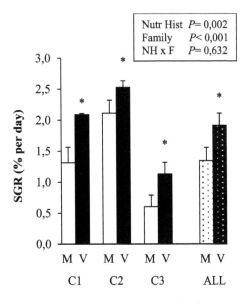

Figure 2. Specific growth rate (SGR) of the trout during the 25-day V-challenge according to the early nutritional history (M or V) and family (C1, C2, C3). Values are means ± SEM (n = 4, except for C3M and C1V with n = 3). Dotted bars represent the effect of nutritional history (M or V) during the V-challenge, averaged over all three families (ALL, means ± SEM, n = 11). The significance of the effects of nutritional history, family (C2>C1>C3) and their interaction (2-way ANOVA) is added in the figure, * indicates a significant effect of nutritional history (V>M, p<0.05).

respectively (Table 3). FI differed between the three families, with family C2 consuming significantly more of diet V than C1 which in turn had higher intakes than C3 (C2>C1>C3). The absence of a significant family*nutritional history interaction shows that differences in FI due to nutritional history were independent of the family effect. *Ad libitum* FI data are detailed in Figure 1 for the three consecutive periods of the V-challenge, *i.e.* days 0–8, 9–17 and 18–25. Three major observations are noteworthy. First, the transition from diet M (and acclimation to diet V) resulted in a huge drop in FI in all groups, as seen during the first period (d0–8) of V-feeding. Secondly, in all periods and for all families (no interaction), V-fish consumed significantly more of diet V than M-fish. This positive effect of early V-exposure on FI was more marked during the first week (V/M ratio of 1.80) than during the last week (V/M ratio leveled off at 1.21). Thirdly, the family effect was significant in all 3 periods with highest FI in C2 groups (C2M, C2V) and lowest FI in C3 groups (C3M, C3V).

Feed efficiency (FE) was significantly affected by both nutritional history and family (Figure 3). With a FE of 0.97±0.11, V-families gained on average 18% more in weight per unit FI than M-families which had a mean FE of 0.84±0.18 (Figure 3). FE in fish of family C3 was lower than that in C1 which was lower than in C2. There was no significant interaction between both factors, though the positive effect of nutritional history V on FE seemed somewhat less pronounced in family C2. The efficiency of protein, lipid and energy retention (gain per unit intake) was, respectively, 11, 36 and 24% higher in V- compared to M-groups (Table 3).

The efficiency of feed utilization during the restricted V-challenge

The 4-week restricted V-challenge confirmed the positive effect of early diet V-exposure (P = 0.02) and the effect of family (P = 0.01, C2>C1) on the utilization efficiency of diet V, without significant interaction between both factors (P = 0.24, Figure 4).

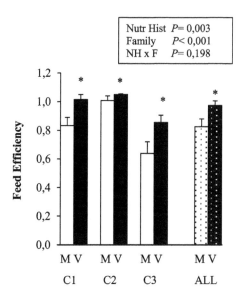

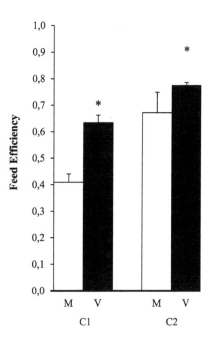

Figure 3. Feed efficiency (FE) of the trout during the 25-day V-challenge according to the early nutritional history (M or V) and family (C1, C2, C3). Values are means ± SEM (n = 4, except for C3M and C1V with n = 3). Dotted bars represent the effect of nutritional history (M or V) during the V-challenge, averaged over all three families (ALL, means ± SEM, n = 11). The significance of the effects of nutritional history, family (C2 = C1>C3) and their interaction (2-way ANOVA) is provided in the figure. * indicates a significant effect of nutritional history (V>M, p<0.05).

Figure 4. Feed efficiency (FE) during the restricted V-challenge. Two families of rainbow trout with nutritional history (M or V) received for 4 weeks diet V at 0.75% of their body weight (restricted feeding). Values are means ± SEM (n = 2). The significance of the effects of nutritional history, family (C2>C1) and interaction (P values, 2-way ANOVA) is provided in the results section, * shows a significant effect of nutritional history (V>M, p<0.05).

Discussion

The juvenile rainbow trout that were confronted during early first-feeding stages with the plant-based diet V displayed better growth when fed this same diet 7 months later (V-challenge) compared to the non-exposed M-fish. The better growth in the V-fish is attributable to a combination of higher *ad libitum* FI together with better feed utilisation. To our knowledge, so far no study in fish has documented an analogous positive long-term effect of short-term early exposure to a plant-based diet.

V-exposed fish display higher feed intakes (FI) during the V-challenge

FI drastically dropped in all groups challenged to eat diet V, devoid of both fishmeal and fish oil, as frequently seen in salmonids fed diets with high levels of plant ingredients [2]. Key to the present study is the finding that the drop in FI was significantly less prominent in V- than in M-trout which had never been confronted before with diet V. Cumulated over the V-challenge and depending on the family, V-exposed fish consumed 20 to 70% more than M-fish. This long-term positive effect needs emphasis, taking into account the over 300-fold increase in fish body weight between the V-challenge (35–40 g fish) and the early V-exposure (~100 mg fry). Further research should assess the persistency of the observed effect as the contrast in FI caused by nutritional history steadily decreased over time.

In any event, the effect of early nutritional history on FI implies that the V-fish were able to 'recognize' diet V, and this 7 months following initial exposure. The capacity of flavour learning has been shown before in teleost fish, albeit over a shorter time span than in the present study and mostly by applying a conditioned aversion paradigm in which flavor is associated with a noxious stimulus. Goldfish has been found capable to learn to avoid

flavored food particles following injections with lithium chloride for periods of 11 days [35] or 47 days [36] after learning. Both gustatory and olfactory systems seem to be involved in the initial learning process [36] whereas the dorsomedial telencephalic pallium seems to play an essential role in memorizing the associative food aversion in analogy with the amygdala in mammals [35]. Likewise, long-lasting aversive associations between the sensory properties of a food and its postingestive consequences have been repeatedly highlighted in generalist herbivores [37]. Such associative memory is considered crucial for aiding the animal to avoid particular plant-toxins during foraging [38,39]. Importantly, the early V-exposure in our study did not result in an aversion to diet V as might be expected in case the trout had associated the early diet V experience to negative postingestive consequences. Instead, the juvenile fish early exposed to diet V ate more during the V-challenge compared to the M-fish. This is, in essence, strongly suggestive of reduced food neophobia, which may be mediated through mechanisms related with sensory flavor acceptance or with reduced susceptibility to specific plant secondary compounds (e.g. enhanced detoxifying capacity).

In mammals, early flavor experiences are important in establishing life-long food flavor acceptances [30] and may render distasteful flavors palatable [28,29]. The ability to retain nutrient flavors transmitted by the mother's diet (amniotic fluid, milk) has been interpreted as a natural mechanism for the safe transmission of predictive dietary signatures from mother to young [40,41]. In fish, knowledge on the effect of early flavor experience on later food flavor acceptance is scarce. In salmonids, olfaction is believed to be more important in guiding feeding behavior than gustation [42]. Moreover, gustatory preferences show low plasticity in fish and have been reported to be independent of previous feeding experience [32]. Of interest, though not directly related with the development of feed flavor acceptance, is the susceptibility of the salmonid olfactory system to imprinting, a mechanism used by adult salmon to find their way back to the natal streams [43–46]. The olfactory imprinting process is assumed to be linked to major physiological processes (e.g. emergence from the gravel, smolt-parr transformation) and external environmental clues (e.g. exposure to novel water) [47]. Olfactory imprinting under laboratory conditions has been shown to work with compounds such as morpholine or phenetyl alcohol [44,48]. More recent studies indicate a role of free amino acids (L-isoforms), probably derived from a variety of living organisms (e.g. plants in and near streams), as guiding substance for salmonids to return to their natal river [46,49,50]. In our study, specific compounds released in the water from diet V during the trout's early-life exposure perhaps provoked an olfactory imprinting, responsible for the reduced neophobia and higher intakes later in life. The prospect of alleviating food flavor neophobia in fish by early short-term flavor exposure certainly warrants further attention.

Food neophobia is also considered as an innate reaction which prevents the animal to ingest potentially harmful unknown substances [38,39]. In this respect, two mechanisms, not mutually exclusive, may underlie the higher FI seen in the V-trout later in life. This is i) the trout fry had learned during early V-feeding that no severe harmful substances were associated with diet V consumption or ii) the early V-exposure stimulated physiological defense mechanisms to deterrent plant substances. In terrestrial herbivores, early exposure to plant secondary compounds can permanently alter critical physiological detoxification systems [51]. Persistent modifications in xenobiotic metabolism by early plant-feeding have not been considered yet in trout or any other fish.

The amount of food eaten during the V-challenge differed significantly between the three isogenic families, as expected from previous results on paternal effects on *ad libitum* FI of plant-based feed using isogenic rainbow trout lines [52]. Such genetic variability in FI in trout fed plant-based diets is considered of particular interest for setting up selective breeding program [14–17,52]. In humans, genetic differences in the sensitivity to taste substances interfere with early experiences in establishing food likes and dislikes [30]. In our study, however, the effect of early V-exposure on later FI appeared consistent for the three families (no statistical interaction).

V-exposed fish display improved feed efficiency (FE) during the V-challenge

Besides FI, also the efficiency of the utilization of diet V was higher in V- relative to M-fish. Analysis of the components in body gain showed that this was associated with better retention efficiencies of both lipid and protein. The improved capacity of the V-fish to utilise diet V for growth appears promising, but requires caution regarding i) the possible confounding effect of FI on the observed FE-response and ii) the diet specificity of the FE-response.

Regarding the first point, FE in fish is known to show a positive quadratic relationship with feeding level [53–56]. Using good quality feed, optimal FE normally occurs at 20 to 25% below the maximum growth response level, whereas it rapidly declines at the lower intake ranges [56]. This relationship which is feed-dependent is often overlooked in nutritional studies where FI and FE are mostly interpreted as independent parameters. Our data do not allow to estimate the impact of the lower intakes in M-relative to V-groups on the observed reductions in FE. This was the reason for undertaking the small-scale feeding trial in which fish of both nutritional histories were fed restrictively at 0.75% of their body weight. The low feeding levels ensured all feed to be consumed but led to lower growth. Nevertheless, the data confirm

the positive effect of early V-exposure on later FE seen during the V-challenge, in this case more than one year after the early exposure.

For the second question, it is important to say that actual FI during the early feed exposure could not be monitored due to the small size of the feed pellets (300–500 μm) and fry (<200 mg). It is hence conceivable that V-fry consumed less than M-fry during these first weeks of exogenous feeding, despite the hourly feed supply provided in slight excess to all groups. In mammals, early-life exposure to a nutrient-limited environment has been reported to lead to hyperphagia and obesity later in life, probably as a result of metabolic dysfunctions programmed by early nutritional deprivation [19–21,57–60]. We therefore conceived the possibility that a general early 'malnutrition' effect, resulting in overcompensation, might explain the superior ad libitum FI and/or dietary utilization in the V-fish during the V-challenge. However, no such compensatory feeding effects were seen in the V-fish fed with diet M during the 7-month intermediate rearing, as also reflected by the similarity in fish body mass at the start of the challenge which was independent of nutritional history. This clearly points toward a directed diet V response in the juvenile V-fish rather than to an overall compensatory sign of early malnutrition in the V-fish.

A wide range of nutritional conditions and compounds has been found to induce specific adult phenotypes in terrestrial animals and man. When encountered during early life, these may provoke long-lasting adaptive changes in preparation to the potential future environment. If the predicted nutritional environment is correct then the organism's metabolism will match, increasing its evolutionary fitness [18–22]. The foresaid literature has particularly dealt with 'mismatches' as in the case of fetal undernutrition and with nutrient-induced programming of genes whose expression is linked to adult disease (diabetes, cancer). In fish nutrition, to our knowledge only three studies clearly explored the concept of early nutritional programming. These aimed to induce persistent metabolic adaptations, advantageous for dealing with high levels of carbohydrates [24] and low levels of long-chain polyunsaturated fatty acids [25,26], both typical of plant-based feed. Juvenile (10 g) rainbow trout when subjected at early life (200 mg) to a short (3-day) hyperglucidic feeding period were found to display upregulated α-amylase and maltase gene expression [24]. Two other studies reported enhanced delta 6-desaturase mRNA levels in juvenile European seabass, only when they had been exposed before, at larval stage, to a dietary deficiency in long-chain n-3 polyunsaturated fatty acids [25,26]. The adaptive responses at the molecular level were however not associated with noticeable changes in growth when fish were challenged to eat a feed rich in carbohydrates [24] or low in long-chain n-3 polyunsaturated fatty acids [25,26]. The strong positive phenotypic response found in the present study is encouraging as a potential strategy to improve the use of plant-based diets in fish. Yet, further work needs to determine which mechanisms mediated the positive effects set forth by the early life exposure to diet V. A possible mechanism by which an organism can produce different phenotypes from a single genome in response to early life events is through altered epigenetic regulation of genes [19,21,22,61,62]. The great interest in the field of nutritional epigenetics is illustrated by the constantly growing list of food-components known to modulate epigenetic mechanisms, an important subset of which are plant compounds [61]. These studies undoubtedly open new perspectives in fish nutrition and other animal nutrition sectors in general.

In summary, our study shows that an early short term exposure of rainbow trout fry to a plant-based diet improves acceptance and utilization of the same diet when given at a later life stage. Progress in understanding the development of possible epigenetic pathways and interference of genetic predispositions in establishing such adaptive mechanism may contribute to strategies for improving the use of plant-based diets in farmed fish.

Supporting Information

Figure S1 Illustration of the experimental design. Rainbow trout swim-up fry were fed for the first 3 weeks of exogenous feeding either with a plant-based diet (diet V) or with a diet containing fishmeal and fish oil as protein and fat source (diet M). This early feeding period is referred to as 'nutritional history V or M'. After a 7-month common rearing period on diet M, both groups were challenged to feed the plant-based diet V during which voluntary FI, growth and nutrient utilisation were monitored (V-challenge).

Figure S2 Growth (body weight, BW) of the fish during the pre-challenge phase (from first-feeding until the first day of V-challenge). The trout fry were fed either diet M or V

during the first 3 weeks of feeding (nutritional history M or V) and then all received diet M during the rest of the 7 month pre-challenge phase. A: Family C1; B: Family C2; C: Family C3. Data represent means from duplicate groups. The fish were transferred at ~2.5 g (week 20) from 7°C to 16.5°C rearing temperature.

Acknowledgments

We thank JP Fourriot, E Lopestéguy and A Surget for all technical help. The excellent assistance of P Maunas and of F Terrier for the fish rearing and of Y Mercier during the feed challenge is highly appreciated.

Author Contributions

Conceived and designed the experiments: IG. Performed the experiments: IG PB SJK FM. Analyzed the data: IG PB SJK FM MNB SP JWS EQ MDN. Contributed reagents/materials/analysis tools: PB IG MDN EQ. Wrote the paper: IG PB MNB JWS EQ MDN SJK SP FM.

References

1. Naylor RL, Hardy RW, Bureau DP, Chiu A, Elliott M, et al. (2009) Feeding aquaculture in an era of finite resources. Proc Natl Acad Sci 106: 15103–15110.
2. Gatlin DM, Barrows FT, Brown P, Dabrowski K, Gaylord TG, et al. (2007) Expanding the utilization of sustainable plant products in aquafeeds: a review. Aquac Res 38: 551–579.
3. Barrows FT, Bellis D, Krogdahl Å, Silverstein JT, Herman EM, et al. (2008) Report of the Plant Products in Aquafeed Strategic Planning Workshop: An integrated, interdisciplinary research roadmap for increasing utilization of plant feedstuffs in diets for carnivorous fish. Rev Fish Sci 16: 449–455.
4. Hua K, Bureau DP (2012) Exploring the possibility of quantifying the effects of plant protein ingredients in fish feeds using meta-analysis and nutritional model simulation-based approaches. Aquaculture 356–357: 284–301.
5. Glencross B, Evans D, Rutherford N, Hawkins W, McCafferty P, et al. (2006) The influence of the dietary inclusion of the alkaloid gramine, on rainbow trout (*Oncorhynchus mykiss*) growth, feed utilisation and gastrointestinal histology. Aquaculture 253: 512–522.
6. Serrano E, Storebakken T, Penn M, Øverland M, Hansen JØ, et al. (2011) Responses in rainbow trout (*Oncorhynchus mykiss*) to increasing dietary doses of lupinine, the main quinolizidine alkaloid found in yellow lupins (*Lupinus luteus*). Aquaculture 318: 122–127.
7. Bureau DP, Harris AM, Young Cho C (1998) The effects of purified alcohol extracts from soy products on feed intake and growth of chinook salmon (*Oncorhynchus tshawytscha*) and rainbow trout (*Oncorhynchus mykiss*). Aquaculture 161: 27–43.
8. Wacyk J, Powell M, Rodnick K, Overturf K, Hill RA, et al. (2012) Dietary protein source significantly alters growth performance, plasma variables and hepatic gene expression in rainbow trout (*Oncorhynchus mykiss*) fed amino acid balanced diets. Aquaculture 356–357: 223–234.
9. van den Ingh TSGAM, Olli JJ, Krogdahl Å (1996) Alcohol-soluble components in soybeans cause morphological changes in the distal intestine of Atlantic salmon, *Salmo salar* L. J Fish Dis 19: 47–53.
10. Krogdahl Å, Bakke-McKellep AM, Baeverfjord G (2003) Effects of graded levels of standard soybean meal on intestinal structure, mucosal enzyme activities, and pancreatic response in Atlantic salmon (*Salmo salar* L.). Aquac Nutr 9: 361–371.
11. Turchini GM, Torstensen BE, Ng W-K (2009) Fish oil replacement in finfish nutrition. Rev Aquac 1: 10–57.
12. Geurden I, Cuvier A, Gondouin E, Olsen R, Ruohonen K, et al. (2005) Rainbow trout can discriminate between feeds with different oil sources. Physiol Behav 85: 107–114.
13. Pettersson A, Johnsson L, Brännäs E, Pickova J (2009) Effects of rapeseed oil replacement in fish feed on lipid composition and self-selection by rainbow trout (*Oncorhynchus mykiss*). Aquac Nutr 15: 577–586.
14. Pierce LR, Palti Y, Silverstein JT, Barrows FT, Hallerman EM, et al. (2008) Family growth response to fishmeal and plant-based diets shows genotype×diet interaction in rainbow trout (*Oncorhynchus mykiss*). Aquaculture 278: 37–42.
15. Le Boucher R, Dupont-Nivet M, Vandeputte M, Kerneïs T, Goardon L, et al. (2012) Selection for adaptation to dietary shifts: towards sustainable breeding of carnivorous fish. Plos One 7: e44898. doi:10.1371/journal.pone.0044898.
16. Le Boucher R, Quillet E, Vandeputte M, Lecalvez JM, Goardon L, et al. (2011) Plant-based diet in rainbow trout (*Oncorhynchus mykiss* Walbaum): Are there genotype-diet interactions for main production traits when fish are fed marine vs. plant-based diets from the first meal? Aquaculture 321: 41–48.
17. Overturf K, Barrows FT, Hardy RW (2013) Effect and interaction of rainbow trout strain (Oncorhynchus mykiss) and diet type on growth and nutrient retention. Aquac Res 44: 604–611.
18. Lucas A (1998) Programming by Early Nutrition: An Experimental Approach. J Nutr 128: 401S–406S.
19. Lillycrop KA, Burdge GC (2012) Epigenetic mechanisms linking early nutrition to long term health. Best Pract Res Clin Endocrinol Metab 26: 667–676.
20. Petry CJ, Ozanne SE, Hales CN (2001) Programming of intermediary metabolism. Mol Cell Endocrinol 185: 81–91.
21. Gluckman PD, Hanson MA (2008) Developmental and epigenetic pathways to obesity: an evolutionary-developmental perspective. Int J Obes 32: S62–S71.
22. Waterland RA, Jirtle RL (2003) Transposable elements: targets for early nutritional effects on epigenetic gene regulation. Mol Cell Biol 23: 5293–5300.
23. Patel MS, Srinivasan M (2002) Metabolic programming: causes and consequences. J Biol Chem 277: 1629–1632.
24. Geurden I, Aramendi M, Zambonino-Infante J, Panserat S (2007) Early feeding of carnivorous rainbow trout (*Oncorhynchus mykiss*) with a hyperglucidic diet during a short period: effect on dietary glucose utilization in juveniles. Am J Physiol - Regul Integr Comp Physiol 292: R2275–R2283.
25. Vagner M, Zambonino Infante JL, Robin JH, Person-Le Ruyet J (2007) Is it possible to influence European sea bass (*Dicentrarchus labrax*) juvenile metabolism by a nutritional conditioning during larval stage? Aquaculture 267: 165–174.
26. Vagner M, Robin JH, Zambonino-Infante JL, Tocher DR, Person-Le Ruyet J (2009) Ontogenic effects of early feeding of sea bass (*Dicentrarchus labrax*) larvae with a range of dietary n-3 highly unsaturated fatty acid levels on the functioning of polyunsaturated fatty acid desaturation pathways. Br J Nutr 101: 1452–1462.
27. Graven S, Browne J (2008) Sensory Development in the Fetus, Neonate, and Infant: Introduction and Overview. Newborn Infant Nurs Rev 8: 169–172.
28. London RM, Snowdon CT, Smithana JM (1979) Early experience with sour and bitter solutions increases subsequent ingestion. Physiol Behav 22: 1149–1155.
29. Youngentob SL, Glendinning JI (2009) Fetal ethanol exposure increases ethanol intake by making it smell and taste better. Proc Natl Acad Sci 106: 5359–5364.
30. Beauchamp GK, Mennella JA (2009) Early flavor learning and its impact on later feeding behavior. J Pediatr Gastroenterol Nutr 48: S25–S30.
31. Hara TJ, Zielinski B (1989) Structural and functional development of the olfactory organ in teleosts. Trans Am Fish Soc 118: 183–194.
32. Kasumyan AO, Døving KB (2003) Taste preferences in fishes. Fish Fish 4: 289–347.
33. National Research Council (2011) Nutrient requirements of fish and shrimp. Washington, DC: National Academies Press.
34. Quillet E, Dorson M, Le Guillou S, Benmansour A, Boudinot P (2007) Wide range of susceptibility to rhabdoviruses in homozygous clones of rainbow trout. Fish Shellfish Immunol 22: 510–519.
35. Martín I, Gómez A, Salas C, Puerto A, Rodríguez F (2011) Dorsomedial pallium lesions impair taste aversion learning in goldfish. Neurobiol Learn Mem 96: 297–305.
36. Manteifel YB, Karelina MA (1996) Conditioned food aversion in the goldfish, *Carassius auratus*. Comp Biochem Physiol A Physiol 115: 31–35.
37. Simões PMV, Ott SR, Niven JE (2012) A long-latency aversive learning mechanism enables locusts to avoid odours associated with the consequences of ingesting toxic food. J Exp Biol 215: 1711–1719.
38. Provenza FD (2006) Post-ingestive feedback as an elementary determinant of food preference and intake in ruminants. J Range Manag 48: 2–17.
39. Kyriazakis I, Anderson DH, Duncan AJ (1998) Conditioned flavour aversions in sheep: the relationship between the dose rate of a secondary plant compound and the acquisition and persistence of aversions. Br J Nutr 79: 55–62.
40. Ueji K, Yamamoto T (2012) Flavor learning in weanling rats and its retention. Physiol Behav 106: 417–422.

41. Hepper PG, Wells DL, Millsopp S, Kraehenbuehl K, Lyn SA, et al. (2012) Prenatal and early sucking influences on dietary preference in newborn, weaning, and young adult cats. Chem Senses 37: 755–766.

42. Hara TJ (2006) Feeding behaviour in some teleosts is triggered by single amino acids primarily through olfaction. J Fish Biol 68: 810–825.

43. Wisby WJ, Hasler AD (1954) Effect of olfactory occlusion on migrating silver salmon (*O. kisutch*). J Fish Res Board Can 11: 472–478.

44. Scholz AT, Horrall RM, Cooper JC, Hasler AD (1976) Imprinting to chemical cues: The basis for home stream selection in salmon. Science 192: 1247–1249.

45. Ueda H (2012) Physiological mechanisms of imprinting and homing migration in Pacific salmon *Oncorhynchus spp*. J Fish Biol 81: 543–558.

46. Yamamoto Y, Hino H, Ueda H (2010) Olfactory imprinting of amino acids in lacustrine sockeye salmon. Plos One 5: e8633. doi:10.1371/journal.pone.0008633.

47. Dittman AH, Quinn TP, Nevitt GA (1996) Timing of imprinting to natural and artificial odors by coho salmon (*Oncorhynchus kisutch*). Can J Fish Aquat Sci 53: 434–442.

48. Cooper JC, Scholz AT, Horrall RM, Hasler AD, Madison DM (1976) Experimental confirmation of the olfactory hypothesis with homing, artificially imprinted coho salmon (*Oncorhynchus kisutch*). J Fish Res Board Can 33: 703–710.

49. Shoji T, Yamamoto Y, Nishikawa D, Kurihara K, Ueda H (2003) Amino acids in stream water are essential for salmon homing migration. Fish Physiol Biochem 28: 249–251.

50. Johnstone KA, Lubieniecki KP, Koop BF, Davidson WS (2011) Expression of olfactory receptors in different life stages and life histories of wild Atlantic salmon (*Salmo salar*). Mol Ecol 20: 4059–4069.

51. Welch KD, Provenza FD, Pfister JA (2012) Do plant secondary compounds induce epigenetic changes that confer resistance or susceptibility to toxicosis in animals? In: Casasús I, Rogošič J, Rosati A, Štokovič I, Gabiña D, editors. Animal farming and environmental interactions in the Mediterranean region. Forages and grazing in horse nutrition, vol 131. Wageningen, The Netherlands: Wageningen Academic Publishers. pp. 33–44. Available: http://www.springerlink.com/content/gm7258437v75l56g/abstract/. Accessed 11 October 2012.

52. Dupont-Nivet M, Médale F, Leonard J, Le Guillou S, Tiquet F, et al. (2009) Evidence of genotype-diet interactions in the response of rainbow trout (*Oncorhynchus mykiss*) clones to a diet with or without fishmeal at early growth. Aquaculture 295: 15–21.

53. Storebakken T, Austreng E (1987) Ration level for salmonids: II. Growth, feed intake, protein digestibility, body composition, and feed conversion in rainbow trout weighing 0.5–1.0 kg. Aquaculture 60: 207–221.

54. Wang Y, Kong L-J, Li K, Bureau DP (2007) Effects of feeding frequency and ration level on growth, feed utilization and nitrogen waste output of cuneate drum (*Nibea miichthioides*) reared in net pens. Aquaculture 271: 350–356.

55. Yuan Y-C, Yang H-J, Gong S-Y, Luo Z, Yuan H-W, et al. (2010) Effects of feeding levels on growth performance, feed utilization, body composition and apparent digestibility coefficients of nutrients for juvenile Chinese sucker, *Myxocyprinus asiaticus*. Aquac Res 41: 1030–1042.

56. Cleveland BM, Burr GS (2011) Proteolytic response to feeding level in rainbow trout (*Oncorhynchus mykiss*). Aquaculture 319: 194–204.

57. Breier B, Vickers M, Ikenasio B, Chan K, Wong WP (2001) Fetal programming of appetite and obesity. Mol Cell Endocrinol 185: 73–79.

58. Langley-Evans SC, McMullen S (2010) Developmental origins of adult disease. Med Princ Pr 19: 87–98.

59. Vickers MH, Breier BH, Cutfield WS, Hofman PL, Gluckman PD (2000) Fetal origins of hyperphagia, obesity, and hypertension and postnatal amplification by hypercaloric nutrition. Am J Physiol - Endocrinol Metab 279: E83–E87.

60. Fernandez-Twinn DS, Ozanne SE, Ekizoglou S, Doherty C, James L, et al. (2003) The maternal endocrine environment in the low-protein model of intra-uterine growth restriction. Br J Nutr 90: 815–822.

61. Milagro FI, Mansego ML, De Miguel C, Martínez JA (2013) Dietary factors, epigenetic modifications and obesity outcomes: Progresses and perspectives. Molecular Aspects of Medicine 34: 782–812.

62. Jiménez-Chillarón JC, Díaz R, Martínez D, Pentinat T, Ramón-Krauel M, et al. (2012) The role of nutrition on epigenetic modifications and their implications on health. Biochimie 94: 2242–2263.

15

Genetic Identification of F1 and Post-F1 Serrasalmid Juvenile Hybrids in Brazilian Aquaculture

Diogo Teruo Hashimoto[1]*, José Augusto Senhorini[2], Fausto Foresti[3], Paulino Martínez[4], Fábio Porto-Foresti[5]

1 Centro de Aquicultura, Universidade Estadual Paulista, (UNESP), Campus de Jaboticabal, Jaboticabal, SP, Brazil, **2** Centro de Pesquisa e Gestão de Recursos Pesqueiros Continentais, Instituto Chico Mendes de Conservação da Biodiversidade, Pirassununga, SP, Brazil, **3** Departamento de Morfologia, Instituto de Biociências, Universidade Estadual Paulista, (UNESP), Campus de Botucatu, Botucatu, SP, Brazil, **4** Departamento de Genética, Universidad de Santiago de Compostela, Facultad de Veterinaria, Lugo, Spain, **5** Departamento de Ciências Biológicas, Faculdade de Ciências, Universidade Estadual Paulista, (UNESP), Campus de Bauru, Bauru, SP, Brazil

Abstract

Juvenile fish trade monitoring is an important task on Brazilian fish farms. However, the identification of juvenile fish through morphological analysis is not feasible, particularly between interspecific hybrids and pure species individuals, making the monitoring of these individuals difficult. Hybrids can be erroneously identified as pure species in breeding facilities, which might reduce production on farms and negatively affect native populations due to escapes or stocking practices. In the present study, we used a multi-approach analysis (molecular and cytogenetic markers) to identify juveniles of three serrasalmid species (*Colossoma macropomum*, *Piaractus mesopotamicus* and *Piaractus brachypomus*) and their hybrids in different stocks purchased from three seed producers in Brazil. The main findings of this study were the detection of intergenus backcrossing between the hybrid ♀ patinga (*P. mesopotamicus* × *P. brachypomus*) × ♂ *C. macropomum* and the occurrence of one hybrid triploid individual. This atypical specimen might result from automixis, a mechanism that produces unreduced gametes in some organisms. Moreover, molecular identification indicated that hybrid individuals are traded as pure species or other types of interspecific hybrids, particularly post-F1 individuals. These results show that serrasalmid fish genomes exhibit high genetic heterogeneity, and multi-approach methods and regulators could improve the surveillance of the production and trade of fish species and their hybrids, thereby facilitating the sustainable development of fish farming.

Editor: Marinus F. W. te Pas, Wageningen UR Livestock Research, Netherlands

Funding: This work was supported by grants from Fundação de Amparo à Pesquisa do Estado de São Paulo (FAPESP) and Conselho Nacional de Desenvolvimento Científico e Tecnológico (CNPq number 481823/2011-5). The funders had no role in study design, data collection and analysis, decision to publish, or preparation of the manuscript.

Competing Interests: The authors have declared that no competing interests exist.

* E-mail: diogo@caunesp.unesp.br

Introduction

In Brazil, approximately 40 native fish species and 6 interspecific hybrids are cultivated on fish farms [1,2]. The representatives of the Serrasalmidae family, *i.e.*, *Colossoma macropomum* (tambaqui), *Piaractus mesopotamicus* (pacu), *Piaractus brachypomus* (pirapitinga or caranha), and their interspecific hybrids tambacu (female tambaqui × male pacu), patinga (female pacu × male pirapitinga), and tambatinga (female tambaqui × male pirapitinga) correspond to native fish with the largest production in Brazilian aquaculture (56.2 million kg per year) according to IBAMA (Instituto Brasileiro do Meio Ambiente e dos Recursos Naturais Renováveis) [3]. Reciprocal hybrids (*e.g.*, female pacu x male tambaqui) are also viable [4], but these individuals are not typically produced or cultivated on fish farms.

The cultivation of serrasalmids varies among different regions of Brazil [2]. In Southern Brazil, the only species produced is pacu. Tambaqui and pirapitinga, and their hybrids, are not produced in this region because these species cannot tolerate the low temperatures of Southern Brazil. In Northern Brazil, serrasalmid hybrids are produced at a lower rate compared with the pure species tambaqui, the main aquaculture resource in this region. In contrast, hybrids are associated with high production rates in the other regions of Brazil, particularly in the Midwest [3], and the hybrid tambacu has a greater economic importance than other serrasalmid hybrids. This fish group is also widely farmed in other Latin American (Colombia, Venezuela, and Cuba) [5] and Asian (China, Myanmar, Thailand, and Vietnam) countries [6,7].

However, the diversity and zootechnical differences among fish are problematic for the aquaculture industry because pure species or their hybrids can be produced or cultivated as a single species. This inaccuracy primarily reflects the morphological similarity between species, particularly in the case of hybrids and parental species in the juvenile stage. Thus, the use of genetic markers is essential to monitor the production and management of fish hybrid, particularly for the trade between seed suppliers and fish farmers, which is a critical point in the production chain [8].

Currently, PCR-RFLP (polymerase chain reaction - restriction fragment length polymorphism) and multiplex-PCR have been characterized as efficient methods for the rapid and inexpensive identification of hybrids [9–11]. For serrasalmid hybrids, these methodologies facilitate diagnoses based on the combination of single nucleotide variants in the mitochondrial genes, Cytochrome

C Oxidase subunit I (*mt-co1*) and Cytochrome b (*mt-cyb*), with nuclear genes, such as α-Tropomyosin (*tpm1*) and Recombination Activating Gene 2 (*rag2*) [12]. Nuclear diagnostic markers are essential to differentiate hybrids between species, but mitochondrial markers, although haploid, identify the maternal origin of hybrids, and this information is crucial for the assessment of hybridization.

Cytogenetic analysis methods have also been described for the identification of serrasalmid hybrids. Through C-banding and fluorescence *in situ* hybridization (18S ribosomal RNA probe), chromosome markers have facilitated the precise identification of the hybrids tambacu and tambatinga, respectively [13,14]. Although cytogenetic methods have limitations of low throughput because of the effort and time required for data analysis and processing [2], these techniques provide important information to verify ploidy level [13], which cannot be directly assessed through molecular markers.

Despite hybrid vigor in some cases, there are problems associated with the inadequate use of interspecific hybrids for aquaculture production. Occasionally, fish farmers have mistakenly used hybrids as broodstock, as reported for tilapia, catfish, and carp [15–17]. Superior performance or desirable characteristics associated with hybrid vigor might be lost in post-F1 individuals because introgressive hybridization reduces the heterosis obtained in F1 hybrids, and particularly because post-F1 hybrids typically show reduced offspring viability due to high mortality rates [18]. These observations have been previously reported in catfish, where hybrids were used as broodstock [17].

In the present study, we used PCR-RFLP, multiplex-PCR, and cytogenetic methods to evaluate the juvenile fish trade between seed suppliers and fish farmers in Brazil, focusing on the genetic identification of F1 and post-F1 serrasalmid hybrids. The novelty of this study was the discovery of fertility in the hybrid patinga (female pacu x male pirapitinga), and its use as broodstock in Brazilian aquaculture. Moreover, mistaken trade of hybrid tambacu was detected in fish farms and one post-F1 hybrid was characterized as triploid.

Materials and Methods

We performed the genetic identification of 924 juvenile individuals from eight stocks of live fish purchased from three private Brazilian aquaculture seed producers (herein referred to as SPS, MGS, and SES) (Table 1). All fish farms assessed in this study represent large companies in Brazil. From fish farm SPS, located in São Paulo State (Southeastern Brazil), we analyzed two commercially available stocks, labeled as hybrids tambacu (SPS1) and patinga (SPS2). From fish farm MGS, located in the Minas Gerais State (Southeastern Brazil), we analyzed three commercially produced stocks, labeled as pure tambaqui (MGS1), pacu (MGS2), and hybrid tambacu (MG3). From fish farm SES, located in Sergipe State (Northeastern Brazil), we analyzed three commercially stocks, labeled as pure tambaqui (SES1), hybrids tambatinga (SES2), and tambacu (SES3). The size of the analyzed fish ranged from 5 to 10 cm. We did not notify the producers that the fish would be used for identification purposes.

This study was conducted in strict accordance with the recommendations of the National Council for Control of Animal Experimentation (Brazilian Ministry for Science, Technology and Innovation). The present study was performed under authorization N° 33435-1 issued through ICMBio (Chico Mendes Institute for the Conservation of Biodiversity, Brazilian Ministry for Environment). Fin fragments were collected from each fish under benzocaine anesthesia and all efforts were made to minimize

Table 1. Molecular identification of juvenile serrasalmid fish stocks purchased from different fish farmers.

Fish farm	Stocks purchased	Stock identification	n
SPS (São Paulo State)	tambacu	SPS1	50
	patinga	SPS2	33
MGS (Minas Gerais State)	tambaqui	MGS1	143
	pacu	MGS2	115
	tambacu	MGS3	133
SES (Sergipe State)	tambaqui	SES1	150
	tambatinga	SES2	150
	tambacu	SES3	150

suffering. DNA was extracted from the fin fragments using the Wizard Genomic DNA Purification Kit (Promega) according the manufacturer's protocol. The DNA concentration was assessed against a molecular marker standard (the Low DNA Mass Ladder, Invitrogen) through electrophoresis on a 1% agarose gel.

The samples were genotyped using two methods and different genes, as previously described [12]: 1) multiplex-PCR based on nuclear α-Tropomyosin (*tpm1*) and mitochondrial Cytochrome C Oxidase subunit I (*mt-co1*) genes; and 2) PCR-RFLP using the nuclear Recombination Activating Gene 2 (*rag2*) and mitochondrial Cytochrome b (*mt-cyb*) genes. Both methods provide diagnostic electrophoretic fragments for each parental species and their interspecific hybrids. Diagnostic sizes of the PCR products or restriction fragments are described in the Table 2. The sequences for the primers and restriction enzymes, PCR reagents, reagent concentrations, and reaction conditions were used as previously described [12]. We used multiplex-PCR, followed by PCR-RFLP in subsequent analyses for confirmation. DNA samples from the pure parental species were used as controls for reaction specificity in all experiments. These samples were previously identified through morphological and molecular analyses [12] and obtained from the stock maintained at the Centro Nacional de Pesquisa e Conservação de Peixes Continentais (CEPTA/ICMBio, Pirassununga, São Paulo State, Brazil).

Cytogenetic analysis was also performed to verify the ploidy level in individuals of the SPS2 stock. Chromosomal preparations were obtained according to the methods of Foresti et al. [19]. The chromosome morphology was determined based on the arm ratio consistent with Levan et al. [20], and the chromosomes were subsequently classified as metacentric (m), submetacentric (sm),

Table 2. Sizes of the PCR products or restriction fragments, according to Hashimoto et al. [12].

Method	Gene	Diagnostic fragment size (bp)		
		pacu	tambaqui	pirapitinga
Multiplex-PCR	*mt-co1*	307	435	307 and 610
	tpm1	269	172	131
PCR-RFLP	*mt-cyb*	152 and 513	261 and 405	665
	rag2	750	357 and 393	250 and 500

subtelocentric (st), and acrocentric (a). Fluorescence *in situ* hybridization (FISH) was performed using the method of Yang et al. [21]. These experiments encompassed all of the genotypes shown below through the molecular identification of the SPS2 stock. The 5S ribosomal RNA (rRNA) gene sequences were PCR amplified from DNA using the primers described by Pendás et al. [22]. To prepare the probe, PCR products of the 5S rRNA gene were labeled with biotin-16-dUTP (Roche) through nick translation (Invitrogen). The chromosomes were counterstained with DAPI (4′,6-diamidino-2-phenylindole, Vector Laboratories). The FISH images were captured and processed using the CytoVision Genus system (Applied Imaging, USA) and a Cohu CCD camera mounted on an Olympus BX-60 microscope.

Results

We obtained the same genotype with all molecular markers in the samples purchased as hybrid tambacu (stocks SPS1, MGS3, and SES3), pure tambaqui (MGS1 and SES1), pacu (MGS2), and hybrid tambatinga (SES2), indicating that these species correspond to hybrid tambacu. The results of the multiplex-PCR analysis of the nuclear marker *tpm1* (Figure 1a) revealed a heterozygous genotype (fragments of 172 and 269 bp), characteristic of hybrid tambacu. Moreover, multiplex-PCR of the mitochondrial marker *mt-co1* (Figure 1b) showed that these hybrids exhibited the genotype of the maternal species tambaqui (fragment of 435 bp), consistent with the identification of these samples as tambacu (♀ tambaqui x ♂ pacu) instead of the reciprocal hybrid paqui (♀ pacu×♂ tambaqui). The results were confirmed through PCR-RFLP using the nuclear *rag2* and mitochondrial *mt-cyb* genes. Thus, the MGS1, MGS2, SES1, and SES2 samples were mislabeled, as these species were actually tambacu.

The results of molecular identification in the SPS2 stock demonstrated that these juveniles likely correspond to post-F1 hybrids, resulting from the backcrossing of the hybrid ♀ patinga (♀ pacu x ♂ pirapitinga) with ♂ tambaqui. Consistent with this hypothesis, all the offspring showed the pacu genotypes for the mitochondrial *mt-co1* (fragment of 307 bp) and *mt-cyb* (fragments of 152 and 513 bp) genes, and segregating genotypes at the nuclear *tpm1* and *rag2* markers, as indicated below (Figures 2 and 3):

- Genotype A (8 individuals): pattern of hybrid tambacu for both nuclear markers *tpm1* (fragments of 172 and 269 bp) and *rag2* (fragments of 357, 393, and 750 bp).
- Genotype B (9 individuals): pattern of hybrid tambatinga for both nuclear markers *tpm1* (131 and 172 bp) and *rag2* (250, 357, 393, and 500 bp).

- Genotype C (6 individuals): nuclear markers of hybrid tambacu for the *tpm1* (172 and 269 bp) gene, and genotype of hybrid tambatinga for the *rag2* (250, 357, 393, and 500 bp) gene.
- Genotype D (9 individuals): opposite nuclear patterns to the genotype C, *i.e.*, genotype of hybrid tambatinga for the *tpm1* (131 and 172 bp) gene, and genotype of hybrid tambacu for the *rag2* (357, 393, and 750 bp) gene.
- Genotype E (1 individual): atypical genotype comprising gene fragments from the three pure species (tambaqui, pacu, and pirapitinga) for the *tpm1* (131, 172, and 269 bp) gene, and genotype of hybrid tambacu for the *rag2* (357, 393, and 750 bp) gene. This unexpected pattern is compatible with a triploid individual, consistent with the cytogenetic results shown below.

According to the Mendelian inheritance, the hypothesized backcross ♀ patinga (♀ pacu x ♂ pirapitinga) x ♂ tambaqui would produce each A–D genotype at 25% among the offspring (Figure 3), and the observed data did not deviate from the null hypothesis, confirmed using the $\chi2$ test (p = 0.86).

The cytogenetic analysis showed a diploid chromosome number of 2n = 54 for most individuals of the SPS2 stock, with chromosomes presenting morphologies of the types m and sm, excluding one specimen comprising 81 chromosomes (50 metaphases with this chromosome number were counted), suggesting a polyploid individual with three chromosome sets (3n) (Figure 4), and corresponding to genotype E. The occurrence of this individual might reflect a likely meiotic segregation pattern, including a crossing-over between the *tpm1* gene locus and the centromere, but not in the *rag2*-bearing chromosomes, during female oogenesis, with the fusion/retention of the second polar body in the egg, resulting in a triploid individual (Figure 5).

The FISH analysis of the specimens in the SPS2 stock revealed 5S rRNA clusters in the subcentromeric region of four chromosomes, some of which were non-homologous, as revealed through differences in morphology/size and the positions of the hybridization signals (Figures 4a and 4b). In two chromosomes, the genes were located on the long arms (major clusters), and in the other two chromosomes, the genes were located on the short arms (minor clusters). We observed different combinations of these chromosomes in the analyzed individuals, further suggesting the occurrence of post-F1 individuals. Moreover, the individual previously identified as triploid was characterized with six 5S rRNA clusters: three chromosomes with FISH signals in the subcentromeric region of the long arms (major clusters), and the other three chromosomes with signals in the short arms (minor clusters) (Figure 4c). No correlation was observed between the different chromosomes carrying 5S rRNA clusters with the

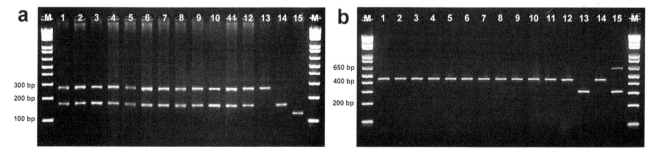

Figure 1. Molecular identification of the samples SPS1, MGS1, MGS2, MGS3, SES1, SES2, and SES3 using the nuclear *tpm1* (a) and mitochondrial *mt-co1* (b) genes in multiplex-PCR. Lanes 1–12, genotypes of hybrid tambacu (♀ tambaqui x ♂ pacu); and lanes 13, 14, and 15, control DNA samples from the pure pacu, tambaqui, and pirapitinga species, respectively; M, 1 Kb Plus DNA Ladder.

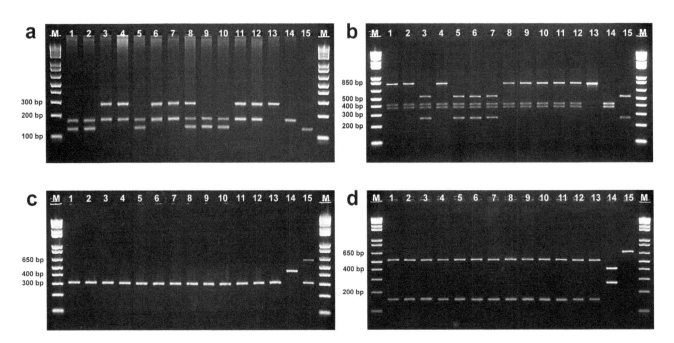

Figure 2. Molecular identification of the SPS2 stock using the nuclear *tpm1* **(a) and** *rag2* **(b) genes, and mitochondrial** *mt-co1* **(c) and** *mt-cyb* **(d) genes.** Lanes 4, 11, and 12, genotype A; lane 5, genotype B; lanes 3, 6, and 7, genotype C; lanes 1, 2, 9, and 10, genotype D; lane 8, genotype E; and lanes 13, 14, and 15, control DNA samples from the pure pacu, tambaqui, and pirapitinga species, respectively; M, 1 Kb Plus DNA Ladder.

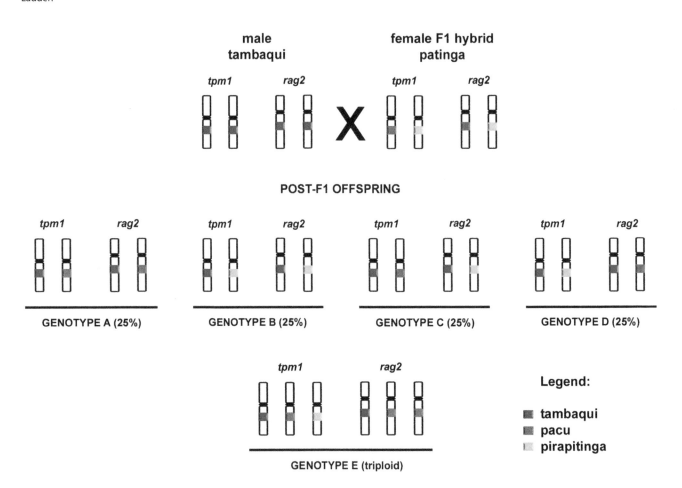

Figure 3. Schematic representation of the backcrossing between ♂ **tambaqui x** ♀ **patinga (**♀ **pacu x** ♂ **pirapitinga), demonstrating** the *tpm1* and *rag2* gene loci and the expected probability of each genotype.

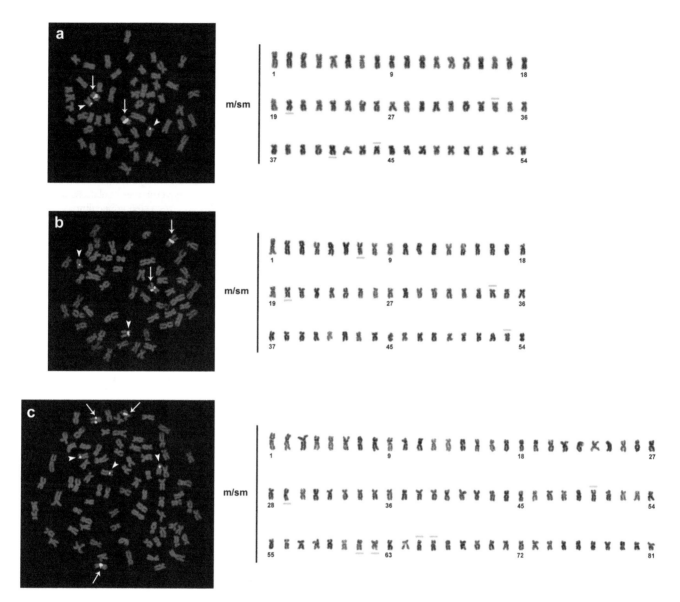

Figure 4. Metaphases and the respective karyotypes of individuals of the SPS2 stock, showing the chromosome location of the 5S rRNA clusters. The arrows and arrowheads indicate the 5S rRNA major and minor clusters, respectively, in metaphase. In the karyotypes, the green bars indicate the chromosomes bearing 5S rRNA genes and their respective locations (long or short arms). (a) an individual of genotype A, (b) an individual of genotype B, and (c) an individual of genotype E.

genotypes of the nuclear markers *tpm1* and *rag2*, thus suggesting independent segregation between these markers.

Discussion

Hybrids between tambaqui and pacu are popular in Brazil, as these hybrids combine the robustness and faster growth rate of tambaqui with the low temperature resistance of pacu [23,24]. The technology required for the reproduction of hybrid tambacu through hormonal induction has been well established and widely used in farming systems, making cultivation easier than with other hybrids and even pure species. This effect might explain the results obtained in the present study, in which the hybrid tambacu was traded as other hybrids and pure species. The admixture of different types of fish is another problem observed on fish farms [12], where stocks of serrasalmid juvenile fish comprised up to five types of fish, including pure species and hybrid individuals.

However, these mislabeling activities represent a fraud to the market and are not productive for cultivation and aquaculture because different hybrids and pure species have specific zootechnical characteristics and economic values [2].

Special attention should be given when hybrids are sold as pure species, as demonstrated in the present study for the stocks of tambaqui (MGS1 and SES1) and pacu (MGS2). The results showed that in addition to fraud, fish farmers are not aware of the potential biological and environmental risks represented by hybrids, whose impact could affect the aquaculture industry and threaten native species, as previously described in other species, such as tilapia, catfish, and trout [25 28]. Moreover, the same problems have been observed for other Brazilian fish farms in several States (São Paulo, Minas Gerais, Piauí, Sergipe, and Pará) [12,29], indicating that could be a common practice in the aquaculture industry.

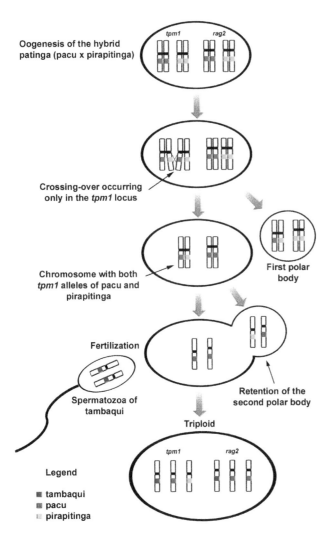

Figure 5. Schematic representation of the events of crossing-over and the fusion/retention of the second polar body, which likely generated the triploid individual of genotype E.

In addition, the results of the molecular characterization in the present study provided clear evidence that individuals of female patinga are fertile, similar to the hybrids tambacu and tambatinga [18,30]. Hence, due to the difficulty of morphological identification, these hybrids can be erroneously used as broodstock on Brazilian fish farms, which is not productive for aquaculture and represents even higher risks [15,31]. The negative effects of post-F1 individuals include the dilution or loss of the desirable characteristics, resulting from hybrid vigor [32], low hatching rate and high mortality level [17,18]. Therefore, the data obtained in the present study show that molecular tools should be applied in breeding facilities to ensure the integrity of pure stocks used on the Brazilian fish farms, as previously demonstrated [18], where even post-F1 hybrids have been observed in the farmed broodstock of catfish.

Serrasalmid hybrids have also been detected in the natural environment and some authors have suggested that these individuals are the consequence of aquaculture activities [12,33,34]. The presence of fertile hybrids on the farms observed in the present study reveals the potential risks of these farming practices to wild populations. Introgressive hybridization poses a threat to the genetic integrity of pure species, which might result in

a single hybrid population, as reported in trout, catfish, and tilapia [27,35–37]. In Brazil, this situation has also been described for hybrids between the Neotropical catfish species *Pseudoplatystoma corruscans* and *Pseudoplatystoma reticulatum* [38]. Genetic analysis of these species have revealed a high frequency of hybrids, including post-F1 generation individuals in the Mogi Guaçu (50%) and Aquidauana (30%) rivers, where the majority of the Brazilian fish farms are located, suggesting that hybrids might be introduced from farmed stocks [38]. The results obtained from the present study should be complemented with the analysis of wild populations to confirm whether hybrids in natural environments result from farm escapees or natural hybridization. This point is critical for the development of government regulations to achieve sustainable and environmentally respected aquaculture.

The combination of molecular markers and cytogenetic techniques was essential for the identification of a triploid individual in the present study and to confirm the hybridization processes. Cytogenetic information is considered a suitable tool to verify ploidy level and analyze parental chromosome sets in hybrids [4,39,40]. In previous studies, pure species of tambaqui, pacu, and pirapitinga were characterized by a diploid chromosome number of 54 m/sm chromosomes [13,14]. The individuals cytogenetically analyzed in this study also showed $2n = 54$ chromosomes, with the exception of the sample corresponding to genotype E, which was characterized by triploidy (3n). This type of event was also described for an individual of the hybrid tambacu [13] and for a post-F1 specimen resulting from backcrossing between ♀ pacu x ♂ tambacu (♀ tambaqui x ♂ pacu), which also generated gynogenetic individuals [18].

Spontaneous triploid fish can be explained by several cytological mechanisms that produce unreduced diploid gametes [41]. Premeiotic endomitosis is characterized by genome doubling without cytokinesis before meiotic division, followed by two quasinormal divisions. In apomixis, meiosis is repressed in the first division, and the oocyte is produced through mitosis, without the recombination and segregation of the homologous chromosomes. However, in both mechanisms, the unreduced gametes are isogenic, *i.e.*, genetically identical to the parent without genetic variation of the resulting eggs [41,42]. Premeiotic endomitosis and apomixis occurs in several fish species [43,44], including interspecific hybrids [45], but cannot explain the triploid event observed in the present study, as genetic variations were detected in the molecular analyses.

Alternatively, these results are consistent with a cytological mechanism similar to automixis, in which meiosis is maintained and diploid gametes are generated after meiosis through the fusion/retention of the polar body. This phenomenon has been well documented in the triploid offspring of poeciliid interspecific hybrids [42]. The resulting triploid products are not genetically identical to the parental genome, as segregation and recombination result in nonidentical homologous chromosomes [42], explaining the atypical genotype E observed herein. However, further studies are needed to evaluate whether this triploid event is due to automixis or whether hybridization facilitates polyploidization, as shown in other species [46].

Consistent with the data obtained in this study, the presence of 5S rRNA clusters in two pairs of chromosomes in diploid individuals is a common characteristic in fish genomes [47,48]. However, the results of the present study demonstrated the occurrence of different combinations of chromosomes bearing 5S rRNA sites, suggesting these combinations were inherited from chromosomes of distinct species.

The molecular and cytogenetic data obtained herein are consistent with the hypothesis of the fertility of hybrid patinga.

However, the presence of additional post-F1 hybrids (F2 or advanced backcrosses) cannot be ruled out at least for some specimens because of the small number of nuclear markers used in this study. According to Boecklen and Horward [49], more than 70 nuclear markers are needed for the reliable differentiation between pure species and advanced hybrid crosses or backcrosses. Thus, the acquisition of additional markers based on nuclear genes is necessary for the identification of serrasalmid hybrids. The combination of next-generation sequencing technologies with restriction enzyme analyses simplifies genome sequencing to obtain deeper coverage at particular sites, thus facilitating the identification of thousands of SNPs (single nucleotide polymorphisms) at low cost [50–52], as Hohenlohe et al. [53] and Amish et al. [54] demonstrated through the identification of thousands of SNPs for the accurate detection of hybrids between *Oncorhynchus mykiss* and *Oncorhynchus clarkii lewisi*.

In Brazil, most aquaculture establishments are not licensed and there is little proposed legislation regulating fish breeding [8]. In contrast, some countries have legislation that specifically addresses issues concerning hybridization: California (USA) has laws prohibiting the hybridization of fish without a proper license [2,55]. Thus, specific laws should be implemented in Brazil to address the problems of the uncontrolled trade and management of the fish hybrids detected in several studies [12,17,29,38]. Moreover, confinement measures are indispensable to avoid the widespread dissemination of fish hybrids [2], particularly physical and reproductive measures required for transgenic fish [56,57].

Consistent with Hashimoto et al. [12], the results obtained in the present study show that genetic tools should be applied to monitor the trade of juvenile fish hybrids, representing a preventive measure for the sustainable development of the aquaculture industry, particularly because serrasalmid hybrids are fertile (*e.g.*, hybrid patinga) and hybrid tambacu can be erroneously traded as pure species. In conclusion, this multi-approach analysis (molecular, cytogenetic, and FISH methods) was useful for the detection of hybridization and the results provided new insights concerning the genome plasticity of serrasalmid species, including the occurrence of intergenus backcrossing between ♀ patinga (♀ pacu x ♂ pirapitinga) x ♂ tambaqui and the presence of an post-F1 hybrid triploid, likely derived from similar mechanisms of unreduced gametes in automixis described for other fish hybrids [42].

Author Contributions

Conceived and designed the experiments: DTH JAS FPF. Performed the experiments: DTH FPF. Analyzed the data: DTH FF PM FPF. Contributed reagents/materials/analysis tools: DTH JAS FF FPF. Wrote the paper: DTH JAS FF PM FPF.

References

1. Godinho HP (2007) Estratégias reprodutivas de peixes aplicadas à aqüicultura: bases para o desenvolvimento de tecnologias de produção. Rev Bras Reprod Anim 31(3): 351–360.
2. Hashimoto DT, Senhorini JA, Foresti F, Porto-Foresti F (2012) Interspecific fish hybrids in Brazil: management of genetic resources for sustainable use. Rev Aquaculture 4: 108–118.
3. IBAMA (2007) Estatística da Pesca 2007: Brasil – Grandes regiões e unidades da Federação. 113 p.
4. Porto-Foresti F, Hashimoto DT, Alves AL, Almeida RBC, Senhorini JA, et al. (2008) Cytogenetic markers as diagnoses in the identification of the hybrid between Piauçu (*Leporinus macrocephalus*) and Piapara (*Leporinus elongatus*). Genet Mol Biol 31(suppl.): 195–202.
5. Flores Nava A (2007) Aquaculture seed resources in Latin America: a regional synthesis. In: Bondad-Reantaso MG, editor. Assessment of freshwater fish seed resources for sustainable aquaculture. FAO Fisheries Technical Paper, No. 501, FAO, Rome. pp. 91–102.
6. Honglang H (2007) Freshwater fish seed resources in China. In: Bondad-Reantaso MG, editor. Assessment of freshwater fish seed resources for sustainable aquaculture. FAO Fisheries Technical Paper, No. 501, FAO, Rome. . pp. 185–199.
7. FAO (2010) The State of World Fisheries and Aquaculture - 2010 (SOFIA). Food and Agriculture Organization of the United Nations - FAO Fisheries and Aquaculture Department, Roma.
8. Suplicy FM (2007) Freshwater fish seed resources in Brazil. In: Bondad-Reantaso MG, editor. Assessment of freshwater fish seed resources for sustainable aquaculture. FAO Fisheries Technical Paper, No. 501, FAO, Rome. . pp. 129–143.
9. Hashimoto DT, Mendonça FF, Senhorini JA, Bortolozzi J, Oliveira C, et al. (2010) Identification of hybrids between Neotropical fish *Leporinus macrocephalus* and *Leporinus elongatus* by PCR-RFLP and multiplex-PCR: tools for genetic monitoring in aquaculture. Aquaculture 298: 346–349.
10. Prado FD, Hashimoto DT, Mendonça FF, Senhorini JA, Foresti F, et al. (2011) Molecular identification of hybrids between Neotropical catfish species *Pseudoplatystoma corruscans* and *Pseudoplatystoma reticulatum*. Aquaculture Res 42: 1890–1894.
11. Porto-Foresti F, Hashimoto DT, Prado FD, Senhorini JA, Foresti F (2013) Genetic markers for the identification of hybrids among catfish species of the family Pimelodidae. J Appl Ichthyol 29: 643–647.
12. Hashimoto DT, Mendonça FF, Senhorini JA, Oliveira C, Foresti F, et al. (2011) Molecular diagnostic methods for identifying Serrasalmid fish (Pacu, Pirapitinga, and Tambaqui) and their hybrids in the Brazilian aquaculture industry. Aquaculture 321: 49–53.
13. Almeida-Toledo LF, Foresti F, Toledo-Filho SA, Bernardino G, Ferrari VA, et al. (1987) Cytogenetic studies in *Colossoma mitrei*, *C. macropomum* and their interspecific hybrid. In: Tiews K, editor. Selection, Hybridization and Genetic Engeneering in Aquaculture. Berlin Heenemann Verlagsgesellshaft mb II vol. 1. pp. 189–195.
14. Nirchio M, Fenocchio AS, Swarça AC, Pérez JE, Granado A, et al. (2003) Cytogenetic characterization of hybrids offspring between *Colossoma macropomum* (Cuvier, 1818) and *Piaractus brachypomus* (Cuvier, 1817) from Caicara del Orinoco, Venezuela. Caryologia 56: 405–411.
15. Mia MY, Taggart JB, Gilmour AE, Gheyas AA, Das TK, et al. (2005) Detection of hybridization between Chinese carp species (*Hypophthalmichthys molitrix* and *Aristichthys nobilis*) in hatchery broodstock in Bangladesh, using DNA microsatellite loci. Aquaculture 247: 267–273.
16. Mair GC (2007) Genetics and breeding in seed supply for inland aquaculture. In: Bondad-Reantaso MG, editor. Assessment of freshwater fish seed resources for sustainable aquaculture. FAO Fisheries Technical Paper, No. 501, FAO, Rome. pp. 519–548.
17. Hashimoto DT, Prado FD, Senhorini JA, Foresti F, Porto-Foresti F (2013) Detection of post-F1 fish hybrids in broodstock using molecular markers: approaches for genetic management in aquaculture. Aquaculture Res 44: 876–884.
18. Almeida-Toledo LF, Bernardino G, Oliveira C, Foresti F, Toledo-Filho SA (1996) Gynogenetic fish produced by a backcross involving a male hybrid (female *Colossoma macropomum*×male *Piaractus mesopotamicus*) and a female *Piaractus mesopotamicus*. Boletim Técnico CEPTA 9: 31–37.
19. Foresti F, Almeida-Toledo LF, Toledo-Filho SA (1981) Polymorphic nature of nucleolus organizer regions in fishes. Cytogenet Cell Genet 31: 137–144.
20. Levan A, Fredga K, Sandberg AA (1964) Nomenclature for centromeric position on chromosomes. Hereditas 52: 201–220.
21. Yang F, O'Brien PC, Milne BS, Graphodatsky AS, Solanky N, et al. (1999) A complete comparative chromosome map for the dog, red fox, and human and its integration with canine genetic maps. Genomics 62(2): 189–202.
22. Pendás AM, Moran P, Freije JP, Garcia-Vazquez E (1994) Chromosomal mapping and nucleotide sequence of two tandem repeats of Atlantic salmon 5S rDNA. Cytogenet Cell Genet 67: 31–36.
23. Calcagnotto D, Almeida-Toledo LF, Bernardino G, Toledo-Filho SA (1999) Biochemical genetic characterization of F1 reciprocal hybrids between neotropical pacu (*Piaractus mesopotamicus*) and tambaqui (*Colossoma macropomum*) reared in Brazil. Aquaculture 174: 51–57.
24. Gomes LC, Simões LN, Araujo-Lima CARM (2010) Tambaqui (*Colossoma macropomum*). In: Baldisserotto B, Gomes LC, editors. Espécies nativas para a piscicultura no Brasil. Universidade Federal de Santa Maria, Santa Maria. pp. 589–606.
25. Bartley DM, Gall AE (1991) Genetic identification of native cutthroat trout (*Oncorhynchus clarki*) and introgressive hybridization with introduced rainbow trout (*O. mykiss*) in streams associated with the Alvord Basin, Oregon and Nevada. Copeia 3: 854–859.
26. Docker MF, Dale A, Heath DD (2003) Erosion of interspecific reproductive barriers resulting from hatchery supplementation of rainbow trout sympatric with cutthroat trout. Mol Ecol 12: 3515–3521.
27. Na-Nakorn U, Kamonrat W, Ngamsiri T (2004) Genetic diversity of walking catfish, *Clarias macrocephalus*, in Thailand and evidence of genetic introgression from introduced farmed *C. gariepinus*. Aquaculture 240: 145–163.

28. Angienda PO, Lee HJ, Elmer KR, Abila R, Waindi EM, et al. (2011) Genetic structure and gene flow in an endangered native tilapia fish (*Oreochromis esculentus*) compared to invasive Nile tilapia (*Oreochromis niloticus*) in Yala swamp, East Africa. Conserv Genet 12: 243–255.

29. Gomes F, Schneider H, Barros C, Sampaio D, Hashimoto DT, et al. (2012) Innovative molecular approach to the identification of the tambaqui (*Colossoma macropomum*) and its hybrids, the tambacu and the tambatinga. An Acad Bras Cienc 84: 89–97.

30. Martino G (2002) Retrocruce de hembras híbridos (F1) (*Colossoma macropomum* × *Piaractus brachypomus*) con machos de las especies parentales. In: CIVA 2002, Comunicaciones y Foros de Discusión, Espanha. pp. 688–693.

31. Pullin RSV (1988) Tilapia Genetic Resources for Aquaculture. International Center for Living Aquatic Resources Management. Manila, Philippines.

32. Toledo-Filho SA, Almeida-Toledo LF, Foresti F, Bernardino G, Calcagnotto D (1994) Monitoramento e conservação genética em projeto de hibridação entre pacu e tambaqui. Cadernos de Ictiogenética 2, CCS/USP, São Paulo.

33. Orsi ML, Agostinho AA (1999) Introdução de peixes por escapes acidentais de tanques de cultivo em rios da Bacia do rio Paraná, Brasil. Rev Bras Zool 16(2): 557–560.

34. Paiva MP, Andrade-Tubino MF, Godoy MP (2002) As represas e os peixes nativos do Rio Grande - Bacia do Paraná - Brasil. Interciência, Rio de Janeiro.

35. Allendorf FW, Leary RF, Spruell P, Wenburg JK (2001) The problems with hybrids: setting conservation guidelines. Trends Ecol Evol 16(11): 613–622.

36. Young WP, Ostberg CO, Keim P, Thorgaard GH (2001) Genetic characterization of hybridization and introgression between anadromous rainbow trout (*Oncorhynchus mykiss irideus*) and coastal cutthroat trout (*O. clarki clarki*). Mol Ecol 10: 921–930.

37. Senanan W, Kapuscinski AR, Na-Nakorn U, Miller LM (2004) Genetic impacts of hybrid catfish farming (*Clarias macrocephalus* × *C. gariepinus*) on native catfish populations in central Thailand. Aquaculture 235: 167–184.

38. Prado FD, Hashimoto DT, Senhorini JA, Foresti F, Porto-Foresti F (2012) Detection of hybrids and genetic introgression in wild stocks of two catfish species (Siluriformes: Pimelodidae): the impacts of artificial hybridisation in Brazil. Fish Res 125: 300–305.

39. Hashimoto DT, Parise-Maltempi PP, Laudicina A, Bortolozzi J, Senhorini JA, et al. (2009) Repetitive DNA probe linked to sex chromosomes in hybrids between Neotropical fish *Leporinus macrocephalus* and *Leporinus elongatus* (Characiformes, Anostomidae). Cytogenet Genome Res 124: 151–157.

40. Hashimoto DT, Laudicina A, Bortolozzi J, Foresti F, Porto-Foresti F (2009) Chromosomal features of nucleolar dominance in hybrids between the Neotropical fish *Leporinus macrocephalus* and *Leporinus elongatus* (Characiformes, Anostomidae). Genetica 137: 135–140.

41. Arai K, Fujimoto T (2013) Genomic constitution and atypical reproduction in polyploid and unisexual lineages of the *Misgurnus* Loach, a Teleost Fish. Cytogenet Genome Res 140: 226–240.

42. Lampert KP, Lamatsch DK, Fischer P, Epplen JT, Nanda I, et al. (2007) Automictic reproduction in interspecific hybrids of poeciliid fish. Curr Biol 17: 1948–1953.

43. Itono M, Okabayashi N, Morishima K, Fujimoto T, Yoshikawa H, et al. (2007) Cytological mechanisms of gynogenesis and sperm incorporation in unreduced diploid eggs of the clonal loach, *Misgurnus anguillicaudatus* (Teleostei: Cobitidae). J Exp Zool A Ecol Genet Physiol 307: 35–50.

44. Lamatsch DK, Stöck M (2009) Sperm-dependent parthenogenesis and hybridogenesis in teleost fishes. In: Schön I, Martens K, van Dijk P, editors. Lost sex: the evolutionary biology of parthenogenesis. Springer, Dordrecht. pp. 399–432.

45. Shimizu Y, Shibata N, Sakaizumi M, Yamashita M (2000) Production of diploid eggs through premeiotic endomitosis in the hybrid medaka between *Oryzias latipes* and *O. curvinotus*. Zool Sci 17: 951–958.

46. Gomelsky B (2003) Chromosome set manipulation and sex control in common carp: a review. Aquat Living Resour 16: 408–415.

47. Martins C, Wasko AP (2004) Organization and evolution of 5S ribosomal DNA in the fish genome. Chapter X. In: Williams CR, editor. Focus on Genome Research. Nova Science Publishers, New York. pp. 289–318.

48. Hashimoto DT, Ferguson-Smith MA, Rens W, Foresti F, Porto-Foresti F (2011) Chromosome mapping of H1 histone and 5S rRNA gene clusters in three species of *Astyanax* (Teleostei, Characiformes). Cytogenet Genome Res 134: 64–71.

49. Boecklen WJ, Howard DJ (1997) Genetic analysis of hybrid zones: numbers of markers and power of resolution. Ecology 78: 2611–2616.

50. Baird NA, Etter PD, Atwood TS, Currey MC, Shiver AL, et al. (2008) Rapid SNP Discovery and Genetic Mapping Using Sequenced RAD Markers. Plos One 3(10): e3376.

51. Davey JW, Hohenlohe PA, Etter PD, Boone JQ, Catchen JM, et al. (2011) Genome-wide genetic marker discovery and genotyping using next-generation sequencing. Nat Rev Genet 12: 499–510.

52. Wang S, Meyer E, McKay JK, Matz MV (2012) 2b-RAD: a simple and flexible method for genome-wide genotyping. Nat Methods 9: 808–810.

53. Hohenlohe PA, Amish SJ, Catchen JM, Allendorf FW, Luikart G (2010) Next-generation RAD sequencing identifies thousands of SNPs for assessing hybridization between rainbow and westslope cutthroat trout. Mol Ecol Res 11: 117–122.

54. Amish SJ, Hohenlohe PA, Painter S, Leary RF, Muhlfeld C, et al. (2012) RAD sequencing yields a high success rate for westslope cutthroat and rainbow trout species-diagnostic SNP assays. Mol Ecol Resour 12(4): 653–60.

55. Bartley DM, Rana K, Immink AJ (2001) The use of inter-specific hybrids in aquaculture and fisheries. Rev Fish Biol Fish 10: 325–337.

56. Mair GC, Nam YK, Solar II (2007) Risk management: reducing risk through confinement of transgenic fish. In: Kapuscinski AR, Hayes KR, Li S, Dana G, editors. Environmental risk assessment of genetically modified organisms, Volume 3: Methodologies for transgenic fish. CABI Publishing, Cambridge. pp. 209–238.

57. Hallerman E (2008) Application of risk analysis to genetic issues in aquaculture. In: Bondad-Reantaso MG, Arthur JR, Subasinghe RP, editors. Understanding and applying risk analysis in aquaculture. FAO Fisheries and Aquaculture Technical Paper. No. 519. FAO, Rome. pp. 47–66.

16

Coupled Ecosystem/Supply Chain Modelling of Fish Products from Sea to Shelf: The Peruvian *Anchoveta* Case

Angel Avadí[1,2]*, Pierre Fréon[2], Jorge Tam[3]

1 Université Montpellier 2– Sciences et Techniques, Montpellier, France, **2** Institut de Recherche pour le Développement (IRD), UMR212 EME IFREMER/IRD/UM2, Sète, France, **3** Instituto del Mar del Perú (IMARPE), Callao, Peru

Abstract

Sustainability assessment of food supply chains is relevant for global sustainable development. A framework is proposed for analysing fishfood (fish products for direct human consumption) supply chains with local or international scopes. It combines a material flow model (including an ecosystem dimension) of the supply chains, calculation of sustainability indicators (environmental, socio-economic, nutritional), and finally multi-criteria comparison of alternative supply chains (e.g. fates of landed fish) and future exploitation scenarios. The Peruvian *anchoveta* fishery is the starting point for various local and global supply chains, especially via reduction of *anchoveta* into fishmeal and oil, used worldwide as a key input in livestock and fish feeds. The Peruvian *anchoveta* supply chains are described, and the proposed methodology is used to model them. Three scenarios were explored: *status quo* of fish exploitation (Scenario 1), increase in *anchoveta* landings for food (Scenario 2), and radical decrease in total *anchoveta* landings to allow other fish stocks to prosper (Scenario 3). It was found that Scenario 2 provided the best balance of sustainability improvements among the three scenarios, but further refinement of the assessment is recommended. In the long term, the best opportunities for improving the environmental and socio-economic performance of Peruvian fisheries are related to sustainability-improving management and policy changes affecting the reduction industry. Our approach provides the tools and quantitative results to identify these best improvement opportunities.

Editor: Howard I. Browman, Institute of Marine Research, Norway

Funding: This work is a contribution to the cooperative agreement between the Instituto del Mar del Peru (IMARPE), the Institut de Recherche pour le Developpement (IRD), and of the LMI DISCOH. Angel Avadí was financially supported by the Direction des Programmes de Recherche et de la formation au Sud (DPF) of the IRD. The funders had no role in study design, data collection and analysis, decision to publish, or preparation of the manuscript.

Competing Interests: The authors have declared that no competing interests exist.

* Email: angel.avadi@ird.fr

Introduction

Sustainability in food systems has several dimensions of concern, including environmental [1,2], socio-economic and food security [3,4], consumption patterns [5], technology [6], information [7] and governance/policy [8]. Moreover, sustainability arises from the complex interrelation among these factors, and thus science should focus on the most significant cause-and-effect relationships and driving forces that shape these interrelations so as to inform and provide tools for management and policy [9].

A recent journal editorial stressed the growing challenges of sustainability in food systems, given the increasing demand for food and the environmental impacts associated with modern food production [10]. The editorial referred to the relevance of trade policy and trade impacts on vulnerable communities, as well as to the need for globally-accepted metrics and policies for sustainability. This kind of narrative is representative of the general interest of the research community in studying and advancing sustainability tools for policy and decision-making. Agricultural and fishfood systems feed the world. We use the term "fishfood" to describe edible products from marine and freshwater fisheries and aquaculture. Despite the relatively small size of the global fishfood economic system in comparison to agriculture, it encompasses complex socio-economic networks with considerable impact of the world's environment. Economically, fishfood products represent about 10% of the value of total agricultural exports, and this percentage is increasing. Nutritionally, fish represent over 20% of animal protein intake in low-income and food-deficient countries [11,12]. Therefore, it is imperative to apply sustainability principles to the design, operation and assessment of fishfood systems.

"Fishfood system" is used here an umbrella term for complex systems producing fish directly consumed by humans, and closely interacting with surrounding aquatic and terrestrial ecosystems. Resource management science and research have produced a variety of approaches for capturing interactions between natural and socio-economic realms in such systems.

An essential feature of all approaches for understanding complex systems is modelling [13]. The ideal level/zone of complexity of modelling has been defined as the level of resolution at which essential real-world dynamics are included and analysis is not too burdensome [14] [15].

In fisheries, ecological processes such as predation, competition, environmental regime shifts, and habitat effects have the potential to impact bio-economic dynamics (e.g. recovery of exploited stocks, surplus production) [16]. These impacts may manifest themselves at an order of magnitude comparable to that exerted by fisheries pressure. Ecological/ecosystem modelling is a rich, well established research field; nonetheless, it is not always included in fisheries modelling and management [16]. Whole ecosystem models try to account for all trophic levels in the ecosystems studied. Some of the most notable examples are ECOPATH [17] and ECOSIM [18]. Currently, the most commonly used whole ecosystem modelling approach is probably Ecopath with Ecosim (EwE), a combination of ECOPATH, ECOSIM and a constantly increasing number of add-ons [19]. A software implementation of EwE is freely available for evaluating ecosystem impacts of fisheries [20,21]. EwE modelling is data-intensive, especially regarding biomasses and diets, and its outputs require interpretation to be used for policy-making support, among other limitations [21].

The "supply chain" is a concept used since the early 1980s to refer to dynamics between firms (value chains) contributing to the provision of a good or service. It encompasses all value chains, integrated or not, along the life cycle of the delivered product [22], as well as material, information and financial flows circulating among these value chains [23]. The supply chain concept is the ideal approach for studying today's economic organisations, immersed in a globalised world and both featuring and lacking vertical integration. Supply chain modelling is performed to understand, analyse and improve the efficiency, effectiveness and sustainability of supply chains. Supply chain modelling theory has been extensively applied to the study of food supply chains. Goals of supply chain modelling in food systems include cost reduction, safety, quality, flexibility and responsiveness, among other aspects [24]. Supply and value chain analysis, as well as modelling approaches, have been applied to fisheries, aquaculture and whole fishfood supply chains, as extensively reviewed in [25]. Non-modelling studies have focused on reducing costs, increasing efficiency and improving product quality, as well as (more recently) in developing or re-shaping existing supply chains [26].

Models oriented toward operations research have diverse objectives, depending on the system under study. In fisheries, aspects such as resource allocation problems, uncertainty management, harvest policy and strategy, harvest timing, quota decisions, experimental management regimes, investment in fleet capacity, and stock switching by fishermen [25] are studied. In aquaculture, trade-offs among alternative activities, strategic planning requirements for emerging technologies, planning and management, optimal harvesting time and other optimal control frameworks, feeding regimes, and risk management are studied [25]. Modelling of whole fishfood supply chains is less common; thus, it has been suggested that future research should focus on optimal production planning, costs associated with additional sorting of raw materials (due to the batch nature of many landed species) and quality aspects [24]. Past research has focused on handling and preservation practices for extended shelf life [26].

Although supply chain analysis and modelling of agrifood systems is quite common in research, modelling of fishfood supply chains is less common. Few models combine ecosystems and (fishfood) supply chains. The few social-ecological systems models applied to fisheries (as listed in [13]) and fisheries bio-economic models (e.g. those listed in [27,28]) are spatially explicit and include fishermen/vessel behaviour and their impact on management systems. Despite these few examples, most fisheries-related modelling research has historically focused on ecological (or ecosystem) modelling, that is to say, on ecosystem-fisheries interactions which do not explore socio-economic aspects.

Combining a fish supply-chain modelling approach with an EwE trophic model to model policy scenarios for stock recovery was first proposed in [29]. This approach was based on an idea later published in [30], in which a social-ecological system model combining ecosystem (using EwE trophic models) and proprietary value-chain modelling approaches was proposed. The model coupling (partial two-way interactions limited to the feedback effect of the producers on the ecosystem) proposed in [30] was eventually implemented as a plug-in for EwE 6.2. The coupled model was recently used in a case study [31]. We borrowed the one-way vs. two-way coupling wording and criteria from ecosystem modelling and used it to define the types of interactions between an ecosystem model and a material flow (supply chain) model (MFM). More details on the classification of modelling tools and justification of the models retained is presented in section A in File S1.

In this article, a sustainability modelling and assessment methodology is proposed and applied to compare several fishfood and agricultural supply chains that compete for Peruvian *anchoveta* (*Engraulis ringens*) resources. These chains generate a variety of impacts on Peruvian ecosystems and society, as well as on the global environment and economy. Therefore, we compare relative environmental and socio-economic performance of products from the chains and analyse alternative exploitation and fish fate (final fishfood product) scenarios. Ultimately, we track the fate of one t of landed *anchoveta* channelled through alternative Peruvian supply chains, now and in the future. The system under study encompasses the supply chains from extraction (fisheries and their impact on the Northern Humboldt Current ecosystem), through reduction activities for fishmeal and fish oil, aquafeed production (taking into account other agricultural inputs to aquafeeds), aquaculture, fishfood industries and, finally, to fishfood products on grocery shelves.

The dynamics of these complex supply chains have never been studied in a holistic, sustainability-imbued way. Understanding these dynamics and impacts to the largest extent possible is the motivation of this research, so that decision makers along the chains are informed and actions are taken to improve sustainability of the *anchoveta*-based fishfood fisheries and industries.

The research topic connects with the wider topic of sustainability assessment of food systems, and its importance derives from the prevalence of Peruvian fishmeal in international food supply chains, as Peru is by far the largest global exporter of fishmeal and fish oil used to supply aquaculture and animal production supply chains, mainly in Asia and Europe [12]. Simultaneously, since Peru is a developing country facing nutritional and social challenges, the fact that most fisheries landings are destined for reduction into fishmeal and fish oil is subject to discussion and multi-disciplinary analysis [31,32].

This article first introduces the Peruvian *anchoveta* supply chains and the sustainability assessment framework. It then presents the results obtained from applying the framework to the Peruvian case study by assessing and comparing the sustainability of supply chains and alternative exploitation scenarios. Finally, the methodology is discussed in relation to the results obtained, and suggestions for improving both the current and (possible) future situations are proposed.

Peruvian *Anchoveta* Supply Chains

The Humboldt Current System

The Northern Humboldt Current System (NHCS) identifies the tropical ocean area off Peru and north of Chile. The NHCS is considered the most productive fishing ground in the world because it produces more fish per area than any other region. Moreover, the NHCS has several unique characteristics that determine its productivity [33]. The NHCS is an eastern boundary upwelling ecosystem, extremely sensitive to climatic dynamics. Temperature anomalies, mainly associated with El Niño-Southern Oscillation (ENSO) and Pacific Ocean regime shifts, have historically produced huge changes in seabird populations and fluctuations in abundance of two numerically dominant species of pelagic fish: *anchoveta* and sardine (*Sardinops sagax*). *Anchoveta* is one of the world's largest exploited fish stocks.

The *anchoveta* fishery

The modern *anchoveta* fishery started in Peru around 1955, parallel with the decline of the previously profitable guano industry. The 1957–58 ENSO event decimated guano-producing seabird populations and coincided with further development of the *anchoveta* fishery. During the 1960s the fleet and the fishery grew continuously until 1970, peaking with the largest historical harvest of 12.3 million t, representing 20% of that year's world catch of all fish [33]. In 1972, the *anchoveta* stock collapsed, probably due to combination of high fishing pressure, a regime shift in the ecosystem and a strong ENSO event, followed by a slow recovery of the *anchoveta* stock and catches as well as changes in fisheries management and legislation [34] (Figure B1 in File S1). From 2000 to 2009, catches were stable compared to historical landings, averaging 7.1 million t per year. In 2010, an ENSO event and management measures reduced landings to 3.4 million t [12,35].

Currently, Peruvian fisheries are ruled by the currently valid Fisheries Act (Decree Law 25977 of 1992) and its applicable by-laws (Supreme Decree 012-2001-PRODUCE, Supreme Decree 005-2012-PRODUCE). The Peruvian purse-seiner fishery is the world's largest mono-specific fishery, both in landings and in number of vessels [33,36,37]. The fleet is heterogeneous. The industrial fleet (vessels with holding capacity >32.6 m^3) includes steel vessels and wooden vessels nicknamed "Vikingas". As of 2012, ~660 industrial steel vessels (operating directly under regime Decree Law 25977) target *anchoveta* for reduction (i.e. for fishmeal plants). Additionally, almost 700 Vikingas (operating under regime Law No. 26920) also target *anchoveta* for reduction. The small-scale fleet includes vessels with holding capacity < 10 m^3, while the medium-scale fleet has vessels with holding capacity of 10–32.6 m^3. Small-scale vessels also differ from medium-scale ones in the level of technology and capture systems used; small-scale vessels are characterised by manual labour and basic technology [38]. In total, the small- and medium-scale (SMS) fleet includes about 850 wooden vessels that by law target *anchoveta* (among other species) only for direct human consumption (DHC), but also illegally for reduction fishmeal plants. As a result, a small percentage of national catches is rendered into seafood products for DHC, according to both official PRODUCE statistics and IMARPE comprehensive data [34] as detailed in [39,40]. PRODUCE is the Peruvian Ministry of Production (www. produce.gob.pe/), while IMARPE is the Peruvian Marine Research Institute (Instituto del Mar del Perú, www.imarpe.pe/ imarpe/), a public institution leading national research on marine resources and the marine environment.

Catches by the steel fleet represent around 81% of the total *anchoveta* catches for reduction, while the Vikingas capture 19%, according to IMARPE statistics (Marilú Bouchon, unpublished data). The industrial fleet landings for indirect human consumption (IHC) (i.e. reduction) represent >99% of total catches, while SMS fleet landings for DHC (fresh, freezing, canning, curing) represent <1% of total catches, according to PRODUCE statistics, as summarised in Table 1.

Overcapitalisation/overcapacity affects the *anchoveta* fleets, largely due to the existence of a semi-regulated open access

Table 1. Statistics for *anchoveta* landings and processing (2001–2011).

Year	2001	2002	2003	2004	2005	2006	2007	2008	2009	2010	2011	Average
Anchoveta landings	6 358 217	8 104 729	5 347 187	8 808 494	8 655 461	5 935 302	6 159 802	6 257 981	5 935 165	3 450 609	7 103 061	6 556 001
Anchoveta for reduction	6 347 600	8 082 897	5 335 500	8 797 100	8 628 400	5 891 800	6 084 700	6 159 387	5 828 600	3 330 400	6 994 051	6 498 221
Fishmeal production[a]	2 034 900	1 562 116	1 416 500	1 807 000	2 067 900	1 367 900	1 284 500	1 585 600	1 584 100	1 119 300	1 235 674	1 551 408
National consumption	91 800	46 686	43 700	53 600	66 400	25 400	20 700	20 800	36 700	33 600		39 944
Exports	1 943 100	1 515 430	1 372 800	1 753 400	2 001 500	1 342 500	1 263 800	1 564 800	1 547 400	1 085 700		1 399 130
Fish oil production	447 200	206 150	267 508	363 000	339 400	346 773	371 600	280 400	335 000	320 800	248 637	320 588
National consumption	131 800	45 245	80 800	78 200	60 600	58 200	65 900	41 800	46 800	69 700		61 731
Exports	315 400	160 905	186 708	284 800	278 800	288 573	305 700	238 600	288 200	251 100		236 253
Anchoveta for DHC	10 617	21 832	11 687	11 394	27 061	43 502	75 102	98 594	106 565	120 209	109 010	57 779
Canning	3 286	13 364	4 823	2 631	14 887	31 000	61 944	78 851	84 957	94 234	84 194	43 106
Freezing	1 137	4 326	655	214	1 405	1 268	5 286	12 265	11 517	15 160	14 680	6 174
Fresh fish	398	9	392	320	348	538	401	336	293	223	44	300
Curing	3 717	4 132	5 806	8 194	10 425	10 658	7 459	7 142	9 762	10 579	10 092	7 997

[a]all species, >90% *anchoveta*. Based on PRODUCE data [48,117,121]. DCH: Direct Human Consumption.

system that existed until the 2008 fishing season (inclusive) and a single national quota (Total Allowable Catch, TAC) that is revised each fishing season. Overcapitalisation is still substantial in Peru; in 2007 the fishing fleet was estimated to be 2.5–4.6 times its optimal size [41].

The Peruvian *anchoveta* fishery operates in two well-defined coastal areas in the South Pacific, as determined by the species habitat and behaviour: the north-central area (from 4°–14° S) and the south area (from 15° to ~18° S, which continues in Chile from parallels ~18° to 24° S). More detailed descriptions of the industrial steel, semi-industrial and SMS fleets are presented in [42], while discussions on their environmental performance are presented in [39].

The reduction industries

Fishmeal plants produce fishmeal as the main product and fish oil as co-product. Inclusion of fishmeal and fish oil in aquafeeds has decreased [43] as alternative protein sources have become available and their effectiveness has been demonstrated. Nonetheless, the demand for fish reduction products has remained constant due to expansion of aquaculture, which consumed 61% of all fishmeal and 74% of all fish oil produced in 2008, and to continuous growth of livestock feed and pet food industries [12,43,44]. Peruvian fishmeal and fish oil represented 40–47% and 34–47% of the world's supply from 2007–2011, respectively [45].

In Peru, more than 98% of fishmeal produced is derived from *anchoveta*. Plants can be classified into conventional, high-protein and residual, according to the technology used and product quality obtained ([46,47]). Peruvian product labels describe "fair average quality" fishmeal (~64% protein), dried with direct heat, "high protein content" fishmeal (67–70% protein), dried with indirect heat (steam, hot air), and residual fishmeal (processing residues, ≤ 55% protein), dried with direct heat. Peru had 160 industrial fishmeal plants in 2012, but not a single registered artisanal fishmeal plant, according to [48]. Fifty percent of plants are concentrated in the northern coastal region, mainly in Chimbote and Chicama [49].

The reduction industry suffers from overcapacity: in 2007 the industry was 3–9 times its optimal size [41]. Although the 1992 General Fisheries Act prohibited further increase in capacity of reduction plants, the overcapacity issue was worsened by privatisation of the sector in the 1990s and many mergers and acquisitions from 2006–2008 that concentrated the sector [41]. A shift towards better technology, and thus better and more lucrative products, is noticeable in the increase in high-protein fishmeal processing capacity and production (fair average quality fishmeal from 37.6% in 2010 to 34.0% in 2011; prime fishmeal from 62.4% in 2010 to 66.0% in 2011) [50,51].

Production and export of fishmeal and fish oil is the main driver for the thriving *anchoveta* industry. Peruvian fishmeal and oil are exported, among other aquaculture-producing countries, to China, Chile and some European countries. The main users of these imports are farms producing shrimp, salmonids, carp, tilapia and other cultivated species. It has been suggested that Chinese carp cultures may be the largest single consumer of fishmeal, despite low inclusion rates in feeds, due to the enormous volume of production [12,52]. Other authors suggest shrimp farming in China as the main consumer (Patrik Henriksson, SEAT, pers. comm., 2012).

The fish-to-fishmeal conversion ratio in the Peruvian industry has increased from more than 5:1 in the early 1990s to ~4.2:1 in recent years. Conversion ratios below 4.2 are considered impossible in the Peruvian context [41]. Table 2 compares several

reported conversion ratios. Fish oil conversion ratios fluctuate greatly because they depend on the lipid content of *anchoveta*, which varies over time. The mean yield from 2001–2011 was 21.3:1, as calculated based on statistics from PRODUCE and [48]. A more detailed discussion about the reduction industry is under preparation by our team (The Anchoveta Supply Chain project, ANCHOVETA-SC, http://anchoveta-sc.wikispaces.com).

The processing industry for food

Peru surpassed 30 million inhabitants in 2012 [53], more than 70% of whom live in urban areas. Annual per capita fish consumption was estimated at ~19 kg in 2005 and 23 kg in 2009. Consumption is notably higher along the coast (seafood) and in Amazonian areas (river fish), while it is much lower in the highlands (industrialised fish products and Andean aquaculture) [54].

The amount of fresh *anchoveta* landed for DHC has increased in the last decade at a mean annual rate of 37%, according to PRODUCE statistics. Nonetheless, DHC of only 1–2% of landings is low in a country with a large percentage of its population suffering from malnutrition [36]. It has been suggested that increased DHC of *anchoveta* could help solve some of the nutritional problems in Peru and the larger region [55].

Peruvian consumption of *anchoveta*, despite its recent increase, is still relatively small (3.3 kg per capita in 2010), yet it represents, on average, >70% of *anchoveta* DHC products. The scarcity of *anchoveta* for DHC is due to a combination of factors, including regulatory limitations (industrial vessels cannot supply the DHC industry), consumer preferences and lack of a cold chain for fish in Peru. Some believe a key factor is the shelf price of *anchoveta* DHC products. Moreover, one of the factors that direct or divert (for SMS captures) most *anchoveta* landings to reduction is the small difference, if any, in prices paid to fishermen per t of fish landed [32]. Fishmeal plants paid more than DHC plants until recently. Additionally, to keep *anchoveta* acceptable for DHC, vessels must carry ice, which reduces their holding capacity by at least 30%. These topics are further analysed in [32]. More detailed discussion of Peruvian *anchoveta* processing for DHC is presented in [56].

Key *anchoveta*-based aquaculture systems in Peru

In Peru, aquaculture has been and is still dominated by scallops (*Argopecten purpuratus*) and shrimp (mainly *Litopenaeus vannamei*) for marine species and by trout (mainly *Oncorhynchus mykiss*), tilapia (*Oreochromis* spp.) and black pacu (*Colossoma macropomum*) for freshwater species [57,58]. Marine aquaculture contributes ~81% of Peruvian cultured fishfood production, while freshwater production represents ~19% [59].

Peruvian aquaculture, mostly represented by small-scale or artisanal practices (~63% of total production in 2010 [59]) has featured continuous growth over the last 20 years. Most trout culturing operations are artisanal yet semi-intensive, especially those in the Puno Department (Lake Titicaca and nearby water bodies), where most national production takes place. Trout farming in Puno department water bodies consist of artisanal wood- or metal-nylon floating cages (800–2000 kg carrying capacity) and larger metal-nylon floating cages (up to 6 000 kg carrying capacity). Trout is destined mainly for export, despite increasing consumption in the producing areas and larger cities of Peru, particularly Lima. Black pacu are cultured mainly in large, semi-intensive artificial pond systems, while tilapia is produced using a variety of methods and operational scales, mostly intensive. Black pacu is almost exclusively cultured in the Amazonia (Loreto and San Martin Departments) and tilapia in the Piura region. Black pacu is mostly consumed locally, mainly because of the

Table 2. Fish to fishmeal and fish oil conversion ratios: national averages from 2001–2006.

Countries	Landings (1 000 t)	Fishmeal (1 000 t)	Fish oil (1 000 t)	FM ratios A	FO ratios	FM ratios B	Species used for reduction (common names)
Chile	3 161	773	157	4.09	20.13	3.93	Jack mackerel, *anchoveta*, sardine
China	2 041	769	-	2.65	-	3.80	Various
Denmark	881.5	327	106	2.70	8.32	3.01	Sandeel, sprat, blue whiting, herring
Iceland	1 262	221	74	5.71	17.05	4.44	Blue whiting, herring, trimmings
Japan	1 141	226	66	5.05	17.29	4.45	Sardine, pilchard
Norway	1 061	203	47.5	5.23	22.34	5.11	Blue whiting, capelin, trimmings
Peru	7 561	1 700	270	4.45	28.00	4.54	*Anchoveta*
Peru (this study)[b]	**6 498.2**	**1 551.4**	**320.6**	**4.21**	**21.30**		*Anchoveta*
United States	909	258	88	3.52	10.33	3.61	Menhaden, Alaska pollock

Notes: Landings, fishmeal (FM) and fish oil (FO) production, FM ratios A and FO ratios were taken from [122]. FM ratios B were taken from [123], for the period 2000–2005. Data for this study were taken from PRODUCE reported landings and production values for the period 2001–2011 [117].

physical isolation of the Amazonian communities that produce it. Tilapia was historically destined for national markets, but over the last decade increasing proportions of production have been exported.

Among these types of culture, shrimp aquaculture is the main consumer of fishmeal, given high percentages of fishmeal (20–50%) in commercial feeds [43,60–62] and production volumes. As in other fish farming systems, a key aspect of Peruvian aquaculture is feed supply. In Peru, both artisanal and commercial feeds are used, but the latter prevail, especially for trout. National production of aquaculture products in Peru was estimated at 89 000 t in 2010, whereas national consumption was estimated at 0.52 kg per capita (~15 000 t for a population of 29 million), yet a growth pattern in consumption of 22% per year has been recorded [59]. A more detailed discussion of Peruvian (freshwater) aquaculture is presented in [63].

Distribution channels

Distribution channels for fisheries for DHC consist of 1) landing in several fishing ports and piers, both private and public; 2) transportation of fish in isothermal trucks, often organised by wholesalers; 3) processing in DHC plants; and 4) distribution to retailers for national consumption and export to foreign markets [64]. Most landing facilities for DHC have never met the requirements set by the sanitary standard for fisheries and aquaculture resources, as established by Supreme Decree 040-2001-PRODUCE [64]. The lack of a cold chain for fish in Peru is a major factor limiting further development of domestic distribution channels.

Peruvian aquaculture products are distributed within Peru by retailers (e.g. distributors, markets) and exported by producers or specialised exporting firms. Wholesaler markets concentrate ~29% of total landings destined for fresh fish, 3.2% of which are not captured by Peruvian vessels but imported from neighbouring countries (mainly jack mackerel, *Trachurus murphyi*). In coastal areas, wholesaler markets supply retailers, supermarkets, restaurants and final consumers, although this does not apply to the scarce supply of fresh *anchoveta*. Lima alone accounts for 32% of national fish consumption.

Regarding canned fish, both processing plants and importers supply wholesalers, who subsequently supply supermarkets and retailers. Five percent of canned fish consumed in Peru is either imported as final product or as frozen fish to be processed in Peru, mainly tuna from Ecuador. Frozen food products are both produced in Peru and imported. Imports, representing ~60% of frozen fish consumed in Peru, largely consist of jack mackerel (when national production of this highly fluctuating resource is too low) from Chile and tuna from Ecuador. Producers and importers supply wholesalers, who subsequently supply restaurants and supermarkets across the country (transported mainly in refrigerated trucks). Cured and salted products are both produced in Peru and imported, notably anchovy from Argentina (18% of national consumption of cured products). Producers and importers directly supply markets across the country.

Fisheries management and policy environment

IMARPE provides the scientific foundation for fisheries management in Peru, which is implemented by PRODUCE [65]. IMARPE struggles between scientific and political considerations for its recommendations due to its relationship with PRODUCE (e.g. IMARPE's Chairman of the Board is a political, rather than technical, position) [66].

IMARPE estimates the *anchoveta* population off Peru and recommends an annual TAC to PRODUCE [55]. This estimate is based on 1) hydro-acoustic data collected since 1975 from 2–3 annual surveys of the entire Peruvian coastline and 2) modelling of *anchoveta* population dynamics as a function of environmental conditions and recruitment levels using Virtual Population Analysis based upon a bio-economic age-structured model [67]. The recommended TAC is related to the Maximum Sustainable Yield. Spawning biomass is calculated using the Egg-Production Method (a meta-review is available in [68]).

Since the north-central stock contains >90% of the *anchoveta* biomass, most regulation and legislation applies only to it, leaving the south stock to be exploited under an open-access regime (featuring closures related to the proportion of juveniles in the total population). Fisheries legislation has been introduced since the early 1990s, and currently fisheries are mostly managed in an adaptive-reactive manner, with mixed effects. For instance, the decrease in catches to 3.4 million t in 2010 was due mostly to management measures applied to protect a large juvenile ratio. Because of that management decision, 2011 catches exceeded those of 2009 [12].

Other effects of legislation are still unfolding in the Peruvian *anchoveta* fishery and reduction industries. For instance, before 2008 legislation introducing individual vessel quotas (IVQ), up to 1200 vessels competed for the TAC in a so-called "Olympic race", reducing the annual fishing season to 50 days [41,69]. A list of key historical legislation governing fisheries in Peru is available in Table B1 in File S1. Fishing companies have reacted to the IVQ regime in various ways. For instance, large vertically integrated companies encompassing fishing and reduction are using their more efficient vessels to harvest their company-wide quotas (since IVQ are transferable within the same company) [41,69]. As intended, this will eventually reduce fleet overcapacity, but has generated several other negative consequences [41,70].

Most legislation regulates the activities of industrial, large-scale vessels, while the SMS fleets are poorly regulated and practically operate in an open-access regime [71]. Regulations on SMS fisheries include the exclusive use of the sea within 5 nautical miles (9.3 km) off shore, holding capacity, length, manual labour, mesh size of nets, prohibition of beach seines, minimum catch sizes for some species, and protection for cetaceans, turtles and seabirds [71,72].

Some researchers consider that legislation related to Peruvian *anchoveta* is either insufficient, ineffective or poorly enforced [35,41,66], a situation affecting all *anchoveta* fleets. Moreover, several issues permeate the enforcement of Peruvian fisheries legislation and management guidelines (based on publications, pers. comm. with various researchers and experts, as well as on journalistic pieces), including the following:

- few data exist for smaller scale operations (Juan Carlos Sueiro, pers. comm., 2013)
- illegal, under-reported and un-regulated (IUU) landings are common [70]
- illegal reduction plants operate profusely, partially supplied by IUU landings (Pablo Echevarría, pers. comm., 2013)
- illegally produced fishmeal is "washed" by brokers
- regulations that mandate proper solid and liquid waste management from fishing vessels and processing plants are generally ignored
- capital and bargaining power are concentrated in a handful of vertically integrated companies
- SMS fisheries pay no fishing rights and have no quota assigned, while the money that industrial operations pay for fishery rights is clearly insignificant compared to their profits

and insufficient to finance fishery regulation, supervision and control [47,70,73–76]

Despite these problems, Peruvian fisheries are generally considered among the most sustainably managed in the world [67,77,78], mostly because of their adaptive and reactive management measures that compensate for deficiencies in the legislation and management system. This management relies mostly on acoustic surveying-based annual quotas and on-demand fishery closures.

Socio-economic dynamics

Fisheries and seafood products, especially exports of fishmeal and fish oil, represent the third largest individual source of foreign income for the Peruvian economy (on average, 8% from 2000–2011) [79]. China and Germany are the largest importers of Peruvian fishmeal, while Denmark and Chile are the main importers of fish oil. Most Peruvian fishmeal, most of which is high-quality, is destined for aquafeeds. In terms of employment, industrial and SMS fisheries, as well as reduction and other fish-processing industries, provide a large number of jobs. It is difficult to isolate the jobs associated exclusively with the extraction and processing of *anchoveta*, other than those in the reduction industries. Nonetheless, [80] estimated the number of jobs directly associated with the *anchoveta* industrial and SMS fleets at 10 000 and 8 000, respectively. Recently, employment in the Peruvian fisheries and processing sector was estimated more comprehensively [31]. These and other socio-economic indicators of *anchoveta* supply chains (gross profit generation, added value) are presented and discussed in [81].

Nutritional value of fishfood products of *anchoveta* supply chains

According to the FAO and the Global Hunger Index [82–84], Peru has advanced in hunger reduction, yet remains one of the few Latin-American countries with moderate hunger. The International Food Policy Research Institute (IFPRI) defines "moderate hunger" as a level of hunger associated with a Global Hunger Index value of 5–10 out of 40. This index is built by combining three equally weighted indicators (undernourishment, child underweight and child mortality; as defined by FAO) [82,84]. According to the FAO, hunger is associated with poverty [85]. Especially in Andean communities, indicators such as chronic malnutrition of children under five, stunting and undernourishment are still elevated [82,85,86], and thus government policies should be (and to some extent are being) oriented to provide these communities with cheaper sources of animal protein and improve access to nutritious food.

Seafood, especially that derived from the thriving *anchoveta* supply chains, has often been suggested as a suitable means to improve nutritional intake of vulnerable communities and people at large. Fishfood products of the *anchoveta*-based supply chains include *anchoveta* products as well as marine and freshwater aquaculture products. *Anchoveta* products are extremely high in beneficial omega-3 fatty acids, mineral salts and essential amino acids [55]. Further discussion on nutritional values of *anchoveta* and other Peruvian fishfood products is presented in [81].

Ecosystem and bio-economic modelling of the Peruvian *anchoveta* fishery

The NHCS ecosystem and its sensitivity to environmental conditions, often emphasising population dynamics/stock assessment of commercially important species (e.g. *anchoveta* [87,88] and Pacific hake (*Merluccius gayi*) [89] an *anchoveta* predator.) or threatened species (e.g. fur seals [90]), has been modelled since the 1970s [91,92]. A preliminary EwE [17,18] trophic model of the NHCS was presented in [93], highlighting that natural predators contribute more to total *anchoveta* mortality than fisheries. On the other hand, hake mortality, for instance, is due mostly to fisheries. A more comprehensive EwE-based trophic model was later presented by [94,95], which discusses trophic and ecosystem dynamics under El Niño and La Niña conditions. The model by [94] was used to apply the ecosystem approach to hake and *anchoveta* fisheries [96,97] and is currently used in the project IndiSeas [98]. Currently, these trophic models are not used for management because they are considered to be under development and to lack comprehensive data. Several bio-economic models have been also developed for the Peruvian *anchoveta* fishery [99], some of which have been used to estimate stock biomass and calculate the TAC. In recent years, new age-structured and integrated assessment models have been used by IMARPE [88].

The proposed framework

The proposed framework is based on a one-way coupled model of the ecosystem and the supply chains that exploit it. It aims to provide tools and rationale for assessing and comparing current and future exploitation strategies of *anchoveta* and *anchoveta* supply chains by means of trophic, biophysical and socio-economic modelling.

A one-way coupled ecosystem/supply chain model

We propose an enlarged framework featuring an integrated ecosystem/supply chain model by combining existing models towards a holistic depiction of the ecosystem/seafood system interactions. This framework depicts flows and stocks of materials and energy occurring through the supply chain (from ecosystem to product retailing) and selected socio-economic elements (Figure 1). The proposed framework follows previous endeavours [29–31] in selecting EwE as a suitable ecosystem modelling platform to be coupled in a one-way or two-way manner with mass/socio-economic models. The frameworks differ in the approach for modelling supply chains. Our approach de-emphasises economic flows and highlights flows associated with the sustainability indicators selected to better describe sustainability performance of the system. The framework intends to assess overall sustainability, yet emphasises its environmental dimension, mainly due to data availability. We consider the proposed coupled model as an example of "ecosystem-based supply-chain modelling". Moreover, the goals of both approaches differ as well: the value chain analysis in [30] accounts for socio-economic benefits of fisheries and subsequent links in the value chain, while our analysis compares the relative sustainability performance of competing fisheries-based supply chains.

In our framework, monetary flows are analysed at the industrial-segment level rather than at the value chain level; that is to say, no individual firms are modelled, but rather whole production sectors (e.g. fisheries, reduction industry, species-specific aquaculture sector) by aggregating and generalising individual firm results.

An EwE trophic model of the marine ecosystem exploited by the modelled supply chain can be used as the base ecosystem model. The outputs of the EwE model would feed a material and energy-flow model, which could be built, for example, with Umberto, a modelling tool specifically designed to study material flow networks [100]. Umberto represents material flow networks as Petri nets, that is to say, in terms of transitions (transformational processes), places (placeholders for materials and energy) and arrows (flows). These are the modelling tools/approaches that we

selected, but almost any combination of a whole-ecosystem model and a MFM would be suitable, especially if the coupling could be established in a dynamic fashion (i.e. models interacting in real time during simulations).

The framework has three main phases (Figure 2): 1) characterisation and modelling of the fishfood system under study, 2) definition and calculation of sustainability indicators, 3a) comparison of competing supply chains, and 3b) definition and comparison of alternative policy-scenarios for the set of all supply chains. Phases 1 and 2 are to a certain extent concurrent, since the selection of sustainability indicators largely determines the direction and complexity of system characterisation (data collection and processing).

In Phase 1, material, energy, nutritional and monetary flows of target supply chains, both short (DHC products) and long (aquaculture), are modelled. In Phase 2, a set of suitable sustainability indicators is compiled to compare the performance of supply chains modelled in Phase 1, as detailed and illustrated for a subset of *anchoveta* supply chain-derived products in [81]. In Phase 3, supply chains are compared and policy-based scenarios for future exploitation and production are defined and contrasted.

Since the main goal of the characterisation stage is to inform sustainability assessment of complex anthropogenic systems directly interacting with ecosystems, the characterisation must include both biophysical and socio-economic flows. The study of biophysical flows illustrates ecosystem/industry interactions and provides data about flows and stocks of materials and energy occurring along the supply chain, including their effects on the environment. In contrast, analysis of socio-economic flows offers insights about social and economic dynamics occurring parallel to the material ones. By understanding the system from at least these three perspectives, sustainability can be evaluated.

Supply chain characterisation and modelling

The biophysical accounting framework used to model supply chains was Life Cycle Assessment (LCA). LCA is a mature approach, and current Life Cycle Impact Assessment (LCIA) methods encompass a great diversity of environmental impact categories. Socio-economic aspects would ideally be assessed by combining life cycle methods and economic analysis frameworks, such as Life Cycle Costing (LCC), Social LCA and cost-benefit analysis. Nonetheless Social LCA is not yet mature, and it is usually difficult to obtain all the data required from fishery and fishfood industries to apply it [101,102]. Not enough data were available for LCC or cost-benefit analyses.

Several LCA studies were required to characterise environmental impacts and resource consumption (including energy use) of components of fish supply chains: fisheries, processing for DHC, reduction into fishmeal and fish oil, aquaculture and distribution. LCAs were performed using the software SimaPro [103], which features integration with the widely used database ecoinvent [104] and various LCIA methods, including CML baseline 2000 [105], ReCiPe [106], Cumulative Energy Demand [107] and USEtox [108]. LCA methodology and results associated with *anchoveta* supply chains are presented in [39,40,42,56,63].

LCA results (including additional and fishfood-specific impact categories and other Life Cycle Inventory-based indicators), EwE outputs and socio-economic performance indicators become inputs for the Umberto modelling environment. Umberto outputs include mass and energy balances and flow diagrams (e.g. Sankey diagrams).

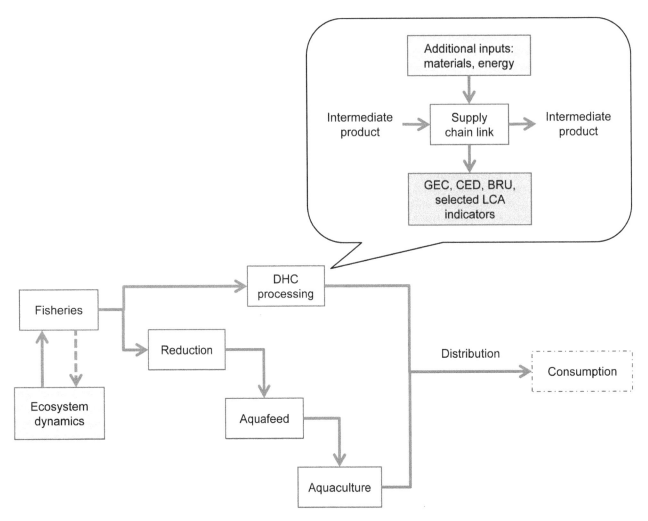

Figure 1. Simplified one-way coupled ecosystem/supply chain model. The zoom view illustrates how industrial processes and sub-processes are detailed within the supply chain. Environmental and socio-economic impacts of a given link of the supply chain are carried to the next link.

Definition and calculation of the indicator set

Once the target supply chains are modelled based upon detailed operational and socio-economic data, a set of sustainability indicators is calculated to assess sustainability and compare alternative supply chains (e.g. DHC vs. IHC chains based on the same fishery).

Several sustainability indicators were selected from the large indicator pool available in the literature so that all aspects of sustainability –especially the environmental dimension, but also energy efficiency, human nutrition and socio-economic factors– were addressed. Main criteria for this selection were historical use in the fishfood research field; purpose (mainly environmental plus key socio-economic aspects); practicability, given data availability; and comparability with other food systems. Table 3 lists the indicator set, introduced and detailed in [81], and expanded in this study with a few IndiSeas ecological indicators [109,110] to compare alternative states of the exploited ecosystem. The indicators "Trophic level of landings", "Proportion of predatory fish" and "Inverse fishing pressure" can be used to measure maintenance of ecosystem structure and functioning, conservation of biodiversity and maintenance of resource potential, respectively (Eq. 1, 2 and 3 [110]):

$$TL_{land} = \sum_{s}(TL_s \cdot Y_s)/Y \qquad (1)$$

where TL is the trophic level, Y is catch and s is species.

$$\textit{Proportion of predatory fish} = \\ \textit{Biomass of predatory fish}/\textit{Total biomass} \qquad (2)$$

where *Total biomass* includes the biomass of demersal, pelagic and commercially relevant invertebrates.

$$\textit{Inverse fishing pressure} = (\textit{Landings} / \textit{Biomass})^{-1} \qquad (3)$$

where *Landings* and *Biomass* refer to those of the species selected. For these three indicators, a larger value represents in principle a healthier ecosystem (but see discussion).

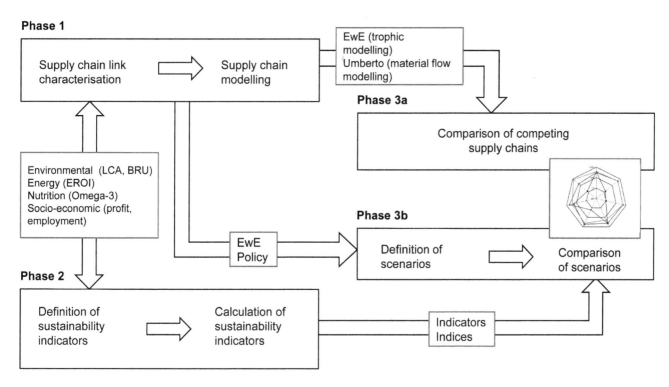

Figure 2. Proposed sustainability assessment framework for seafood supply chains.

Table 3. The sustainability indicators proposed, per dimension of sustainability addressed.

Sustainability dimension	Indicator (unit)	Reference publications	Calculation
Ecological	$I_{BNR,sp}$ (years)	[124]	Manual
	$I_{BNR,eco}$ (years)		
	TL_{land}	[110]	
	Proportion of predatory fish		
	Inverse fishing pressure		
Ecological/environmental	BRU (g C kg^{-1})	[125]	Manual
	BRU-based discard assessment	[126,127]	
Environmental	LCA/ReCiPe (Pt)	[128]	LCIA methods
	LCA/CED (MJ)	[107]	
	LCA/CML [USES-LCA] (kg 1,4-DB eq)	[105,129]	
	LCA/USEtox (CTU)	[108]	
Nutritional	GEC (MJ kg^{-1})	[130]	Manual
	Nutritional profile	[131]	
Energy efficiency	Gross edible EROI (%)	[130,132,133]	Manual
	Edible protein EROI (%)		
Socio-economic	Production costs (USD)	[134]	Manual
	Employment (USD)		
	Value added (USD)		
	Gross profit generation (USD)	Accounting concept	

Abbreviations: BRU: Biotic Resource Use, CED: Cumulative Energy Demand, CTU: comparative toxic units, EROI: Energy Return On Investment, GEC: Gross Energy Content, $I_{BNR,sp}$: impacts on Biotic Natural Resources at the species level, $I_{BNR,eco}$: impacts on Biotic Natural Resources at the ecosystem level, LCA: Life Cycle Assessment, LCIA: Life Cycle Impact Assessment, TL_{land}: Trophic level of landings. Modified from [81].

Definition and comparison of policy-based scenarios

In the context of fishfood research, comparing the sustainability of competing or alternative exploitation scenarios could inform decision making. Figure 3 illustrates proposed scenarios for comparing sustainability of fishfood supply chains, using the typology discussed in [111] (see also section A in File S1).

By integrating the ecosystem compartment in the supply chain model, it is possible to predict, for instance, changes in stock related to changes in exploitation regimes. Changes in stock (e.g. stock recovery) are not only linked to fishing pressure, but also to ecological processes [16]. The EwE model features biological processes such as respiration and predation. It can represent environmental regime shifts and ENSO events as different scenarios (e.g. states of the NHCS in an El Niño and non-El Niño year [94]), although this was not explored here. The integration can also help estimating overall environmental impacts associated with alternative fates of landed fish.

Supply chains and policy-based scenarios are compared based on functional units, typically one t of fish (live weight) produced or processed. Supply-chain-wide flow analyses and product comparisons by means of the sustainability indicator set are the comparison tools. Visualisation devices include mass and energy balances, tables, Sankey diagrams [112,113] and graphs.

Data sources and methods

Establishing inventory data for LCAs was the most data-intensive endeavour in this study. Most background processes had been previously modelled in ecoinvent and reference publications. Data were collected in Peru from 2008–2013 in the ANCHO-VETA-SC project, in cooperation with PRODUCE, IMARPE, the Research Institute of the Peruvian Amazonia [114], a trout development project from the Puno regional government [115], Peruvian universities, various large fishing and reduction enterprises –organised into the National Fisheries Society [51]–, as well as many confidential and anonymous sources. Detailed statistics and operational data about all key links in the complex *anchoveta*-based supply chains were gathered. Moreover, experts and analysts of the *anchoveta* industries were also approached, and

historical datasets obtained from them, some including data from a large enterprise no longer in operation but whose vessels were operated by other companies. Surveys were extensively used to obtain data, particularly from industrial and SMS fisheries. Field visits included fishing ports, fishmeal plants, fish processing plants, aquaculture farms and shipyards. Details about all data sources used are presented in [39,40,42,56,63,81].

A screening-level LCA (Life Cycle Screening, LCS) of the industrial hake fleet was performed using literature data and landings statistics from PRODUCE and IMARPE (R. Castillo, pers. comm., 2013; R. Adrien, pers. comm., 2013). This screening relied heavily on assumptions, since detailed data on Peruvian hake fisheries was not available. Based on these uncertain data, sustainability indicators were calculated so as to compare the fishery of this carnivorous fish with those of *anchoveta* and another carnivorous fish (farmed trout), as well as their respective products.

The ecosystem model used is based on the above-mentioned EwE trophic models of the NHCS by [94,95]. The model domain extends from $4°–16°$ S and 60 nautical miles (111 km) offshore, covering an area of $\sim$165 000 km^2 and including 32 living functional groups. The model was fitted to historical time-series data of biomass and catch of main fishery resources from 1995–2003. After the historical period, scenario simulations were run for the period 2004–2033. A key feature of the EwE scenarios was the behaviour of *anchoveta* and hake biomasses. Observed and fitted *anchoveta* biomasses decreased during El Niño in 1997–1998, recovered in 2000, and fluctuated until stabilising around 70 t km^{-2}. On the other hand, hake biomasses also decreased during El Niño, but recovered more slowly in 2006 and stabilised around 1 t km^{-2}.

Figure 4 lists the alternative exploitation scenarios derived from the EwE simulation. These scenarios were recommended in [32]. Two types of scenarios seemed suitable, both policy-induced: 1) changes in fish fates (DHC vs. IHC) and 2) changes in landings and landing composition. Therefore, three alternative exploitation scenarios were derived from the EwE model, projecting the reference year (2011) into the future:

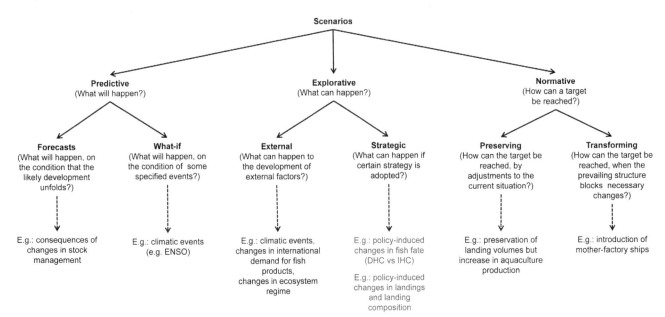

Figure 3. Types of scenarios suitable for seafood sustainability research. Based on [111]. Examples in red represent the preferences of this research. DHC: direct human consumption; IHC: indirect human consumption (i.e. reduction).

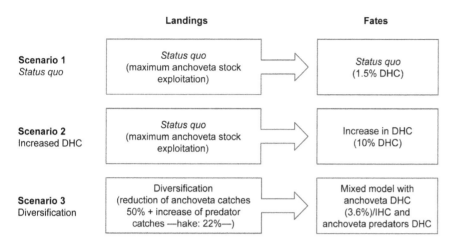

Figure 4. Alternative exploitation scenarios.

- Scenario 1 (S1) - *Status quo*. This is an extrapolation of the current situation (2011) in which the *anchoveta* fishery is fully developed and landings oriented to DHC remain low, varying from 1.5% in the reference year 2011 to 3.6% in 2021. The increase in the percentage of DHC represents an extrapolation of the current slightly increasing trend. After the historical period, *anchoveta* and hake fishing mortality are set constant and equal to the last historical value.

- Scenario 2 (S2) - Increased DHC. The same fully developed *anchoveta* fishery as in Scenario 1, but 10% of the landings are oriented to DHC. *Anchoveta* and hake fishing mortalities are the same as in Scenario 1.

- Scenario 3 (S3) - Diversification. In this scenario, *anchoveta* exploitation decreases and exploitation of hake, increases. *Anchoveta* landings are oriented both to DHC and IHC, and hake landings are oriented to DHC. From the end of the historical period onwards, *anchoveta* fishing mortality was linearly decreased to 50% over the next ten years (2013), then hake fishing mortality was linearly increased to 22% over the next ten years (2023); afterwards, fishing mortalities were kept constant for 10 more years (2033) to stabilise EwE outputs.

EwE modelling provides the ecosystem perspective of these scenarios, while LCA-derived and other indicators based on a functional unit can easily be scaled up or down to varying production volumes. The one-way coupling between the EwE model and the MFM, built with Umberto [100], is mono-directional, since dynamic linking was not feasible. EwE outputs are inputs to the MFM model, but changes in the MFM model cannot influence the EwE model directly. Therefore, the one-way coupled model was used to model the current situation and alternative fish-exploitation scenarios. Nonetheless, the MFM model can be used alone, as a supply chain modelling tool to explore variations within a defined scenario (e.g. changes in relative production volumes of aquaculture products or *anchoveta* DHC products). The Umberto project file containing the MFM model is available upon request to the corresponding author.

For the alternative exploitation scenarios, changes in the percentage of *anchoveta* landings destined to DHC and in aquaculture production were modelled for future years by extrapolating historical landing and production data [116,117] using statistically representative trend lines. Operational costs and prices were not extrapolated due to a lack of detailed annual data. Eventual changes in captures per unit effort (CPUE), which is

accepted to be proportional to changes in biomass and fish catchability (affecting fuel-use intensity), were considered, in such a way that all environmental modelling in this study is based on CPUE-adjusted fuel use intensities.

The coupled trophic/supply chain model is fed from several models: the EwE trophic model of the NHCS, LCAs of each link in the *anchoveta* supply chain, and additional sustainability and nutrition indicators (Table 4).

Results

Comparison of current supply chains

The proposed ecosystem/supply chain model produced an overview of the sustainability of the entire *anchoveta* supply chain. The MFM is presented in Figure B2 in File S1. All studied products were ranked (Figure 5), including distribution at the national level of fisheries-DHC and aquaculture products. Fresh *anchoveta* and low energy-intensive *anchoveta* products perform better from a sustainability perspective than other products. [81] presents a more detailed comparison of *anchoveta* DHC and aquaculture products, representing the current status of these supply chains.

When including national distribution of DHC products (using refrigerated chains when necessary), overall environmental performance (represented by the ReCiPe single score and toxicity indicators) increase, with a wide range of values (from 3% for canned products to 250% for frozen products). Nonetheless, the relative environmental ranking of studied products does not change significantly, because distribution contributes relatively little to total impacts (Table 5).

The fate of one t of Peruvian *anchoveta*, from sea to plant or farm gate (and to port gate for fresh *anchoveta* for DHC) was calculated (Figure 6). DHC products have markedly higher yields of products (and are directly edible by humans) than reduction products. Aquaculture products are not directly comparable because they also require agricultural inputs.

Alternative exploitation scenarios

In S1 and S2, *anchoveta* biomass (Figure C3 in File S1) and hake biomass (Figure C4 in File S1) remained stable in the simulation based on historical values because no further changes were introduced. However, in the diversification scenario (S3), due to the decrease in *anchoveta* landings, *anchoveta* biomass increased by 21% and stabilised around 85 t·km^{-2} (Figure C3 in File S1).

Table 4. Modelled sub-systems of the Peruvian anchoveta supply chain.

Sub-models →	EwE outputs	Biophysical indicators				
		LCA	LCS	Other environ-mental indicators	Nutrition/energy indicators	Socio-economic indicators
Supply chain links ↓						
Fisheries						
Industrial *anchoveta* fleet	X	X		X	X	X
Vikinga (*anchoveta*) fleet	X	X		X	X	X
Small- and medium-scale (SMS) *anchoveta* fleet/average landed *anchoveta* for IHC	X	X		X	X	X
Average landed *anchoveta* for reduction (weighted mean of industrial and Vikinga fleets)	X	X		X	X	X
Ice plants supplying SMS fisheries			X			
Industrial hake fishery	X		X	X	X	
Direct Human Consumption						
Canned *anchoveta*		X		X	X	X
Frozen *anchoveta*		X		X	X	X
Salted/cured *anchoveta*		X		X	X	X
Indirect Human Consumption (reduction)						
Prime fishmeal		X		X	X	X
Fair Average Quality fishmeal		X		X	X	X
Residual fishmeal			X	X	X	X
Aquafeeds						
Artisanal feeds, Peru		X		X	X	X
Commercial feeds, Peru			X	X	X	X
Commercial feeds international (ingredients and energy use)			X	X	X	X
Aquaculture						
Tilapia: artisanal/commercial feeds, Peru			X	X	X	X
Black pacu: artisanal/commercial feeds, Peru		X		X	X	X
Trout: artisanal/commercial feeds, Peru		X		X	X	X

Abbreviations. LCA: Life Cycle Assessment, LCS: Life Cycle Screening, IHC: Indirect Human Consumption.

Consequently, hake biomass increased by 18% and stabilised around 1.2 t•km^{-2} (Figure C4 in File S1). It is noteworthy that biomasses of other predators also increased in this scenario (e.g. other piscivorous fish such as Eastern Pacific bonito (*Sarda chiliensis chiliensis*), seabirds and pinnipeds), yet hake is the most commercially interesting species among them. EwE outputs for 2011 and simulation scenarios, including fish biomasses, are presented in section C in File S1. A key input datum for the hake fisheries LCS is mean fuel-use intensity, estimated at 84 kg fuel per landed tonne, mass-allocated between hake and by-catch (93% of landings were hake, according to detailed landing records for the hake fleet in 2010; IMARPE, unpublished data).

From the main masses of products in the three scenarios in the reference future year 2021 (Figure 7), conclusions about masses of target seafood products and the total biomass of all commercial species in the marine ecosystem can be drawn: the former increases by 1% in S2 and decreases by 40% in S3, while the latter does not change in S2 and increases by 8% in S3. Sankey diagrams [112,113] of the main masses (biomass and other materials) and energy flows were produced for the supply chains in

the reference year 2011 and for the three scenarios in 2021 (Figures B3 to B5 in File S1).

Graphical comparison of the scenarios according to other dimensions of analysis (e.g. ecological and socio-economic) is presented in Figure 8 (and detailed per product in Figures B6 to B10 in File S1). The results depicted in these figures refer only to the fishfood products studied.

Comparative gross economic benefits are expressed as gross profit (revenues – production costs). Gross profit of the fishfood-product supply chains studied increases by 12% in S2 but decreases by 36% in S3. Detailed mass and economic balances, as well as detailed data for other dimensions of analysis (environmental impacts, biotic resource use, nutritional value) are shown in Tables B2 and B3 in File S1. Employment related to the fishfood-product supply chains studied increases by 18% in S1, which was expected due to the increase in job-intensive production of DHC products. In S2 the increase in employment reaches 53%, while in S3 employment decreases by 6%.

Environmental impacts, as expressed by the ReCiPe single score, increase by 10% in S1 and by 54% in S2, associated with increased production of energy-intensive processed

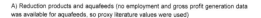

A) Reduction products and aquafeeds (no employment and gross profit generation data
was available for aquafeeds, so proxy literature values were used)

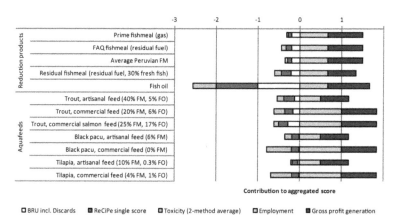

B) Fishfood products

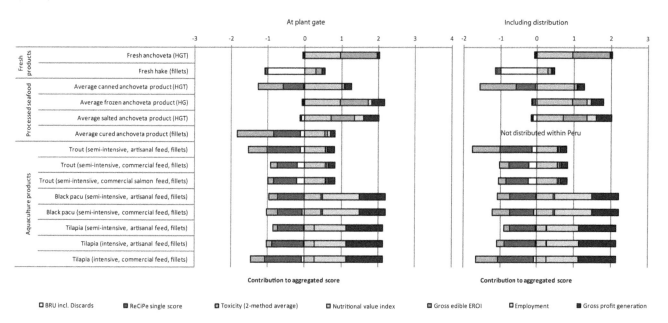

Figure 5. Ranking of DHC products studied from *anchoveta* supply chains according to the proposed indicator set. Per t of fish in product. Shorter negative bars and longer positive bars represent better performance: the range of values on the x-axis represents the maximum positive and negative scores possible for each product. In bottom graphs, all units have the same length, even though the x-axis of the right-side graph was shortened for convenience. Only five indicators are shown in order to limit redundancy between indicators, simplify the diagram, and increase balance among indicators from the three pillars of sustainability (impacts are cumulative and no weighting factor was used). Species names: anchoveta (*Engraulis ringens*), black pacu (*Colossoma macropomum*), hake (*Merluccius gayi*), trout (*Oncorhynchus mykiss*), tilapia (*Oreochromis* spp.).

seafood products. In S3, environmental impacts decrease by 32% due to the large decrease in *anchoveta* landings. The biotic resource use subtotal decreases by only 3% in S1 and 4% in S2, but decreases by 40% in S3, also due to the large decrease in *anchoveta* landings.

The sum of available protein (a proxy for the nutritional value of each scenario) of target products increases in all scenarios from 2011 to 2021. In S1, the 112% increase in the available protein of target products is associated with increasing landings for DHC, while in S2 the increase is 434%. In S3, the increase, by comparison, appears moderate (53%) but substantial due to the increase in hake landings for DHC. It is worth noting that the sum of available protein of other commercial species such as Peruvian sea catfish (*Galeichthys*

peruvianus), fine flounder (*Paralichthys adspersus*) and Eastern Pacific bonito, display a different pattern from that for target products, with small decreases of 4% from 2011 in S1 and S2 but a 47% increase in S3. When this increase is expressed as an absolute value (2 039 Mt) it overcompensates the lower available protein subtotal in S3 compared to those of S1 and S2 (−2.2 Mt).

Among the ecosystem level indicators chosen (Figure 9), a higher value for $I_{BNR,sp}$ represents lower ecosystem health, while the higher values for all IndiSeas indicators represent a healthier ecosystem. Results of $I_{BNR,sp}$ among scenarios (the same amount of biomass is extracted in S1 and S2) show progressive improvement for *anchoveta* and worsening for hake. Applying IndiSeas indicators to EwE outputs of all commercial species results in an increase in

Table 5. Comparison of environmental performance of fisheries and aquaculture direct-human-consumption products, at plant gate and after distribution, per t of fish in product.

Product group	Products	At plant gate			Including distribution			Percentage increase	
		ReCiPe single score (Pt)	Toxicity (CML, kg 1,4-DB eq)	Ranking (1 = best)	ReCiPe single score (Pt)	Toxicity (CML, kg 1,4-DB eq)	Ranking (1 = best)	ReCiPe single score	Toxicity (CML)
Fresh products	Fresh *anchoveta* (HGT)	31	51 918	1	51	75 829	1	68%	46%
	Fresh hake (fillets)	111	129 869	4	205	241 039	4	85%	86%
Processed seafood	Average canned *anchoveta* product (HGT)	866	3 229 195	6	893	3 260 146	5	3%	1%
	Average frozen *anchoveta* product (HG)	38	60 272	2	132	171 443	3	250%	184%
	Average salted *anchoveta* product (HGT)	46	103 633	3	62	122 566	2	36%	18%
Aquaculture products	Trout (semi-intensive, artisanal feed, fillets)	1 045	1 783 975	8	1 140	1 895 146	8	9%	6%
	Trout (semi-intensive, commercial feed, fillets)	849	1 151 958	5	943	1 263 129	6	11%	10%
	Trout (semi-intensive, commercial salmon feed, fillets)	980	1 170 740	7	1 074	1 281 910	7	10%	9%
	Black pacu (semi-intensive, artisanal feed, fillets)	1 052	1 158 268	10	1 146	1 269 438	10	9%	10%
	Black pacu (semi-intensive, commercial feed, fillets)	1 045	1 121 131	9	1 140	1 232 301	9	9%	10%
	Tilapia (semi-intensive, artisanal feed, fillets)	1 105	1 017 474	11	1 200	1 128 644	11	9%	11%
	Tilapia (intensive, artisanal feed, fillets)	1 355	1 178 435	12	1 450	1 289 605	12	7%	9%
	Tilapia (intensive, commercial feed, fillets)	1 573	1 653 337	13	1 667	1 764 507	13	6%	7%

Abbreviations. HGT: headed, gutted, tailed.

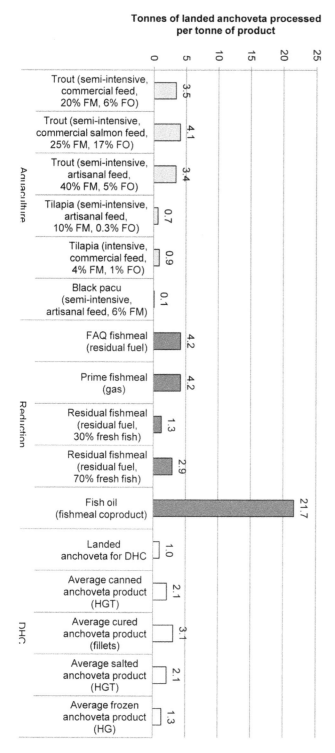

Figure 6. Alternative fates of 1 t of landed *anchoveta.* Excluding other agricultural inputs to aquafeeds and DHC products, expressed as tonnes of landed anchoveta processed into 1 t of final product; HGT: headed, gutted, tailed; FM: fish oil. Species names: anchoveta (*Engraulis ringens*), black pacu (*Colossoma macropomum*), trout (*Oncorhynchus mykiss*), tilapia (*Oreochromis* spp.).

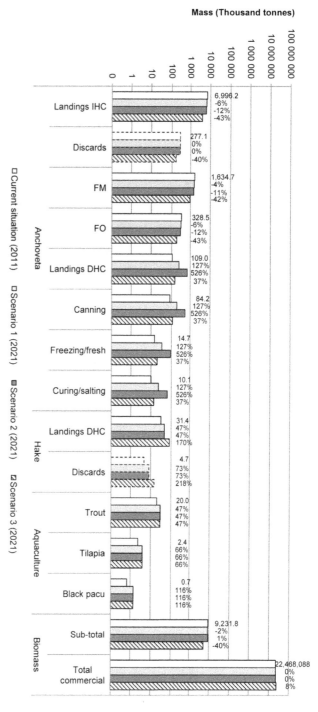

Figure 7. Mass outputs associated with alternative exploitation scenarios. Per key product on a $\log_{10}$ scale. Percentages represent variation from the current situation. Species names: anchoveta (*Engraulis ringens*), black pacu (*Colossoma macropomum*), hake (*Merluccius gayi*), trout (*Oncorhynchus mykiss*), tilapia (*Oreochromis* spp.).

Discussion

Methodological choices

The use of trophic level (TL)-based ecological indicators is suitable for Peru because its fisheries are fully- or over-exploited. TL-based indicators, under the fishing-down-the-food-web concept [118], represent a measure of ecosystem structure and functioning and thus can be used to measure state and trends.

the trophic level of landings, from 2.53 in S1 and S2 to 2.61 in S3. They also show an increase in inverse fishing pressure from 2.51 to 4.07, while the proportion of predatory fish decreases slightly in S3, from 0.19 to 0.18.

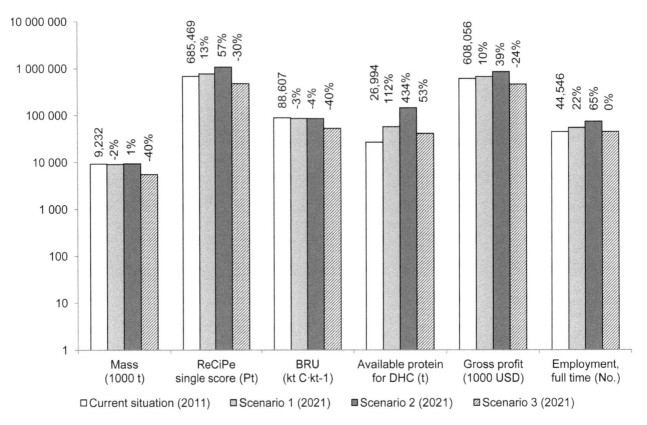

Figure 8. Comparison of alternative exploitation scenarios. In terms of product masses, environmental score, biotic resource use, human nutritional protein availability, gross profit and employment per key product on a $\log_{10}$ scale. Percentages represent variation from the current situation.

These indicators would be less useful in situations in which exploitation is still developing.

Both 1) the proportion of *anchoveta* landings destined for DHC and 2) aquaculture production are expected to grow. Complete historical annual data was available until 2011, and both DHC and aquaculture production datasets indicated a growing trend until 2010. Nonetheless, in 2011 total *anchoveta* landings for DHC were lower than those in 2010, but it was not possible to predict a decreasing trend based on a single "low-catch" year. Aquaculture output, on the other hand, shows continuous growth since 2001.

If future scenarios had been built assuming no growth, relative results would not differ significantly. Particularly in the case of S3, in which total *anchoveta* landings dramatically decrease, we simulated the fate of *anchoveta* landings maintaining the trend of DHC of the reference situation (2011). That is to say, the landing ratios of S1 (~3.6% to DHC) were kept. Since reduction and canning industries are highly vertically integrated and have overcapacity, it is likely that a shortage of *anchoveta* would severely constrain fish reduction, leading firms to prioritise their most recent investment: processing of *anchoveta* for DHC, especially canning. The fact that operational costs and prices were not extrapolated is not a major issue, since our approach is comparative.

Another fundamental decision for scenario modelling was that the reference situation (2011) was modelled (in the MFM) using biomasses from PRODUCE statistics rather than from EwE predictions. Differences are minor, but we preferred the more realistic depiction of the reference situation. For future scenarios, total catches (resulting from the fishing mortality rate) for *anchoveta* and hake were taken from EwE predictions, as previously described, but the uncertainty in predictions of this kind of model [21,119] was not considered. As a result, predictions for future scenarios must be used with caution and regarded as tentative indications of trends. Reduction efficiencies were not altered (we assumed that the technical optimum has been reached), nor were aquaculture data (e.g. inter-species production ratios, general trends in feed compositions).

The current situation: could it be better?

In the current situation, a variety of *anchoveta*-based products are produced. The fishmeal industry has improved its technical performance over the years, and the current state-of-the-art mainly involves use of natural gas and an indirect drying process. Prime-quality fishmeal produced at gas-based indirect drying plants has the best sustainability performance according to the set of sustainability indicators applied. Nonetheless, the legal production of residual fishmeal remains necessary, not only from a socio-economic standpoint, but also from the environmental standpoint, to make the best use of fish resources.

As for DHC products, optimum sustainability would come from landing, processing and distributing fresh/chilled/frozen *anchoveta* products; however, salted and canned products currently provide certain vulnerable communities with fish products. Freshwater aquaculture products could play a larger socio-economic role in Peru if an adequate distribution chain is established and current landing infrastructure for the SMS fleet is improved and enlarged. Among cultured species, black pacu has higher sustainability performance. Moreover, production of black pacu (and by extension other Amazonian species) seems promising for Peru,

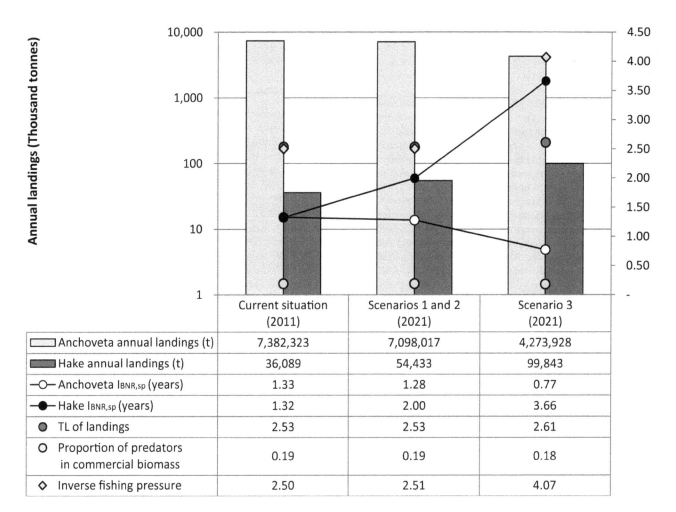

	Current situation (2011)	Scenarios 1 and 2 (2021)	Scenario 3 (2021)
Anchoveta annual landings (t)	7,382,323	7,098,017	4,273,928
Hake annual landings (t)	36,089	54,433	99,843
—O— Anchoveta $I_{BNR,sp}$ (years)	1.33	1.28	0.77
—●— Hake $I_{BNR,sp}$ (years)	1.32	2.00	3.66
● TL of landings	2.53	2.53	2.61
O Proportion of predators in commercial biomass	0.19	0.19	0.18
◇ Inverse fishing pressure	2.50	2.51	4.07

Figure 9. Indicators of ecosystem impacts. Impacts on Biotic Natural Resources (I_{BNR}) at the species level, mean trophic level (TL) of landings, proportion of predatory fish in commercial biomass, and inverse fishing pressure under the alternative exploitation scenarios. The Maximum Sustainable Yield (MSY) of *anchoveta* was estimated at over 5 million t [135]; thus, a 5-year mean of total landings (5.5 million t) was used as a proxy. The MSY of hake was estimated at ~27,000 t until its stock fully recovers [89]. Species names: anchoveta (*Engraulis ringens*), hake (*Merluccius gayi*).

but again, it would depend on a currently non-existent distribution chain.

It is necessary to improve monitoring throughout the supply chains to increase compliance with management measures (e.g. satellite monitoring of SMS vessels; monitoring of diseases, discards, and juveniles). To improve the quality of fish (especially *anchoveta*) landed for DHC, it would be advisable to increase the awareness of fishermen and the personnel who inspect landing points about food-safety issues to reduce in-plant discards.

Policy measures should also be adopted to improve production of *anchoveta* DHC products, such as establishing a quota system for SMS fleets and/or allowing all fleets to land fish for either DHC or IHC as long as minimum requirements (e.g. fish preservation) for each are fulfilled [32]. These measures would also help reduce in-plant discards, rationalise pricing for raw *anchoveta* and reduce IUU.

Scenarios 1 and 2: *Anchoveta* for reduction or for food?

S1 represents the status quo, that is to say the management strategy of the reference year (2011) retained after the beginning of the simulation (2004), and extrapolated into the future. This scenario is sustainable from the perspective of managing *anchoveta* stock but sub-optimal in socio-economic aspects. Indeed, the

current redistribution of the fish processing industry profit is limited, due to several factors [32], and does not provide enough income to the lowest economic classes of the Peruvian population to alleviate their hunger and nutritional issues. S2 would improve sustainability in a variety of ways. For instance, by extracting nearly the same amount of biomass without reducing the mean TL of landings or the proportion of predatory fish in the ecosystem (Figure 6), gross profit would increase by a factor of 1.2 due to increased activity of DHC processing industries. Similarly, employment would increase by a factor of 1.5 and available protein for consumers by a factor of 2.5. The environmental costs of these improvements represent a 1.4-fold increase compared to S1. The implications of S2 are complex: for instance, gross profit would be generated by more firms than at present, and national distribution chains would have to be developed. Moreover, because it is unlikely that Peruvian consumers will consume all the additional *anchoveta* production (which increases by a factor of 2.8 in whole-fish equivalents), an export market must be found, which remains uncertain. But if demand for exported canned Peruvian *anchoveta* became dominant, prices in the domestic market could increase [32]. Nonetheless, S2 would be more sustainable at the national level than S1, especially if profit becomes more evenly distributed within the DHC sector.

Scenario 3: *Anchoveta* today or hake tomorrow?

The goal of S3 is tempting: allow over- or fully-exploited stocks to increase to the point that they can be exploited again (hopefully more sustainably than in the past). Decreasing *anchoveta* fishing mortality to 50% over at least 10 years would increase other NHCS stocks, notably hake (by 18% in biomass). An associated increase in hake catches (by a factor of ~1.4) would thus be possible, and a similar increase is predicted for stocks of other predators –e.g. conger (*Ophichthus remiger*), flatfish, horse mackerel (*Trachurus murphyi*), pinnipeds and seabirds–, whereas a decrease of a few species that compete with *anchoveta* (e.g. other small pelagic species) and of cetaceans is predicted. The implications of such a dramatic change in resource exploitation are diverse but underestimated due to considering only species biomass: total biomass, biotic resource use, total gross profit, employment and environmental impacts would all decrease (by factors of 0.4, 0.3, 0.3, 0.2 and 0.4, respectively). Moreover, the mean TL of landings slightly increases due to the change in the proportions of *anchoveta* and hake landed. The proportion of predators (fish and others) in the ecosystem slightly decreases (from 0.19 to 0.18) in this scenario because biomass of *anchoveta* increases slightly more than that of predators. The inverse fishing pressure increases due to the drastic reduction in total landings (Figure 9). The available amount of protein of *anchoveta* and hake for Peruvian consumers would also decrease (by a factor of 0.3), but is likely to be partly compensated (if not overcompensated) by an increase in landings of other species caught for DHC (Figure 8; Figure C2 in File S1).

Overall, according to these indicators, S3 seems less preferable than S1 and S2 despite some ecological and environmental improvements. Moreover, obtaining the national consensus required to decrease exploitation of the *anchoveta* stock so dramatically would be a daunting endeavour, to say the least. Nonetheless, this scenario deserves more in-depth study, varying the exploitation rates less drastically and taking into account all species in the ecosystem that are exploited or potentially exploitable by fisheries or for tourism. Furthermore, how changes in volumes of fish landings could affect fishing costs and prices of each species should be considered in scenarios. Expected changes in fishing costs were already taken into account through our CPUE-adjusted fuel use intensities. Only minor additional changes can be expected because the major fisheries are already large and mature industrial ones, preventing major changes due to economy of scale. In contrast, the existing increasing trend in fishmeal prices may be exacerbated by a decrease in Peruvian catches of *anchoveta* in S3. Indeed, the Peruvian share of this commodity is over 40%, its global production is decreasing and its sustained demand is price-inelastic [32,120]. However, the concomitant increase in Peruvian hake production (by a factor of 1.4) will not result in lower hake prices because the Peruvian share of this commodity is too low (3–5% [45]) to impact its global market. As a result, the decrease in gross profit of the Peruvian fishing industry should be lower than predicted in S3.

Conclusions

The proposed framework, as illustrated with the Peruvian case study, provides a multi-criteria toolset for decision-making to improve fishfood supply chain dynamics. Scenario analysis confirmed previous speculations that an increase in the proportion of *anchoveta* destined for DHC would positively contribute to Peru's sustainable development (S2). It also indicated that a dramatic reduction in *anchoveta* landings would not be, in general, positive for the country (S3), although this scenario deserves deeper investigation (e.g. consideration of all species, sensitivity analysis of more realistic changes in exploitation rates to estimate optimal levels). The preservation of ecosystem services should also be considered in more detail.

Due to the huge size of the reduction industry and its supplier fisheries, results per functional unit do not align with absolute results per industry (DHC vs. IHC vs. aquaculture). Indeed, in absolute terms, the activities in Peru with most impact at present are those related to capturing and reducing *anchoveta* into fishmeal and fish oil. As a result, the best opportunities for improving the environmental and socio-economic performance of Peruvian *anchoveta* supply chains would be related to management and policy changes that improve the sustainability of the reduction industry and its suppliers.

Acknowledgments

This work, carried out by members of the Anchoveta Supply Chain (ANCHOVETA-SC) project (?http://anchoveta-sc.wikispaces.com), is a contribution to the International Joint Laboratory "Dynamics of the Humboldt Current System" (LMI-DISCOH) coordinated by the Institut de Recherche pour le Développement (IRD) and the Instituto del Mar del Perú (IMARPE), which groups together several other institutions. The authors are extremely grateful to Dr. Michael Corson (INRA, France) for his contribution to the language improvement of this manuscript.

Author Contributions

Conceived and designed the experiments: AA PF JT. Performed the experiments: AA PF JT. Analyzed the data: AA PF JT. Contributed reagents/materials/analysis tools: AA PF JT. Wrote the paper: AA PF.

References

1. Power A (1999) Linking ecological sustainability and world food needs. Environ Dev Sustain 1: 185–196. Available: http://www.springerlink.com/index/H6211R3816488154.pdf. Accessed 18 April 2012.
2. Ingram J, Ericksen P, Liverman D, editors (2010) Food security and global environmental change. London: Earthscan. Available: http://www.publish.csiro.au/nid/21/pid/6702.htm. Accessed 18 April 2012.
3. SOFA (2011) The state of food and agriculture 2010–11. Rome: Food and Agriculture Organization of the United Nations (FAO). Available: http://www.fao.org/docrep/013/i2050e/i2050e.pdf.
4. Nellemann C, MacDevette M, Manders T, Eickhout B, Svihus B, et al, editors (2009) The environmental food crisis: the environment's role in averting future food crises: a UNEP rapid response assessment. Available: http://www.grida.no/publications/rr/food-crisis/. Accessed 18 April 2012.
5. Tukker A, Goldbohm RA, de Koning A, Verheijden M, Kleijn R, et al. (2011) Environmental impacts of changes to healthier diets in Europe. Ecol Econ: 1–13. doi:10.1016/j.ecolecon.2011.05.001.
6. Spiertz H (2010) Food production, crops and sustainability: restoring confidence in science and technology. Curr Opin Environ Sustain 2: 439–443. doi:10.1016/j.cosust.2010.10.006.
7. Wognum PM, Bremmers H, Trienekens JH, van der Vorst JGaJ, Bloemhof JM (2011) Systems for sustainability and transparency of food supply chains – Current status and challenges. Adv Eng Informatics 25: 65–76. doi:10.1016/j.aei.2010.06.001.
8. McMichael P (2011) Food system sustainability: Questions of environmental governance in the new world (dis) order. Glob Environ Chang 21: 804–812. doi:10.1016/j.gloenvcha.2011.03.016.
9. Dahl AL (2012) Achievements and gaps in indicators for sustainability. Ecol Indic 17: 14–19. doi:10.1016/j.ecolind.2011.04.032.
10. Food Policy (2011) The challenge of global food sustainability (Editorial). Food Policy 36: S1–S2. doi:10.1016/j.foodpol.2011.01.001.

11. SOFIA (2010) The state of world fisheries and aquaculture 2010. Rome: FAO Fisheries and Aquaculture Department. Available: http://www.fao.org/docrep/013/i1820e/i1820e.pdf. Accessed 17 May 2011.

12. SOFIA (2012) The State of World Fisheries and Aquaculture 2012. Rome: FAO Fisheries and Aquaculture Department. Available: http://www.fao.org/docrep/016/i2727e/i2727e.pdf.

13. Schlüter M, Mcallister R, Arlinghaus R, Bunnefeld N, Eisenack K, et al. (2012) New horizons for managing the environment: A review of coupled social-ecological systems modeling. Nat Resour Model 25: 219–272. doi:10.1111/j.1939-7445.2011.00108.x.

14. Levin S (1992) The problem of pattern and scale in ecology. Ecology 6: 1943–1967.

15. Grimm V, Revilla E, Berger U, Jeltsch F, Mooij WM, et al. (2005) Pattern-oriented modeling of agent-based complex systems: lessons from ecology. Science 310: 987–991. doi:10.1126/science.1116681.

16. Link J (2002) Ecological considerations in fisheries management: when does it matter? Fisheries 27. Available: http://www.tandfonline.com/doi/abs/10.1577/1548-8446(2002)027<0010:ECIFM>2.0.CO; 2. Accessed 3 November 2012.

17. Christensen V, Pauly D (1992) ECOPATH II - a software for balancing steady-state ecosystem models and calculating network characteristics. Ecol Modell 61: 169–185. Available: http://www.sciencedirect.com/science/article/pii/0304380092900168. Accessed 30 April 2012.

18. Walters C, Christensen V, Pauly D (1997) Structuring dynamic models of exploited ecosystems from trophic mass-balance assessments. Rev Fish Biol Fish 7: 139–172.

19. Travers M, Shin Y-J, Jennings S, Cury P (2007) Towards end-to-end models for investigating the effects of climate and fishing in marine ecosystems. Prog Oceanogr 75: 751–770. doi:10.1016/j.pocean.2007.08.001.

20. Pauly D, Christensen V, Walters C (2000) Ecopath, Ecosim, and Ecospace as tools for evaluating ecosystem impact of fisheries. ICES J Mar Sci 57: 697–706. Available: http://icesjms.oxfordjournals.org/cgi/doi/10.1006/jmsc.2000.0726.

21. Christensen V, Walters CJ (2004) Ecopath with Ecosim: methods, capabilities and limitations. Ecol Modell 172: 109–139. Available: http://linkinghub.elsevier.com/retrieve/pii/S030438000300365X. Accessed 3 November 2012.

22. Jain J, Dangayach G, Agarwal G (2010) Supply Chain Management: Literature Review and Some Issues. J Stud Manuf 1: 11–25. Available: http://www.hypersciences.org/JSM/Iss.1-2010/JSM-2-1-2010.pdf. Accessed 25 April 2012.

23. Kasi V (2005) Systemic Assessment of SCOR for Modeling Supply Chains. Proceedings of the 38th Annual Hawaii International Conference on System Sciences. Ieee, Vol. 00. p. 87b–87b. Available: http://ieeexplore.ieee.org/lpdocs/epic03/wrapper.htm?arnumber = 1385413.

24. Jensen TK, Nielsen J, Larsen EP, Clausen J (2010) The Fish Industry–Toward Supply Chain Modeling. J Aquat Food Prod Technol 19: 214–226. doi:10.1080/10498850.2010.508964.

25. Bjørndal T, Lane DE, Weintraub A (2004) Operational research models and the management of fisheries and aquaculture: A review. Eur J Oper Res 156: 533–540. doi:10.1016/S0377-2217(03)00107-3.

26. Howieson J, Lawley M (2010) Implementing Whole of Chain Analyses for the Seafood Industry: A Toolbox Approach. In: Ballantine P, Finsterwalder J, editors. Australian and New Zealand Marketing Academy (ANZMAC) Annual Conference 2010, Nov 29 2010. Christchurch, New Zealand: University of Canterbury. 1–8.

27. Prellezo R, Accadia P, Andersen JL, Little A, Nielsen R, et al. (2009) Survey of existing bioeconomic models. Final report. Sukarrieta: AZTI-Tecnalia.

28. Prellezo R, Accadia P, Andersen JL, Andersen BS, Buisman E, et al. (2012) A review of EU bio-economic models for fisheries: The value of a diversity of models. Mar Policy 36: 423–431. doi:10.1016/j.marpol.2011.08.003.

29. Khan A (2009) A fish chain analysis of Northern Gulf cod recovery options: exploring EwE modeling approaches for policy scenarios. In: Palomares MLD, Morissette L, Cisneros-Montemayor A, Varkey D, Coll M, et al., editors. Ecopath 25 Years Conference Proceedings: Extended Abstracts. Vancouver: Fisheries Centre, University of British Columbia, Vol. 17. Available: http://www.fisheries.ubc.ca/webfm_send/141. Accessed 9 December 2011.

30. Christensen V, Steenbeek J, Failler P (2011) A combined ecosystem and value chain modeling approach for evaluating societal cost and benefit of fishing. Ecol Modell 222: 857–864. doi:10.1016/j.ecolmodel.2010.09.030.

31. Christensen V, de la Puente S, Sueiro JC, Steenbeek J, Majluf P (2013) Valuing seafood: the Peruvian fisheries sector. Mar Policy 44: 302–311. doi:10.1016/j.marpol.2013.09.022.

32. Fréon P, Sueiro JC, Iriarte F, Miro Evar OF, Landa Y, et al. (2013) Harvesting for food versus feed: a review of Peruvian fisheries in a global context. Rev Fish Biol Fish 24: 381–398. doi:10.1007/s11160-013-9336-4.

33. Chavez FP, Bertrand A, Guevara-Carrasco R, Soler P, Csirke J (2008) The northern Humboldt Current System: Brief history, present status and a view towards the future. Prog Oceanogr 79: 95–105. doi:10.1016/j.pocean.2008.10.012.

34. Arias M (2012) The evolution of legal instruments and the sustainability of the Peruvian anchovy fishery. Mar Policy 36: 78–89. doi:10.1016/j.marpol.2011.03.010.

35. Tveteras S, Paredes C, Peña-Torres J (2011) Individual Fishing Quotas in Peru: Stopping the Race for Anchovies. Marine Resources Foundation. Available: http://fen.uahurtado.cl/wp/wp-content/uploads/2010/07/I263.pdf. Accessed 5 September 2011.

36. Fréon P, Bouchon M, Domalain G, Estrella C, Iriarte F, et al. (2010) Impacts of the Peruvian anchoveta supply chains: from wild fish in the water to protein on the plate. GLOBEC Int Newsl: 27–31.

37. Fréon P, Bouchon M, Estrella C (2010) Comparación de los impactos ambientales y aspectos socioeconómicos de las cadenas de producción de anchoveta (Comparison of environmental impacts and socio-economical aspects of the Peruvian anchovy supply chains). Boletín Inst del Mar del Perú 25: 63–71.

38. Alvarado F (2009) Diagnóstico social sobre el trabajo y el empleo en el sector pesquero de Ecuador y Perú (Social diagnosis on work and employment in the fisheries sector of Ecuador and Peru). Madrid: Organización Internacional del Trabajo (OIT). Available: http://www.ilo.int/public/spanish/region/eurpro/madrid/download/peruecuador.pdf.

39. Avadí A, Vázquez-Rowe I, Fréon P (2014) Eco-efficiency assessment of the Peruvian anchoveta steel and wooden fleets using the LCA+DEA framework. J Clean Prod (in press). doi:10.1016/j.jclepro.2014.01.047.

40. Fréon P, Avadí A, Marín W, Negrón R (2014) Environmentally-extended comparison table of large- vs. small- and medium-scale fisheries: the case of the Peruvian anchoveta fleet. Can J Fish Aquat Sci (accepted).

41. Paredes C (2010) Reformando el Sector de la Anchoveta Peruana: Progreso Reciente y Desafíos Futuros (Reform of the Peruvian anchoveta sector: Recent progress and challenges for the future). Available: http://www.institutodelperu.org.pe/descargas/Publicaciones/DelInstitutodelPeru/DOC/contenido_carlos_paredes_-_reforma_de_la_presqueria_anchoveta_peru.pdf.

42. Fréon P, Avadí A, Amelia R, Chavez V (2014) Life cycle assessment of the Peruvian industrial anchoveta fleet: boundary setting in life cycle inventory analyses of complex and plural means of production. Int J Life Cycle Assess (in press). doi:10.1007/s11367-014-0716-3.

43. Tacon A, Hasan M, Metian M (2011) Demand and supply of feed ingredients for farmed fish and crustaceans: trends and prospects. Rome. Available: http://www.fao.org/docrep/015/ba0002e/ba0002e.pdf.

44. De Silva SS, Turchini GM (2008) Towards Understanding the Impacts of the Pet Food Industry on World Fish and Seafood Supplies. J Agric Environ Ethics 21: 459–467. doi:10.1007/s10806-008-9109-6.

45. FAO (2014) FishStatJ-software for fishery statistical time series. In: FAO Fisheries and Aquaculture Department [online]. Rome. Updated 28 November 2013. Available: http://www.fao.org/fishery/statistics/software/fishstatj/en.

46. Jiménez F, Gómez C (2005) Evaluación nutricional de galletas enriquecidas con diferentes niveles de harina de pescado (Nutritional evaluation of cookies enriched with various levels of fishmeal). Lima: Red Peruana de Alimentación y Nutrición (r-PAN). Available: http://www.rpan.org/publicaciones/pv010.pdf.

47. Paredes C, Gutiérrez ME (2008) La industria anchovetera Peruana: Costos y Beneficios. Un Análisis de su Evolución Reciente y de los Retos para el Futuro (The Peruvian anchoveta industry: Costs and benefits. An analysis of recent evolution and challenges for the future). Available: http://institutodelperu.org.pe/descargas/informe_de_la_industria_de_anchoveta.pdf. Accessed 19 May 2011.

48. INEI (2012) Perú: Compendio estadístico 2012. Pesca (Peru: Statistical compendium 2012. Fisheries). Lima: Instituto Nacional de Estadística e Informática (INEI). Available: http://www.inei.gob.pe/biblioineipub/bancopub/Est/Lib1055/cap13/ind13.htm.

49. Centrum (2009) Reporte Sectorial: Sector Pesca. Lima: Noviembre 18, 2009.

50. SNP (2010) Memoria Anual 2010. Lima: Sociedad Nacional de Pesquería. Available: http://snp.org.pe/wp/?cat = 25.

51. SNP (2011) Memoria anual 2011. Lima: Sociedad Nacional de Pesquería. Available: http://snp.org.pe/wp/?cat = 25.

52. Deutsch L, Gräslund S, Folke C, Troell M, Huitric M, et al. (2007) Feeding aquaculture growth through globalization: Exploitation of marine ecosystems for fishmeal. Glob Environ Chang 17: 238–249. doi:10.1016/j.gloenvcha.2006.08.004.

53. INEI (2013) Población y vivienda (Population and housing). Available: http://www.inei.gob.pe/estadisticas/indice-tematico/poblacion-y-vivienda/.

54. INEI (2012) Perú: Consumo per cápita de los principales alimentos 2008–2009 (Peru: Per capita consumption of main staple foods 2008–2009). Available: http://www.inei.gob.pe/biblioineipub/bancopub/Est/Lib1028/index.html.

55. Sánchez N, Gallo M (2009) Status of and trends in the use of small pelagic fish species for reduction fisheries and for human consumption in Peru. In: Hasan M, Halwart M, editors. Fish as feed inputs for aquaculture: practices, sustainability and implications. FAO Fisheries and Aquaculture Technical Paper No. 518. 325–369.

56. Avadí A, Fréon P, Quispe I (2014) Environmental assessment of Peruvian anchoveta food products: is less refined better? Int J Life Cycle Assess (in press). doi:10.1007/s11367-014-0737-y.

57. PRODUCE (2009) Plan Nacional de Desarrollo Acuícula (Aquaculture Development National Plan). Lima: Ministerio de la Producción. Available: http://portal.andina.com.pe/EDPFiles/EDPWEBPAGE_plan-acuicola.pdf.

58. Mendoza D (2013) Situación del extensionismo acuícola en el perú (Situation of aquacultural extensionism in Peru). Lima: Dirección de Acuicultura, Ministerio de la Producción-PRODUCE. Available: http://rnia.produce.gob.pe/images/stories/archivos/pdf/publicaciones/informe_extensionismo_peru.pdf.

59. Mendoza D (2011) Panorama de la acuicultura mundial, América Latina y el Caribe y en el Perú (Overview of global aquaculture, Latin America and the

Caribean, and Peru). Lima: Dirección de Acuicultura, Ministerio de la Producción - PRODUCE. Available: http://www.racua.org/uploads/media/informe_acuicultura_mundo_al_peru.pdf.

60. Amaya Ea, Davis DA, Rouse DB (2007) Replacement of fish meal in practical diets for the Pacific white shrimp (Litopenaeus vannamei) reared under pond conditions. Aquaculture 262: 393–401. doi:10.1016/j.aquaculture.2006.11.015.

61. Sun W (2009) Life Cycle Assessment of Indoor Recirculating Shrimp Aquaculture System University of Michigan Ann Arbor. Available: http://deepblue.lib.umich.edu/handle/2027.42/63582. Accessed 25 May 2011.

62. Tacon A (2002) Thematic review of feeds and feed management practices in shrimp aquaculture. Available: http://library.enaca.org/Shrimp/Case/Thematic/FinalFeed.pdf. Accessed 6 February 2013.

63. Avadí A, Pelletier N, Aubin J, Ralite S, Núñez J, et al. (2014) Comparative environmental performance of artisanal and commercial feed use in Peruvian freshwater aquaculture. Aquaculture (in review).

64. Rokovich JA (2009) Estudios para el desarrollo de clusters en la actividad pesquera industrial y artesanal de anchoveta y pota-Informe final. Lima: ICON-INSTITUT GmbH.

65. IMARPE (2012) Instituto del Mar del Perú (Peruvian Institute of the Sea). Available: http://www.imarpe.pe/imarpe/index.php?id_idioma = EN.

66. De la Puente O, Sueiro JC, Heck C, Soldi G, de la Puente S (2011) La Pesquería Peruana de Anchoveta-Evaluación de los sistemas de gestión pesquera en el marco de la certificación a cargo del Marine Stewardship Council (The Peruvian anchoveta fishery-Assessment of the fishery management systems in the framework of th. Universidada Peruana Cayetano Heredia, Centro para la Sostenibilidad Ambiental. Available: http://www.csa-upch.org/pdf/lapesqueriaperuana.pdf.

67. FishSource (2012) Anchoveta-Peruvian northern-central stock. FishSource: Status and environmental performance of fisheries worldwide. Available: http://www.fishsource.com/fishery/identification?fishery = Anchoveta+-+Peruvian+northern-central+stock.

68. Bernal M, Somarakis S, Witthames PR, van Damme CJG, Uriarte A, et al. (2012) Egg production methods in marine fisheries: An introduction. Fish Res 117–118: 1–5. doi:10.1016/j.fishres.2012.01.001.

69. Aranda M (2009) Developments on fisheries management in Peru: The new individual vessel quota system for the anchoveta fishery. Fish Res 96: 308–312. doi:10.1016/j.fishres.2008.11.004.

70. Paredes CE (2012) Eficiencia y equidad en la pesca peruana: La reforma y los derechos de pesca (Efficiency and equity in Peruvian fisheries: Reform and fishing rights). Lima: Instituto del Perú.

71. Alfaro-Shigueto J, Mangel JC, Pajuelo M, Dutton PH, Seminoff Ja, et al. (2010) Where small can have a large impact: Structure and characterization of small-scale fisheries in Peru. Fish Res 106: 8–17. doi:10.1016/j.fishres.2010.06.004.

72. Estrella C, Swartzman G (2010) The Peruvian artisanal fishery: Changes in patterns and distribution over time. Fish Res 101: 133–145. doi:10.1016/j.fishres.2009.08.007.

73. Paredes C (2013) Eficiencia y equidad en la pesca local: la reforma y los derechos. El Comer: B6.

74. Paredes C (2013) La anchoveta en su laberinto. Gestión: 23.

75. USMP (2013) MESA REDONDA ¿El marco legal está propiciando la pesquería formal en el país? Gestión: 15–18.

76. Paredes CE, Letona Ú (2013) Contra la corriente: La anchoveta peruana y los retos para su sostenibilidad. Lima: World Wildlife Fund (WWF) and Universidad de San Martín de Porres (USMP). Available: http://awsassets.panda.org/downloads/anchoveta_version_final.pdf.

77. Alder J, Pauly D (2008) A comparative assessment of biodiversity, fisheries and aquaculture in 53 countries' Exclusive Economic Zones. Alder J, Pauly D, editors Fisheries Centre, University of British Columbia, Canada.

78. Schreiber M, Halliday A (2013) Uncommon among the Commons? Disentangling the Sustainability of the Peruvian Anchovy Fishery. Ecol Soc 18: 15. Available: http://www.ecologyandsociety.org/vol18/iss2/art12/ES-2012-5319.pdf. Accessed 21 May 2013.

79. SUNAT (2012) National Customs and Tax Administration. Available: http://www.sunat.gob.pe/.

80. Sueiro JC (2008) La actividad pesquera peruana: Características y retos para su sostenibilidad. Lima: CooperAcción, Acción Solidaria para el Desarrollo. Available: http://www.ibcperu.org/doc/isis/10736.pdf.

81. Avadí A, Fréon P (2014) A set of sustainability performance indicators for seafood: direct human consumption products from Peruvian anchoveta fisheries and freshwater aquaculture. Ecol Indic (in review).

82. FAO (2000) FAO - Perfiles nutricionales por países - Peru (FAO - Nutritional profiles by country - Peru). Rome: Food and Agriculture Organisation of the United Nations. Available: http://www.bvsde.paho.org/texcom/nutricion/permap.pdf.

83. IFPRI (2006) The Challenge of Hunger. Global Hunger Index: Facts, determinants and trends. Bonn/Washington: International Food Policy Research Institute (IFPRI) and Deutsche Welthungerhilfe (DWHH). Available: http://www.ifpri.org/sites/default/files/publications/ghi06.pdf.

84. IFPRI (2012) 2012 Global hunger index: the challenge of hunger: Ensuring sustainable food security under land, water, and energy stresses. Bonn/Washington, DC/Dublin: International Food Policy Research Institute (IFPRI), Concern Worldwide and Welthungerhilfe and Green Scenery.

Available: http://www.ifpri.org/sites/default/files/publications/ghi12.pdf. Accessed 13 February 2013.

85. FAO (2011) Country Profile: Food Security Indicators. Country: Peru. Available: http://www.fao.org/fileadmin/templates/ess/documents/food_security_statistics/country_profiles/eng/Peru_E.pdf.

86. INEI (2011) Encuesta Demográfica y de Salud Familiar 2011 (Demographic and family health survey 2011). Lima: Instituto Nacional de Estadística e Informática (INEI). Available: http://proyectos.inei.gob.pe/endes/2011/.

87. Pauly D, Muck P, Mendo I, Tsukayama J, editors (1989) The Peruvian Upwelling Ecosystem: Dynamics and Interactions. ICLARM Conference Proceedings 18. Instituto del Mar del Peru (IMARPE), Callao, Peru, Deutsche Gesellschaft für Technische Zusammenarbeit (GTZ), GmbH, Eschbom, Federal Republic of Germany; and International Center for Living Aquatic Resources Management (ICLARM), Manila, Philippines. p. 438. Available: http://www.anchoveta.info/Documentostecnicos/Perubook2.pdf.

88. IMARPE (2010) Boletín Instituto del Mar del Perú. Quinto panel internacional de expertos en evaluación de la anchoveta peruana (Engraulis ringens Jenyns. Callao, 10–14 agosto 2009. Lima: Instituto del Mar del Perú.

89. IMARPE (2009) Boletín Instituto del Mar del Perú. Informe del Tercer Panel Internacional de Expertos de Evaluación de la merluza peruana Merluccius gayi peruanus Ginsburg. Manejo Precautorio de la Merluza Peruana. Callao 24–28 de marzo 2008. Lima: Instituto del Mar del Perú.

90. Cárdenas-Alayza S (2012) Prey abundance and population dynamics of South American fur seals (Arctocephalus australis) in Peru Vancouver: The University of British Columbia.

91. Hertlein W (1995) A Simulation Model of the Dynamics of Peruvian Anchoveta (Engraulis ringens). NAGA, ICLARM Q: 41–46. Available: http://worldfish.catalog.cgiar.org/naga/na_2236.pdf. Accessed 27 September 2012.

92. Taylor MH, Wolff M (2007) Trophic modeling of Eastern Boundary Current Systems: a review and prospectus for solving the "Peruvian Puzzle." Rev peru biol 14: 87–100.

93. Tam J, Blaskovic V, Alegre A (2005) Modelo ecotrófico del ecosistema de afloramiento peruano durante El Niño 1997–98. In: Wolff M, Milessi A, Mendo J, editors. Modelación ecotrófica multiespecífica e indicadores cuantitativos ecosistémicos para la administración de pesquerías. CURSO INTERNACIONAL. Lima: Universidad Agraria La Molina.

94. Tam J, Taylor MH, Blaskovic V, Espinoza P, Michael Ballón R, et al. (2008) Trophic modeling of the Northern Humboldt Current Ecosystem, Part I: Comparing trophic linkages under La Niña and El Niño conditions. Prog Oceanogr 79: 352–365. doi:10.1016/j.pocean.2008.10.007.

95. Taylor MH, Tam J, Blaskovic V, Espinoza P, Michael Ballón R, et al. (2008) Trophic modeling of the Northern Humboldt Current Ecosystem, Part II: Elucidating ecosystem dynamics from 1995 to 2004 with a focus on the impact of ENSO. Prog Oceanogr 79: 366–378. doi:10.1016/j.pocean.2008.10.008.

96. Tam J, Jarre A, Taylor M, Wosnitza-Mendo C, Blaskovic V, et al. (2009) Modelado de la merluza en su ecosistema con interacciones tróficas y forzantes ambientales. Boletín Inst del Mar del Perú 24: 27–32.

97. Tam J, Blaskovic V, Goya E, Bouchon M, Taylor M, et al. (2010) Relación entre anchoveta y otros componentes del ecosistema. Boletín Inst del Mar del Perú 25: 31–37.

98. Shin Y-J, Bundy A, Shannon LJ, Blanchard JL, Chuenpagdee R, et al. (2012) Global in scope and regionally rich: an IndiSeas workshop helps shape the future of marine ecosystem indicators. Rev Fish Biol Fish 22: 835–845. doi:10.1007/s11160-012-9252-z.

99. Csirke J, Gumy A (1996) Análisis bioeconómico de la pesquería pelágica peruana dedicada a la producción de harina y aceite de pescado. Boletín Inst del Mar del Perú 19: 25–68.

100. IFU (2005) Umberto – A Software Tool for Life Cycle Assessment and Material Flow Analysis – User Manual. Hamburg: Institut für Umweltinformatik Hamburg GmbH and ifeu – Institut für Energie und Umweltforschung Heidelberg GmbH.

101. Zamagni A, Pesonen H-L, Swarr T (2013) From LCA to Life Cycle Sustainability Assessment: concept, practice and future directions. Int J Life Cycle Assess 18: 1637–1641. doi:10.1007/s11367-013-0648-3.

102. Klöpffer W (2008) Life cycle sustainability assessment of products. Int J Life Cycle Assess 13: 89–95. doi:10.1065/lca2008.02.376.

103. PRé (2012) SimaPro by Pré Consultants. Available: http://www.pre-sustainability.com/content/simapro-lca-software/.

104. Ecoinvent (2012) The ecoinvent Centre portal. Available: http://www.ecoinvent.ch/.

105. Guinée JB, Gorrée M, Heijungs R, Huppes G, Kleijn R, et al. (2002) Handbook on life cycle assessment. Operational guide to the ISO standards. I: LCA in perspective. IIa: Guide. IIb: Operational annex. III: Scientific background. Dordrecht: Kluwer Academic Publishers.

106. Goedkoop M, Heijungs R, Huijbregts M, Schryver A De, Struijs J, et al. (2012) ReCiPe 2008. A life cycle impact assessment method which comprises harmonised category indicators at the midpoint and the endpoint level. First edition (revised) Report I: Characterisation. Ministry of Housing, Spatial Planning and the Environment (VROM) and Centre of Environmental Science - Leiden University (CML).

107. Hischier R, Weidema BP, Althaus H, Bauer C, Doka G, et al. (2010) Implementation of Life Cycle Impact Assessment Methods. ecoinvent report No. 3 (v2.2). Dübendorf: Swiss Centre for Life Cycle Inventories.

108. Rosenbaum RK, Bachmann TM, Gold LS, Huijbregts MaJ, Jolliet O, et al. (2008) USEtox–the UNEP-SETAC toxicity model: recommended characterisation factors for human toxicity and freshwater ecotoxicity in life cycle impact assessment. Int J Life Cycle Assess 13: 532–546. doi:10.1007/s11367-008-0038-4.

109. Shin YJ, Shannon LJ (2010) Using indicators for evaluating, comparing, and communicating the ecological status of exploited marine ecosystems. 1. The IndiSeas project. ICES J Mar Sci 67: 686–691. doi:10.1093/icesjms/fsp273.

110. Shin Y-J, Shannon LJ, Bundy A, Coll M, Aydin K, et al. (2010) Using indicators for evaluating, comparing, and communicating the ecological status of exploited marine ecosystems. 2. Setting the scene. ICES J Mar Sci 67: 692–716. doi:10.1093/icesjms/fsp294.

111. Börjeson L, Höjer M, Dreborg K-H, Ekvall T, Finnveden G (2006) Scenario types and techniques: Towards a user's guide. Futures 38: 723–739. doi:10.1016/j.futures.2005.12.002.

112. Schmidt M (2008) The Sankey Diagram in Energy and Material Flow Management. Part I. J Ind Ecol 12: 82–94. doi:10.1111/j.1530-9290.2008.00004.x.

113. Schmidt M (2008) The Sankey Diagram in Energy and Material Flow Management. Part II. J Ind Ecol 12: 173–185. doi:10.1111/j.1530-9290.2008.00015.x.

114. IIAP (2012) Instituto de Investigaciones de la Amazonía Peruana (Research Institute of the Peruvian Amazonia). Available: http://www.iiap.org.pe/.

115. PETT (2012) Regional Government of Puno, Proyecto Especial Truchas Titicaca (Especial Project Trout Titicaca Lake). Available: http://pett.regionpuno.gob.pe/.

116. PRODUCE (2012) Ministry of Production - Aquaculture statistics. Available: http://www.produce.gob.pe/index.php/estadistica/acuicultura.

117. PRODUCE (2012) Ministry of Production - Fisheries statistics. Available: http://www.produce.gob.pe/index.php/estadistica/desembarque.

118. Pauly D, Christensen V, Dalsgaard J, Froese R, Torres F Jr (1998) Fishing Down Marine Food Webs. Science (80-) 279: 860–863. doi:10.1126/science.279.5352.860.

119. Essington TE (2007) Evaluating the sensitivity of a trophic mass- balance model (Ecopath) to imprecise data inputs: 628–637. doi:10.1139/F07-042.

120. Tveterås S, Tveterås R (2010) The Global Competition for Wild Fish Resources between Livestock and Aquaculture. J Agric Econ 61: 381–397. doi:10.1111/j.1477-9552.2010.00245.x.

121. PRODUCE (2010) Anuario Estadístico 2010-Sub sector Pesca. Lima: Ministerio de la Producción, Oficina General de Tecnología de la Información y Estadística.

122. Péron G, François Mittaine J, Le Gallic B (2010) Where do fishmeal and fish oil products come from? An analysis of the conversion ratios in the global fishmeal industry. Mar Policy 34: 815–820. doi:10.1016/j.marpol.2010.01.027.

123. Mullon C, Mittaine J-F, Thébaud O (2009) Modeling the global fishmeal and fish oil markets. Nat Resour Model 22: 564–609. doi:10.1111/j.1939-7445.2009.00053.x/full.

124. Langlois J, Fréon P, Delgenes J-P, Steyer J-P, Hélias A (2014) New methods for impact assessment of biotic-resource depletion in LCA of fishery: theory and application. J Clean Prod. doi:10.1016/j.jclepro.2014.01.087.

125. Pauly D, Christensen V (1995) Primary production required to sustain global fisheries. Nature 374: 255–257. Available: http://www.nature.com/doifinder/10.1038/374255a0.

126. Hornborg S, Nilsson P, Valentinsson D, Ziegler F (2012) Integrated environmental assessment of fisheries management: Swedish Nephrops trawl fisheries evaluated using a life cycle approach. Mar Policy 36: 1–9. doi:10.1016/j.marpol.2012.02.017.

127. Hornborg S, Emanuelsson A, Sonesson U, Ziegler F (2012) D.1.3b: By-catch methods in Life Cycle Assessment of seafood products. LC-IMPACT.

128. Goedkoop M, Heijungs R, Huijbregts M, Schryver A De, Struijs J, et al. (2009) ReCiPe 2008. A life cycle impact assessment method which comprises harmonised category indicators at the midpoint and the endpoint level. Report I: Characterisation. Ministry of Housing, Spatial Planning and Environment (VROM). Available: www.lcia-recipe.info.

129. Van Zelm R, Huijbregts MaJ, Meent D (2009) USES-LCA 2.0–a global nested multi-media fate, exposure, and effects model. Int J Life Cycle Assess 14: 282–284. doi:10.1007/s11367-009-0066-8.

130. Tyedmers P (2000) Salmon and Sustainability: The biophysical cost of producing salmon through the commercial salmon fishery and the intensive salmon culture industry The University of British Columbia. Available: https://circle.ubc.ca/handle/2429/13201. Accessed 16 May 2011.

131. Drewnowski A, Fulgoni III V (2008) Nutrient profiling of foods: creating a nutrient-rich food index. Nutr Rev 66: 23–39. doi:10.1111/j.1753-4887.2007.00003.x.

132. Tyedmers P, Watson R, Pauly D (2005) Fueling global fishing fleets. Ambio 34: 635–638. Available: http://www.ncbi.nlm.nih.gov/pubmed/16521840.

133. Hall CaS (2011) Introduction to Special Issue on New Studies in EROI (Energy Return on Investment). Sustainability 3: 1773–1777. doi:10.3390/su3101773.

134. Kruse S, Flysjö A, Kasperczyk N, Scholz AJ (2008) Socioeconomic indicators as a complement to life cycle assessment–an application to salmon production systems. Int J Life Cycle Assess 14: 8–18. doi:10.1007/s11367-008-0040-x.

135. Csirke J, Guevara-Carrasco R, Cárdenas G, Ñiquen M, Chipollini A (1996) Situación de los recursos anchoveta (Engraulis ringens) y sardina (Sardinops sagax) a principios de 1994 y perspectivas para la pesca en el Perú, con particular referencia a las regiones Norte y Centro de la costa peruana. Callao: Instituto del Mar del Perú.

17

Baltic Salmon, *Salmo salar,* from Swedish River Lule Älv Is More Resistant to Furunculosis Compared to Rainbow Trout

Lars Holten-Andersen[1,2]*****, **Inger Dalsgaard**[2], **Kurt Buchmann**[1]

1 Department of Veterinary Disease Biology, Faculty of Life Sciences, University of Copenhagen, Frederiksberg C, Denmark, **2** Division of Veterinary Diagnostics and Research, National Veterinary Institute, Technical University of Denmark, Copenhagen, Denmark

Abstract

Background: Furunculosis, caused by *Aeromonas salmonicida*, continues to be a major health problem for the growing salmonid aquaculture. Despite effective vaccination programs regular outbreaks occur at the fish farms calling for repeated antibiotic treatment. We hypothesized that a difference in natural susceptibility to this disease might exist between Baltic salmon and the widely used rainbow trout.

Study Design: A cohabitation challenge model was applied to investigate the relative susceptibility to infection with *A. salmonicida* in rainbow trout and Baltic salmon. The course of infection was monitored daily over a 30-day period post challenge and the results were summarized in mortality curves.

Results: *A. salmonicida* was recovered from mortalities during the entire test period. At day 30 the survival was 6.2% and 34.0% for rainbow trout and Baltic salmon, respectively. Significant differences in susceptibility to *A. salmonicida* were demonstrated between the two salmonids and hazard ratio estimation between rainbow trout and Baltic salmon showed a 3.36 higher risk of dying from the infection in the former.

Conclusion: The finding that Baltic salmon carries a high level of natural resistance to furunculosis might raise new possibilities for salmonid aquaculture in terms of minimizing disease outbreaks and the use of antibiotics.

..

Editor: Josep V. Planas, Universitat de Barcelona, Spain

Funding: This study was funded by the Danish Research Council for Strategic Research (Grant #2101-07-0086). The funders had no role in study design, data collection and analysis, decision to publish, or preparation of the manuscript.

Competing Interests: The authors have declared that no competing interests exist.

* E-mail: lhoa@life.ku.dk

Introduction

The bacterial disease furunculosis caused by *Aeromonas salmonicida* is one of the main concerns in European salmonid mariculture due to high mortality rates and significant economic losses [1,2]. Vaccination programs have kept the problem under some control. However, side effects following oil-adjuvanted i.p. vaccination have raised a series of ethical and welfare questions related to the use of vaccines [3,4,5]. Hence, inherent resistance in the fish against furunculosis would be preferable in order to reduce medication and side effects from immunoprophylactic procedures. Recent studies have shown that the isolated salmon stock in the Baltic possesses genes conferring resistance towards the extremely pathogenic parasite *Gyrodactylus salaris* [6,7,8]. This salmon stock comprises numerous sub-populations homing to rivers in Sweden, Finland, Russia, Latvia, Lithuania, Estonia, Poland and Germany draining into the Baltic Sea [9]. Baltic salmon from rivers Ume älv and Lule älv present a clear protective immune response a few weeks after infection with *G. salaris* [8,10,11,12]. In contrast, East-Atlantic salmon (Norwegian, Scottish, and Danish) are very susceptible to *G. salaris* and show no effective immune response during infection [6,7,8,10,13]. These differences between salmon

stocks regarding protective immunity against the very pathogenic *G. salaris* pose the question whether a comparable difference might exist when it comes to infection with the bacterium *A. salmonicida*. Positive correlation between resistance to furunculosis and infectious salmon anaemia (viral) in farmed Atlantic salmon have previously been reported [14]. On the other hand, a successful breeding program for increased resistance in brook trout (*Salvelinus fontinalis*) to furunculosis also led to higher susceptibilities to *Gyrodactylus* sp., bacterial gill disease, and *Chilodonella* sp. infections [15]. Although several studies have investigated the potential difference between various salmonids with regard to inherent resistance to *A. salmonicida* [16,17,18] a comparison between Baltic salmon and rainbow trout has not previously been carried out. In the present study, a population of East-Atlantic salmon naturally infected with *A. salmonicida* was used as disease carriers in a cohabitation study to test for differences in susceptibility to furunculosis between Baltic salmon and rainbow trout. Here, we present evidence that the Baltic salmon stock compared to rainbow trout carries a high level of natural resistance against *A. salmonicida*. Further, the study confirmed a previously reported high susceptibility in East-Atlantic salmon.

Results

Development of furunculosis in East-Atlantic salmon

The course of furunculosis in the group of East-Atlantic salmon started as a low-grade infection that developed into wide spread disease with a significant increase in mortalities from day 10 (Fig. 1). At day 30 the numbers of fish alive in the two replicate tanks were 14 and 15, respectively (9.7% in total). Bacteria isolated from the kidney of dead fish were identified as *A. salmonicida* subsp. *salmonicida*.

Difference between Baltic salmon and rainbow trout in natural resistance against *A. salmonicida*

The cohabitation infection model proved to be effective in terms of disease transmission. At day four and six post exposure (transfer of 50 infected East-Atlantic salmon) mortality was recorded in rainbow trout and Baltic salmon, respectively (Fig. 2). Bacteriological examination confirmed that mortalities resulted from infection with biochemically identical *A. salmonicida*, thus verifying transmission of disease from East-Atlantic salmon to Baltic salmon and rainbow trout.

The median survival time for Baltic salmon and rainbow trout was 19 and eight days, respectively (Table 1). A chi-square test for independence demonstrated a significant difference between the survival curves of Baltic salmon and rainbow trout (Table 1). Calculating the hazard ratio showed that the relative risk of dying from infection with *A. salmonicida* is 3.36 times higher in rainbow trout compared to Baltic salmon.

Linear regression was performed for the linear sections of the mortality curves representing the three salmonids. The analysis generated three slopes that are presented in Table 2.

Rainbow trout showed the steepest decline (highest mortality rate) followed by East-Atlantic salmon and Baltic salmon, in that order. In concordance with these results the fraction of survivors (30 day survival) in the three salmonid species was 6.2% in rainbow trout, 9.7% in East-Atlantic salmon, and 34.0% in Baltic salmon (Table 2). No mortality was found in any control group.

Discussion

Susceptibility to *A. salmonicida* infection was in a direct comparison demonstrated to differ between Baltic salmon and rainbow trout. A significantly higher survival (34%) was found in the Baltic salmon populations over a 30-day infection course

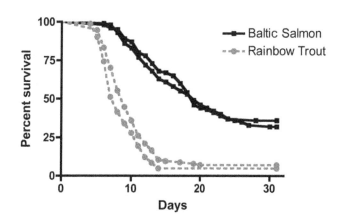

Figure 2. Mortality curves for Baltic salmon and rainbow trout infected through cohabitation (day 0) with *A. salmonicida* infected East-Atlantic Salmon. Curves in black and grey color summarize mortality in 200 fish from duplicate tanks with 100 Baltic salmon or rainbow trout per tank. Non-infected control groups showed zero mortality over the course of the experiment.

compared to rainbow trout (6.2%). A chi-square test and hazard ratio estimation between rainbow trout and Baltic salmon confirmed their difference in susceptibility. East-Atlantic salmon could not be included in these estimations due to their status as a naturally infected population and hence their role as infected cohabitants. That means, the East-Atlantic salmon group should be considered a 100% infected population as opposed to the 33% infection level in the cohabitant groups at the beginning of the experiment. This difference might have affected the kinetics of the infections. Nonetheless, the 30-day survival data (Table 2) indicated a clear trend for a higher susceptibility of East-Atlantic salmon compared to Baltic salmon. Rainbow trout is normally considered more resistant to furunculosis compared to other salmonids (e.g. Atlantic salmon, brown trout, and brook trout) [18,19,20,21]. However, the susceptibility of salmon stocks from the Baltic to *A. salmonicida* have not previously been tested and the

Table 1. Descriptive statistics for differences in resistance to *A. salmonicida* infection in Baltic salmon and rainbow trout.

	Median survival	Chi-square	Hazard ratio[a]
Baltic salmon	19 days	128.5 (P<0.0001)	3.36 (4.08–6.94)
Rainbow trout	8 days		

[a]Hazard ratio with 95% confidence interval.

Table 2. Slope and 30 day survival in East-Atlantic salmon, Baltic salmon and rainbow trout.

	East-Atlantic salmon	Baltic salmon	Rainbow trout
Slope[a]	1.5	0.75	2.4
30 day survival	9.7%	34.0%	6.2%

[a]Slope estimated by linear regression analysis for the linear sections of the mortality curves in Fig. 1 and Fig. 2.

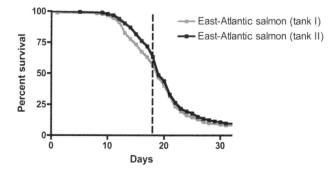

Figure 1. Mortality curves for naturally infected East-Atlantic Salmon. The data summarizes mortality in 300 salmon from duplicate tanks each with 150 fish/tank. The stippled line shows the time-point for randomly picking batches of infected East-Atlantic salmon as cohabitants from parallel tanks (cohab-tanks) with comparable mortalities.

present study is the first to compare resistance to furunculosis in Baltic salmon and rainbow trout.

The mechanisms responsible for the observed difference in resistance between the three salmonids were not investigated in this study. Yet, there are several reports describing factors that might influence resistance to infection in salmonid species. As a first line of defense natural barriers of the skin and the mucus with anti-bacterial properties have been suggested to play a major role [17,20,22]. In the present study, the later onset of mortalities (and confirmed disease) in the Baltic salmon compared to rainbow trout could indicate that the Baltic salmon carries a more resistant exterior as described above. However, a likely entry route besides skin and gills for *A. salmonicida* to the fish is crossing the intestinal lining [23] and systemic disease was confirmed in mortalities of all three salmonids. Thus, the higher survival at the end of the test period in the Baltic salmon compared to both rainbow trout and East-Atlantic salmon points to other defense mechanisms in addition to external barriers. It suggests that means to control systemic disease are present in the Baltic salmon. In this regard, both innate and adaptive anti-bacterial mechanisms are likely to be involved. Moreover, the extreme polymorphisms found at some major histocompatibility complex (MHC) loci and the existence of high- and low-resistance MHC alleles in Atlantic salmon with regard to *A. salmonicida* [24,25,26,27] calls for outbred populations when testing for inherent disease resistance in a given species. Hence, it should be stressed that the population of Baltic salmon used for the present study originated from eggs collected from 100 female fish and fertilized with individual males to eliminate or heavily reduce the effect of individual family differences. As part of the innate response a range of circulating proteins act through neutralizing bacteria or activate downstream effecter mechanisms [28,29,30,31]. During early stages of an *A. salmonicida* infection Atlantic salmon react with a strong and specific humoral response, which during chronic infection is substituted for a less effective response dominated by unspecific natural antibodies [32]. Additionally, opsonization followed by phagocytic clearance [33,34], natural antitoxins [35,36,37], and production of immune complexes [38] are all described as essential parts of an effective anti-bacterial defense in salmonids. Whether these mechanisms or other elements are the reason for the observed survival in the Baltic salmon remains to be investigated.

Differential susceptibility to other pathogens between Baltic salmon, rainbow trout and East-Atlantic salmon has been reported previously. A well-described example is the clear difference in susceptibility between these salmonids to infections with the pathogenic ectoparasitic monogenean *G. salaris* [6,7,10,11,39]. An additional report presented a difference in susceptibility between these fish to infestations with another monogenean species, *Gyrodactylus derjavinoides* [40]. The exact mechanisms responsible for these differences have only been partly elucidated but seem to include variations in expression patterns of a series of cytokine and effector molecules [7,39]. Susceptibility to bacterial kidney disease (BKD) caused by the gram-positive bacterium *Renibacterium salmoninarum* also differs significantly among salmonids, with Pacific salmon species being the most susceptible and rainbow trout the least [41,42]. Baltic salmon were not included in these studies. A comparison between Atlantic salmon and several *Oncorhynchus* spp. in regard to their relative resistance to infectious salmon anaemia (ISA) showed a significantly higher susceptibility and mortality in the Atlantic salmon [43]. Additionally, heritability estimates of susceptibility among Chinook salmon (*Oncorhynchus tshawytscha*) and Atlantic salmon indicated that the heritability component is more pronounced for BKD than for some other bacterial diseases, including furunculosis [16,44].

In salmonid aquaculture the infection pressure with *A. salmonicida* can periodically be substantial [1]. To prevent furunculosis, caused by this bacterium, fish farmers vaccinate their fish and use antibiotics in case of disease outbreak. However, currently used vaccines may cause problematic side effects in the fish [3,4,5]. Moreover, reducing antibiotic treatment remains a goal of aquaculture producers in order to avoid the outlet of antimicrobial residues and the development of resistance in the bacteria [45,46]. In this light, the increased resistance of Baltic salmon to furunculosis shown in the present study may have a series of important implications for future salmon farming since inherent resistance to pathogens could be a means to reduce the need for medication. Moreover, vaccination studies showed that a single vaccination of Baltic salmon smolt eliminated mortality during a four month net-pen period and increased recapture rates significantly after stocking [47,48]. Hence, the combination of improved vaccines with diminutive side effects and use of disease resistant fish stock in the production line may further reduce the need for medication in mariculture. In this regard, the possible use of at least some sub-populations of Baltic salmon should be further investigated. However, the choice of species can obviously not rely solely on one parameter, e.g. resistance to *A. salmonicida*, but needs to take into consideration the differences in susceptibilities to other pathogens in addition to level of domestication, feed conversion rate, and growth potential [49] of the individual species.

Materials and Methods

Ethics statement

The Committee for Animal Experimentation, Ministry of Justice, Copenhagen, Denmark, approved the study including the fish rearing and experimentation (license number 2006/561-1204), which was performed following the ethical guidelines listed in the license.

Fish and rearing conditions

Baltic salmon eggs were collected from 100 wild female spawners and fertilized with sperm from individual wild males all of certified stocks of Baltic salmon (*Salmo salar*, River Lule älv, Vattenfall AB, Umeå, Sweden). The river Lule älv strain is considered an original Baltic salmon strain kept isolated from East-Atlantic salmon stocks for thousands of years [9]. Rainbow trout eggs were obtained from Fousing Trout Farm, Jutland, Denmark. The eggs were collected from more than 50 female spawners (Fousing strain) and fertilized with sperm from six males. To secure a high diversity in the population the egg pool were mixed following incubation. Salmon and rainbow trout eggs were brought to the hatchery and disinfected using iodophore (Actomar K30). Subsequently, they were hatched and fish reared under pathogen-free conditions for three months in recirculated water (Bornholm Salmon Hatchery, Denmark). Hereafter, the fish were brought to our experimental fish keeping facility. The pathogen-free status of the fish was tested and confirmed before the experiment was initiated. In addition, East-Atlantic salmon (River Skjern å, Denmark) carrying a natural infection with *A. salmonicida* were brought to our facility from a salmon hatchery in Jutland, Denmark, for use as infected cohabitants. These fish came from an egg pool based on four female and three male East-Atlantic salmon from River Skjern å, Denmark. The River Skjern å salmon strain is considered to be an original ancient Danish stock, which is currently used for re-stocking of rivers in western Denmark. All fish were acclimated for two weeks and kept in 200 L tanks (200 fish/tank) with bio-filters (Eheim, Germany). Fish were maintained at a 12 h light and 12 h dark cycle in aerated (100% oxygen saturation) tap water at 13°C. All fish were selected for similarity in size (weight 4–5 g). In addition to the above mentioned permission granted by the Committee for Animal Experimentation, Ministry of Justice, Copenhagen, Denmark (see ethics statement), the fish rearing was approved as part of the current restocking program for Baltic salmon in River Lule älv (no file number).

Bacteria

A. salmonicida was isolated from the natural infected East-Atlantic salmon in October 2010, and the infected fish was used for the cohabitation challenge experiment. Isolation was performed on blood agar (blood agar base CM55, Oxoid, supplemented with 5% citrated calf blood) at 20°C for 48 h and the bacteria was identified by the following criteria ([50]): haemolysis, pigment production, cytochrome oxidase, motility, degradation of glucose, arginine dihydrolase, lysine and ornithine decarboxylase, indole and aesculin. Dead and moribund fish were examined bacteriologically to confirm cause of death.

Experimental design

Six groups were established with duplicate tanks for each group: a) Baltic salmon control (no infection; 150 fish/tank), b) rainbow trout control (no infection; 150 fish/tank), c) Baltic salmon+rainbow trout control (no infection; 75+75 fish/tank), d) East-Atlantic salmon (infected, 150 fish/tank), e) Baltic salmon+infected East-Atlantic salmon (100+50 fish/tank), f) rainbow trout+infected East-Atlantic salmon (100+50 fish/tank).

Infection procedure

Fish (Baltic salmon and rainbow trout) were infected through cohabitation with *A. salmonicida*-carrying East-Atlantic salmon. Infected salmon used for cohabitation were tagged (fin-clipped) in order to differentiate these from Baltic salmon and rainbow trout. The use of infected fish as cohabitants provided a natural disease transmission. Initially, disease development and mortality in the infected East-Atlantic salmon was recorded. In addition to the

duplicate tanks described above for this group (d) two additional tanks (cohab-tanks) each holding 200 naturally infected East-Atlantic salmon were set up to produce the cohabitants for effective disease transmission. Dead fish were removed and counted on a daily basis from group (d) during course of infection. When a stable infection was established (Fig. 1) batches of 50 fish were randomly picked from the parallel cohab-tanks and transferred to groups (e) and (f) for infection of Baltic salmon and rainbow trout, respectively. This time-point was day 0 for group (a), (b), (c), (e) and (f). Again, dead fish were removed and counted on a daily basis.

Statistics

The Prism© software package (version 4.0 for Macintosh, GraphPad Software, Inc.) was used to manage data and for statistical analyses. Death from infection was summarized in mortality curves and slopes at the linear section of each curve were estimated by linear regression analysis for comparison between East-Atlantic salmon, Baltic salmon and rainbow trout. The chi-square test for independence was used to test for difference in survival between Baltic salmon and rainbow trout. The hazard ratio (here describing the relative risk of dying from infection) between Baltic salmon and rainbow trout is presented with the 95% confidence interval (CI) [51]. The significance level was set at 0.05.

Acknowledgments

This study was performed in association with the Danish Fish Immunology Research Centre and Network (www.dafinet.dk) and the scientific collaborative FiVac – better fish vaccines (www.fivac.net).

Author Contributions

Conceived and designed the experiments: LHA KB. Performed the experiments: LHA. Analyzed the data: LHA KB. Contributed reagents/materials/analysis tools: LHA KB ID. Wrote the paper: LHA KB ID.

References

1. Pedersen K, Skall HF, Lassen-Nielsen AM, Nielsen TF, Henriksen NH, et al. (2008) Surveillance of health status on eight marine rainbow trout, Oncorhynchus mykiss (Walbaum), farms in Denmark in 2006. J Fish Dis 31: 659–667.
2. Press CM, Lillehaug A (1995) Vaccination in European salmonid aquaculture: a review of practices and prospects. Br Vet J 151: 45–69.
3. Berg A, Bergh Ø, Fjelldal PG, Hansen T, Juell JE, et al. (2006) Animal welfare and fish vaccination - effects and side-effects. Fisken og Havet 9.
4. Koppang EO, Bjerkas I, Haugarvoll E, Chan EK, Szabo NJ, et al. (2008) Vaccination-induced systemic autoimmunity in farmed Atlantic salmon. J Immunol 181: 4807–4814.
5. Koppang EO, Haugarvoll E, Hordvik I, Aune L, Poppe TT (2005) Vaccine-associated granulomatous inflammation and melanin accumulation in Atlantic salmon, Salmo salar L., white muscle. J Fish Dis 28: 13–22.
6. Bakke TA, Jansen PA, Hansen LP (1990) Differences in host resistance of Atlantic salmon, Salmo salar L., stocks to the monogenean Gyrodactylus salaris Malmberg, 1957. J Fish Biol 37: 577–587.
7. Kania PW, Evensen O, Larsen TB, Buchmann K (2010) Molecular and immunohistochemical studies on epidermal responses in Atlantic salmon Salmo salar L. induced by Gyrodactylus salaris Malmberg, 1957. J Helminthol 84: 166–172.
8. Lindenstrøm T, Sigh J, Dalgaard MB, Buchmann K (2006) Skin expression of IL-1beta in East Atlantic salmon, Salmo salar L., highly susceptible to Gyrodactylus salaris infection is enhanced compared to a low susceptibility Baltic stock. J Fish Dis 29: 123–128.
9. Nilsson J, Gross R, Asplund T, Dove O, Jansson H, et al. (2001) Matrilinear phylogeography of Atlantic salmon (Salmo salar L.) in Europe and postglacial colonization of the Baltic Sea area. Mol Ecol 10: 89–102.
10. Dalgaard MB, Nielsen CV, Buchmann K (2003) Comparative susceptibility of two races of Salmo salar (Baltic Lule river and Atlantic Conon river strains) to infection with Gyrodactylus salaris. Dis Aquat Organ 53: 173–176.
11. Heinecke RD, Martinussen T, Buchmann K (2007) Microhabitat selection of Gyrodactylus salaris Malmberg on different salmonids. J Fish Dis 30: 733–743.

12. Kania P, Larsen TB, Ingerslev HC, Buchmann K (2007) Baltic salmon activates immune relevant genes in fin tissue when responding to Gyrodactylus salaris infection. Dis Aquat Organ 76: 81–85.
13. Bakke TA, MacKenzie K (1993) Comparative susceptibility of native Scottish and Norwegian stocks of Atlantic salmon, Salmo salar, to Gyrodactylus salaris Malmberg: Laboratory experiments. Fish Res 17: 69–85.
14. Ødegård J, Olesen I, Gjerde B, Klemetsdal G (2007) Positive genetic correlation between resistance to bacterial (furunculosis) and viral (infectious salmon anaemia) diseases in farmed Atlantic salmon (Salmo salar). Aquaculture 271: 173–177.
15. Hayford CO, Embody GC (1930) Further progress in the selective breeding of brook trout at the New Jersey State Hatchery. Trans Am Fish Soc 60: 109–113.
16. Beacham TD, Evelyn TPT (1992) Genetic-Variation in Disease Resistance and Growth of Chinook, Coho, and Chum Salmon with Respect to Vibriosis, Furunculosis, and Bacterial Kidney-Disease. Trans Am Fish Soc 121: 456–485.
17. Cipriano RC, Ford LA, Jones TE (1994) Relationship between resistance of salmonids to furunculosis and recovery of Aeromonas salmonicida from external mucus. J Wildl Dis 30: 577–580.
18. Ellis AE, Stapleton KJ (1988) Differential susceptibility of salmonid fishes to furunculosis correlates with differential serum enhancement of Aeromonas salmonicida extracellular protease activity. Microb Pathog 4: 299–304.
19. Cipriano RC (1982) Furunculosis in brook trout: infection by contact exposure. Prog Fish Cult 44: 12–14.
20. Cipriano RC, Heartwell CM (1986) Susceptibility of Salmonids to Furunculosis - Differences between Serum and Mucus Responses against Aeromonas-Salmonicida. Transactions of the American Fisheries Society 115: 83–88.
21. Mccarthy DH (1983) An Experimental-Model for Fish Furunculosis Caused by Aeromonas-Salmonicida. J Fish Dis 6: 231–237.
22. Hjelmeland K, Christie M, Raa J (1983) Skin Mucus Protease from Rainbow-Trout, Salmo-Gairdneri Richardson, and Its Biological Significance. J Fish Biol 23: 13–22.
23. Ringø E, Jutfelt F, Kanapathippillai P, Bakken Y, Sundell K, et al. (2004) Damaging effect of the fish pathogen Aeromonas salmonicida ssp. salmonicida

on intestinal enterocytes of Atlantic salmon (Salmo salar L.). Cell Tissue Res 318: 305–311.

24. Kjøglum S, Larsen S, Bakke HG, Grimholt U (2008) The effect of specific MHC class I and class II combinations on resistance to furunculosis in Atlantic salmon (Salmo salar). Scand J Immunol 67: 160–168.

25. Grimholt U, Larsen S, Nordmo R, Midtlyng P, Kjøglum S, et al. (2003) MHC polymorphism and disease resistance in Atlantic salmon (Salmo salar); facing pathogens with single expressed major histocompatibility class I and class II loci. Immunogenetics 55: 210–219.

26. Lohm J, Grahn M, Langefors A, Andersen Ø, Storset A, et al. (2002) Experimental evidence for major histocompatibility complex-allele-specific resistance to a bacterial infection. Proc Biol Sci 269: 2029–2033.

27. Langefors A, Lohm J, Grahn M, Andersen Ø, von Schantz T (2001) Association between major histocompatibility complex class IIB alleles and resistance to Aeromonas salmonicida in Atlantic salmon. Proc Biol Sci 268: 479–485.

28. Hoover GJ, el-Mowafi A, Simko E, Kocal TE, Ferguson HW, et al. (1998) Plasma proteins of rainbow trout (Oncorhynchus mykiss) isolated by binding to lipopolysaccharide from Aeromonas salmonicida. Comp Biochem Physiol B Biochem Mol Biol 120: 559–569.

29. Ottinger CA, Johnson SC, Ewart KV, Brown LL, Ross NW (1999) Enhancement of anti-Aeromonas salmonicida activity in Atlantic salmon (Salmo salar) macrophages by a mannose-binding lectin. Comp Biochem Physiol C Pharmacol Toxicol Endocrinol 123: 53–59.

30. Hoover GJ, Simko E, El-Mowafi A, Ferguson HG, Hayes MA (1998) Lipopolysaccharide-binding lectins and pentraxins in plasma of salmonids genetically resistant to Aeromonas salmonicida. Faseb Journal 12: A807–A807.

31. Smith VJ, Fernandes JM, Jones SJ, Kemp GD, Tatner MF (2000) Antibacterial proteins in rainbow trout, Oncorhynchus mykiss. Fish Shellfish Immunol 10: 243–260.

32. Magnadottir B, Gudmundsdottir S, Gudmundsdottir BK (1995) Study of the humoral response of Atlantic salmon (Salmo salar L.), naturally infected with Aeromonas salmonicida ssp. achromogenes. Vet Immunol Immunopathol 49: 127–142.

33. Griffin BR (1983) Opsonic effect of rainbow trout (Salmo gairdneri) antibody on phagocytosis of Yersinia ruckeri by trout leukocytes. Dev Comp Immunol 7: 253–259.

34. Michel C, Gonzalez R, Avrameas S (1990) Opsonizing Properties of Natural Antibodies of Rainbow-Trout, Oncorhynchus-Mykiss (Walbaum). Journal of Fish Biology 37: 617–622.

35. Cipriano RC (1983) Resistance of Salmonids to Aeromonas-Salmonicida - Relation between Agglutinins and Neutralizing Activities. Trans Am Fish Soc 112: 95–99.

36. Freedman SJ (1991) The Role of Alpha 2-Macroglobulin in Furunculosis - a Comparison of Rainbow-Trout and Brook Trout. Comparative Biochemistry and Physiology B-Biochemistry & Molecular Biology 98: 549–553.

37. Lee KK, Ellis AE (1991) Interactions between Salmonid Serum Components and the Extracellular Hemolytic Toxin of Aeromonas-Salmonicida. Diseases of Aquatic Organisms 11: 207–216.

38. Huntly PJ, Coleman G, Munro ALS (1988) A Comparative-Study of the Sera of Brown Trout (Salmo-Trutta) and Atlantic Salmon (Salmo-Salar L) by Western Blotting against Aeromonas-Salmonicida Exoproteins. Biochemical Society Transactions 16: 999–999.

39. Lindenstrom T, Sigh J, Dalgaard MB, Buchmann K (2006) Skin expression of IL-1beta in East Atlantic salmon, Salmo salar L., highly susceptible to Gyrodactylus salaris infection is enhanced compared to a low susceptibility Baltic stock. J Fish Dis 29: 123–128.

40. Buchmann K, Uldal A (1997) Gyrodactylus derjavini infections in four salmonids: Comparative host susceptibility and site selection of parasites. Dis Aquat Organ 28: 201–209.

41. Sakai M, Atsuta S, Kobayashi M (1991) Susceptibility of 5 Salmonid Fishes to Renibacterium-Salmoninarum. Fish Pathology 26: 159–160.

42. Bruno DW (1986) Histopathology of Bacterial Kidney-Disease in Laboratory Infected Rainbow-Trout, Salmo-Gairdneri Richardson, and Atlantic Salmon, Salmo-Salar L, with Reference to Naturally Infected Fish. Journal of Fish Diseases 9: 523–537.

43. Rolland JB, Winton JR (2003) Relative resistance of Pacific salmon to infectious salmon anaemia virus. Journal of Fish Diseases 26: 511–520.

44. Gjedrem T, Gjoen HM (1995) Genetic variation in susceptibility of Atlantic salmon, Salmo salar L., to furunculosis, BKD, and cold water vibriosis. Aquaculture Research 26: 129–134.

45. McIntosh D, Cunningham M, Ji B, Fekete FA, Parry EM, et al. (2008) Transferable, multiple antibiotic and mercury resistance in Atlantic Canadian isolates of Aeromonas salmonicida subsp. salmonicida is associated with carriage of an IncA/C plasmid similar to the Salmonella enterica plasmid pSN254. J Antimicrob Chemother 61: 1221–1228.

46. Miller RA, Reimschuessel R (2006) Epidemiologic cutoff values for antimicrobial agents against Aeromonas salmonicida isolates determined by frequency distributions of minimal inhibitory concentration and diameter of zone of inhibition data. Am J Vet Res 67: 1837–1843.

47. Buchmann K, Dalsgaard I, Nielsen ME, Pedersen K, Uldal A, et al. (1997) Vaccination improves survival of Baltic salmon (Salmo salar) smolts in delayed release sea ranching (net-pen period). Aquaculture 156: 335–348.

48. Buchmann K, Larsen JL, Therkildsen B (2001) Improved recapture rate of vaccinated sea-ranched Atlantic salmon, Salmo salar L. J Fish Dis 24: 245–248.

49. Gjedrem T, Gunnes K (1978) Comparison of Growth-Rate in Atlantic Salmon, Pink Salmon, Arctic Char, Sea Trout and Rainbow-Trout under Norwegian Farming Conditions. Aquaculture 13: 135–141.

50. Dalsgaard I, Madsen L (2000) Bacterial pathogens in rainbow trout Oncorhynchus mykiss reared at Danish freshwater farms. J Fish Dis 23: 199–209.

51. Ødegård J, Olesen I, Gjerde B, Klemetsdal G (2006) Evaluation of statistical models for genetic analysis of challenge test data on furunculosis resistance in Atlantic salmon (Salmo salar): Prediction of field survival. Aquaculture 259: 116–123.

18

Emergence of Epizootic Ulcerative Syndrome in Native Fish of the Murray-Darling River System, Australia: Hosts, Distribution and Possible Vectors

Craig A. Boys[1]*, **Stuart J. Rowland**[2], **Melinda Gabor**[3], **Les Gabor**[3¤], **Ian B. Marsh**[3], **Steven Hum**[3], **Richard B. Callinan**[4]

1 New South Wales Department of Primary Industries, Port Stephens Fisheries Institute, Nelson Bay, New South Wales, Australia, **2** New South Wales Department of Primary Industries, Grafton Fisheries Centre, Grafton, New South Wales, Australia, **3** State Veterinary Diagnostic Laboratory, NSW Department of Primary Industries, Elizabeth Macarthur Agricultural Institute, Menangle, Narellan, New South Wales, Australia, **4** Faculty of Veterinary Science, University of Sydney, Camden, New South Wales, Australia

Abstract

Epizootic ulcerative syndrome (EUS) is a fish disease of international significance and reportable to the Office International des Epizootics. In June 2010, bony herring *Nematalosa erebi*, golden perch *Macquaria ambigua*, Murray cod *Maccullochella peelii* and spangled perch *Leiopotherapon unicolor* with severe ulcers were sampled from the Murray-Darling River System (MDRS) between Bourke and Brewarrina, New South Wales Australia. Histopathology and polymerase chain reaction identified the fungus-like oomycete *Aphanomyces invadans*, the causative agent of EUS. Apart from one previous record in *N. erebi*, EUS has been recorded in the wild only from coastal drainages in Australia. This study is the first published account of *A. invadans* in the wild fish populations of the MDRS, and is the first confirmed record of EUS in *M. ambigua*, *M. peelii* and *L. unicolor*. Ulcerated carp *Cyprinus carpio* collected at the time of the same epizootic were not found to be infected by EUS, supporting previous accounts of resistance against the disease by this species. The lack of previous clinical evidence, the large number of new hosts (n = 3), the geographic extent (200 km) of this epizootic, the severity of ulceration and apparent high pathogenicity suggest a relatively recent invasion by *A. invadans*. The epizootic and associated environmental factors are documented and discussed within the context of possible vectors for its entry into the MDRS and recommendations regarding continued surveillance, research and biosecurity are made.

Editor: Brian Gratwicke, Smithsonian's National Zoological Park, United States of America

Funding: Fish were collected during routine sampling carried out as part of the Bourke to Brewarrina Demonstration Reach Project, funded under the Native Fish Strategy Program of the Murray-Darling Basin Authority. The funders had no role in study design, data collection and analysis, decision to publish, or preparation of the manuscript.

Competing Interests: The authors have declared that no competing interests exist.

* E-mail: craig.boys@dpi.nsw.gov.au

¤ Current address: Pre Clinical Safety, Novartis Animal Health Australia, Yarrandoo, Kemps Creek, New South Wales, Australia

Introduction

The emergence and spread of aquatic freshwater diseases are a major conservation concern [1]. One aquatic disease implicated in mass mortalities of cultured and wild fish in many countries is epizootic ulcerative syndrome (EUS) [2]. Also known as mycotic granulomatosis, red spot disease and ulcerative mycosis, EUS is caused by the fungus-like oomycete *Aphanomyces invadans* (= *A. piscicida*) and can cause significant ulceration of the skin, necrosis of muscle with extension to subjacent structures including abdominal cavity and cranium, and leading to mortality in many cases [2–4]. Originally described in cultured ayu *Plecoglossus altivelis* in Japan [5], within three decades EUS had been reported in more than 100 fish species [6] in both freshwater and estuarine environments throughout south, south-eastern and western Asia [7–9], the east coast of North America [10,11], in distinct regions of Australia: including New South Wales (NSW), Northern Territory, Queensland and Western Australia [3] and recently in Africa [12]. Due to concern over its potential impact on cultured and wild fisheries,

EUS is officially recognised as a reportable disease by the Network of Aquaculture Centres in Asia-Pacific (NACA) and internationally by the World Organisation for Animal Health (Office International des Epizooties or OIE) [13].

Little is known about the infectious diseases of native fish in the Murray-Darling River System (MDRS), which drains inland catchments, west of the Great Dividing Range in south-eastern Australia. Until recently (2001 in an aquaculture facility), EUS had not been reported from the MDRS, and within Australia was generally considered endemic only to coastal drainages. Originally referred to as Bundaberg Fish Disease, EUS was first reported in Australia in 1972 [14], and subsequently there have been numerous outbreaks reported in wild freshwater and estuarine fishes in the eastern, northern and western coastal drainages [15–18].

EUS has been reported in silver perch *Bidyanus bidyanus* (Mitchell, 1838), a species endemic to the MDRS, being farmed in coastal drainages in northern NSW and south-eastern Queensland [19]. More recently, there have been reports of the presence of EUS within the MDRS, although little is known of its pathogenicity, distribution or susceptibility of species to infection in wild populations in this river system. In May 2008, *A. invadans* was isolated from bony herring *Nematalosa erebi* (Günther, 1868) immediately upstream of Bourke town weir on the Barwon-Darling River, representing the first confirmed, reported diagnosis of EUS in the MDRS, and significantly extending the known range of this pathogen within Australia (J. Go, Elizabeth Macarthur Agricultural Institute, unpubl. data).

During routine sampling of fish in the same section of the Barwon-Darling River in June 2010, dermal lesions and ulcers were observed in six species of fish. Many of the lesions and ulcers appeared to be characteristic of EUS, raising the concern of an epizootic of this reportable disease. As a member country of the OIE, Australia is obliged to report any significant new incursions of EUS, whether it is in a new species or a new area. This paper is the first published account of EUS in four native species: bony herring, golden perch *Macquaria ambigua* (Richardson, 1845), Murray cod *M. peelii* and spangled perch *Leiopotherapon unicolor* (Günther, 1859). The objectives are three-fold. Firstly to document the 2010 epizootic and report its prevalence and extent compared to the time of its first report in *N. erebi* in 2008. Secondly, to document some of the environmental factors associated with the epizootic. Finally, potential explanations of the route of introduction of *A. invadans* into the MDRS are suggested and discussed, with recommendations made for future surveillance, research and biosecurity.

Results

Diagnosis of EUS

In June 2010, *N. erebi*, *M. ambigua*, *M. peelii*, *L. unicolor*, carp *C. carpio* and goldfish *C. auratus* caught from a 200 km section of the Barwon-Darling River between Bourke and Brewarrina weirs (Figure 1) had raised lesions and mild to severe ulcers characteristic of the invasive, tissue-destructive stages of EUS (Figure 2A–F). In some fish the caudal peduncle, caudal fin or dorsal fins were severely eroded (Figure 2D,F), and in others, deep ulcers penetrated into and exposed the peritoneal cavity (Figure 2E).

Aphanomyces invadans was detect as being present in *N. erebi*, *M. ambigua*, *M. peelii* and *L. unicolor* using histopathology, and the diagnosis was further confirmed in *N. erebi* and *M. ambigua* using PCR. Although a small number of *C. carpio* and *C. auratus* were sampled with distinct haemorrhagic, dermal lesions, histopathology and PCR (PCR performed on *C. carpio* only) confirmed that these were not consistent with the case definition of *A. invadans* [2,6,20]. PCR was not performed on *L. unicolor* and *M. peelii*.

Gross observations were consistent in all of the confirmed cases, with focal to multifocal cutaneous ulceration of varying degrees of severity and distribution. Histologically, there was extensive necrosis and ulceration of the epidermis with adjacent epithelial hyperplasia. Subjacent myofibres were, in most cases, severely necrotic, with extensive myofibrillar liquefaction. The endomysium was infiltrated with moderate numbers of histiocytes, lymphocytes and plasma cells, with lesser granulocytes. Distinct, thin sheaths of macrophages (linear granulomas) were frequently observed, surrounded by a narrow band of lymphocytes and plasma cells (Figure 3A). In all cases, myriad fungal hyphae were associated with the ulcers, infiltrating into the myofibres and surrounding connective tissue. Cross sectioned hyphae varied from 10–35 μm in width, non-septate, thick-walled, with occasional branching. The morphology of hyphae was accentuated with GMS staining (Figure 3B). Typically, abundant fibrinocellular

debris was associated with the eroded surfaces. In some *L. unicolor* and *N. erebi*, there were multiple infiltrative granulomas associated with fungal hyphae within internal organs, including kidneys, abdominal adipose tissue, ovary and swim bladder.

Nucleic acid detection of A. invadans

PCR performed on tissue samples detected *A. invadans* DNA in three of three *N. erebi* and three of four *M. ambigua* samples. A negative PCR result was obtained for one *C. carpio* analysed.

Difference in the prevalence of ulcerated fish between 2008 and 2010

In 2008, ulcerated fish representing four species were sampled at 13 of the 30 locations over 200 km from Bourke Weir to Brewarrina Weir, although only N. erebi were submitted for histopathology testing and subsequently confirmed to have EUS (Jeffery Go, unpublished data). By comparison, in 2010 six species with ulcers were sampled from 29 of the 30 locations (Table 1; Figure 4). The prevalence of cutaneous lesions and/or ulcers was 2% (of all fish sampled) in 2008, but 10% in 2010 (Table 1). No *N. erebi*, *L. unicolor* and *M. ambigua* <60 mm had lesions or ulcers (Figure 5). Ulcerated *L. unicolor* ranged in length from 70–170 mm (n = 50), *N. erebi* 60–320 mm (n = 216), and *M. ambigua* 120–480 mm (n = 40). Most ulcerated *L. unicolor* were in the range 70–130 mm, and a majority of *N. erebi* over 140 mm were ulcerated (Figure 5).

Environmental conditions

The epizootics occurred in autumn (2008) and winter (2010) with water temperatures below 16°C and decreasing (Figure 6). In both years, detection of ulcerated fish occurred within two months of significant within-channel flow events (equivalent to between four and eight percentile flows at the Bourke gauge), after an extended period of low-flow conditions (Figure 6). Water quality variables monitored at the time of fish sampling (i.e. once lesions/ulcers were already established) were: temperature 12.7–16.3°C; pH 7.3–8.7; electrical conductivity 350–600 μs.cm^{-1} (electrical conductivity was not significantly different from the five year average obtained from the Bourke gauge; Student *t*-test, d.f. 79, p = 0.593); and dissolved oxygen 4.2–10.6 mg.L^{-1}. Dissolved oxygen concentrations were not below acceptable trigger values for aquatic ecosystems for lowland rivers in south eastern Austrlaia [21].

Discussion

Hosts and distribution

Gross observations and histopathology identified EUS as the cause of the recent epizootic in the Barwon-Darling River in 2010. This is the first published account of EUS in *N. erebi* and first confirmed case of the disease in the native species *M. ambigua*, *M. peelii* and *L unicolor* in the wild. The findings are consistent with the earlier unpublished reports of EUS in *N. erebi* in the Barwon-Darling River in 2008 (Jeffery Go, unpublished data). This increases the number of native fish in the MDRS known to be susceptible to EUS to five, with *B. bidyanus* previously being shown to carry the disease under some culture conditions [19]. Additionally, there has recently (January 2010) been unpublished confirmation of EUS in farmed *M. peelii* at a facility on the Murray River (Brett Ingram and Tracey Bradley, pers. comm). Prior to this, the disease had never been reported in wild or cultured *M. peelii* [22].

Carasius auratus are known to be susceptible to EUS [23], however, we found no histopathological evidence that *A. invadans*

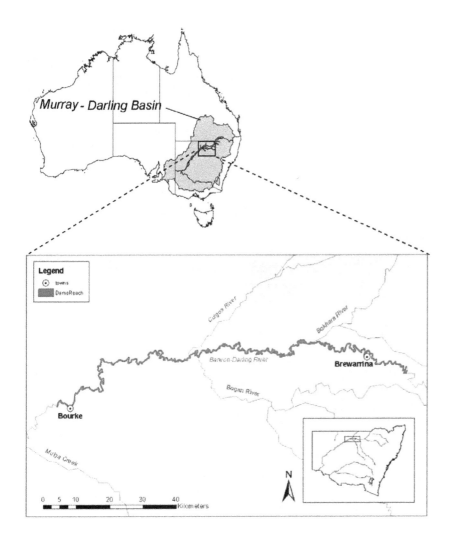

Figure 1. Section of the Barwon-Darling River between Bourke and Brewarrina weirs where fish with EUS were collected.

was the causative pathogen in the two ulcerated *C. auratus* sampled during the recent epizootic. Similarly, despite the abundance of *C. carpio* (n = 953) sampled in the study area during the epizootic, and the observation of seven ulcerated individuals, there was similarly no evidence that *A. invadans* was the causative pathogen for the individuals examined. This is consistent with current evidence through Asia and Europe that *C. carpio* is resistant to EUS [6,24,25].

Prior to 2008, all previous reports of EUS in Australia were from coastal drainages from Bundaberg in Queensland to the Hawkesbury River in NSW [6,14,16,18,19,26]. Following the first report of red spot disease in Australia near Bundaberg in Queensland in 1972 [14], numerous outbreaks followed during the 1980s, and the distribution of the disease expanded south to include the Clarence and Richmond rivers in NSW [16,18,26]. Elsewhere in the world, EUS has spread rapidly throughout south-east Asia and extended deep into the Indian sub-continent since the mid 1980s [27]. These and more recent reports from Africa, Zambia and the USA [12,28–30] demonstrate the extent and ease at/with which this disease can spread.

EUS is a very invasive disease, and when it first occurs in an area, there are high levels of mortality over a very short time, many susceptible species are affected, and individual fish can have numerous lesions and ulcers [27,31]. The severity and prevalence

of this recent epizootic and lack of previous reporting suggests that EUS may be a recent incursion into the MDRS. More than 100 fish species have been reported to be affected by EUS [6], and susceptibility varies between species, with some, including tilapia *Oreochromis* spp. (Castelnau, 1861) resistant to infection by *A. invadans* [24,25,32,33]. Resistance to EUS in tilapia is of significance to the Murray-Darling River System as this species currently poses a significant invasion risk in northern catchments [34]. The severity of ulceration that we observed in *N. erebi*, *M. ambigua* and *M. peelii* suggest that these species are very sensitive to infection by *A. invadans*.

The current geographic distribution of EUS in the MDRS remains largely unclear. The epizootic reported in this study covers a 200 km section of the Barwon-Darling River in the northern region of the Murray-Darling Basin. During publication of this paper, EUS has also been confirmed (histopathologically) in two *M. ambigua* sampled (in July 2011) below Lock 7 in the Murray River, a site downstream of the confluence with the Darling River (M. Gabor and D. Gilligan, NSW Department of Primary Industries, unpubl. data). This, and the previously mentioned report of infected *M. peelii* on a fish farm near the Murray River, confirms that this disease is now widespread in the Murray-Darling River system.

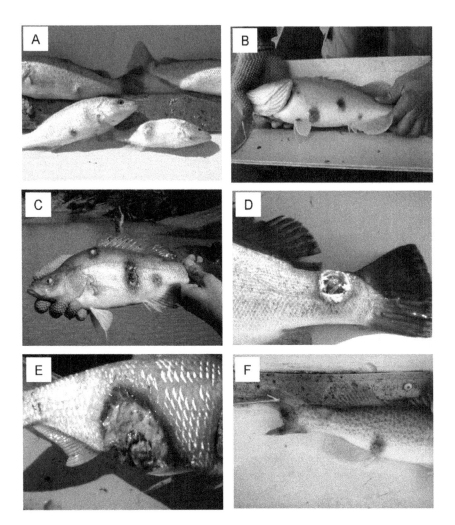

Figure 2. Diseased fish collected from the Barwon-Darling River between Bourke and Brewarrina weirs in June 2010. Panels show A) *L. unicolor*, B) *M. peelii* with raised lesions, C,D) *M. ambigua* and F) *M. peelii* with deep ulceration and muscle or fin necrosis, and E) *N erebi* showing severe ulceration and tissue necrosis exposing the peritoneal cavity and internal organs.

Predisposing environmental factors

Exposure to *A. invadans* spores is a key factor causing EUS [2], and the incidence and transmission of this pathogen throughout the MDRS will largely determine the potential range this disease. Once a pathogen is in an area, subsequent outbreaks of infectious diseases in fishes are closely linked to environmental conditions, particularly temperature and other water quality variables through their effects on stress and the immune system [35–37]. The prevalence of EUS in four species in the Barwon-Darling River in 2010 suggests that conditions were conducive to initial infection and transmission of the pathogen within and between species [2]. Caution must be exercised when interpreting the significance of the water quality measurements taken during this study, as they were taken only at the time of sampling, when fish were ulcerated and the epizootic well advanced. Nevertheless, it is prudent to document the environmental variables associated with all epizootics, to facilitate the development of causal links should more data become available from future outbreaks.

Both outbreaks in 2008 and 2010 were detected at relatively low water temperatures (<16.3°C) following periods of very high flow and flooding. This is consistent with most outbreaks of EUS, which tend to be associated with low and declining water temperatures and high rainfall [6,9,25,33]. Virgona [16] reported significant

correlation between rainfall and the prevalence of early stage lesions in sea mullet *Mugil cephalus* (L., 1758), and found that progression to later stage ulcers occurred after the high flows. Outbreaks of EUS in cultured *B. bidyanus* at the Grafton Aquaculture Centre occur only when water is pumped from the Clarence River during high flows or floods, and when fish in the river are known to have EUS [38].

Temperature is a critical factor determining the severity of EUS outbreaks and most mortalities occur when water temperatures are relatively low [39]. The findings in this paper suggest that high flows and low temperatures may have been predisposing factors to the outbreaks in 2008 and 2010. Low water temperatures (<16°C) and rapid decreases in temperature are immunosuppressive and induce changes to the epidermis, including loss of mucus that predispose fish to infection [37,40]. Outbreaks of the fungal disease saprolegniosis during winter can cause significant mortalities in many species of freshwater fish in the wild and under culture conditions, including *N. erebi* and *B. bidyanus* [38,41].

Other water quality variables including low pH, low dissolved oxygen, decreasing alkalinity, hardness and conductivity have been implicated in outbreaks of EUS [6,9,33]. Although low pH (<6) has been associated with some EUS outbreaks [12,17], Rowland [42] reported an outbreak of EUS in *B. bidyanus* in

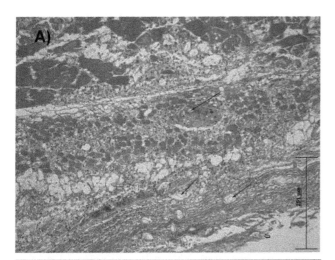

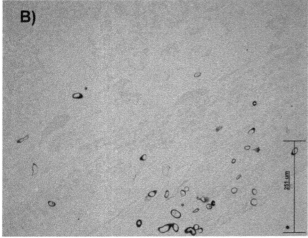

Figure 3. A) *N. erebi*, skin and underlying muscle. Photo micrograph of developing linear granulomas (thin arrow) surrounding faintly eosinophilic fungal hyphae (*). The overlying epithelium is ulcerated (⇔) H & E. (X200). B) *N. erebri*, skin and underlying muscle. Photo micrograph of black staining longitudinal and cross sectional fungal hyphae (*) against green stained tissue. GMS. (X200).

earthen ponds following a bloom of the blue-green algae *Microcystis* and a rapid rise in afternoon pH values to 9.4 and unionised ammonia to 0.39 mg/L. These findings suggest that rapid changes in pH and possibly other water quality variables, and not necessarily absolute values, may initiate changes to the skin which allow attachment of *A. invadans* spores and subsequent invasion of underlying tissue as suggested by Callinan *et al.* [18]. Outbreaks of EUS in the Richmond River in Australia, as well as in the Philippines, Bangladesh and Zambia have been attributed to exposure to acidic runoff draining from acid sulphate soils following heavy rainfall [9,12,43,44]. Sulphidic sediments are not uncommon in floodplain wetlands of the MDRS, potentially being caused by hydrological change brought about by river regulation [45]. Although the risk of wetland acidification appears to be lowest in the Darling River within the vicinity of the recent EUS outbreak, areas of the Murray River appear to contain sulfidic sediments at concentrations which could pose an acidification risk [45]. It is plausible that recent record low flows in the MDRS, combined with the reinstatement of wetting and drying regimes to wetlands may alter pH and provide conditions that predispose fish to infection by *A. invadans*. Extensive

cyanobacterial blooms are known to occur in the Barwon-Darling River [46], but there were no reports coinciding with either of the epizootics.

Possible vectors for introduction of EUS into the MDRS

Controlling the spread of infectious diseases through cultured fish has been a serious problem in many countries, including Australia [47,48]. It is unclear how *A. invadans* has entered the MDRS, but the translocation of cultured *B. bidyanus*, *M. peelii* and *M. ambigua* may have provided a vector for its introduction. In the past, fingerlings have been translocated from Government and commercial native fish farms in eastern drainages in southern Queensland and north-eastern NSW to the western drainage for stock enhancement in impoundments and for commercial aquaculture. In 2001, *B. bidyanus* with EUS were translocated from a commercial hatchery in north-eastern NSW to a fish farm on the Murray River, and although quarantine procedures were implemented, it is unsure if pathogens escaped from the farm (R. Callinan and S. Rowland, unpubl. data). A hatchery quality assurance program and biosecurity measures in NSW now prevent the translocation of native fish fingerlings from eastern drainages to the MDRS, and all fingerlings leaving each hatchery must be free of pathogens and signs of diseases [49]H. Such measures should now be considered for all within-MDRS translocations, including the interstate movement of fish. Until the aspects of distribution and hosts of EUS are better known, no fish should be transported from the Barwon-Darling River to hatcheries for use in breeding programs unless they can be certified free of pathogens.

The extensive, east to west migration of waterbirds from coastal drainages may be a potential vector for the translocation of *A. invadans* to the MDRS that warrants further investigation. Waterbirds are known to disperse a range of aquatic organisms [50]H, and fish-eating birds have been implicated in the spread of some infectious diseases in fish [51–53]. Cormorants *Phalacrocorax* spp., pelicans *Pelecanus conspicillatus*, ibis *Threskiornis* spp., and various species of ducks, including grey teal *Anas gracilis* are commonly found in both coastal and inland drainages, at times aggregating in large numbers on *B. bidyanus* farms where they can introduce pathogens from the wild and move pathogens from pond to pond [38,42]; Jeff Guy, pers. comm.).

It is plausible that *A. invadans* may have been carried into the MDRS from coastal drainages in boats or other equipment. Boats and fishing equipment have been implicated in the transportation of larval and adult stages of some aquatic organisms (in live wells, bilges, bait buckets and engines and the transmission of whirling disease in trout in the USA [54–56]. Whilst there is no evidence that this is responsible for the movement of *A invadans* into the MDRS, it cannot be discounted because is not uncommon for research boats involved in the State and Basin-wide monitoring programs to frequently move between coastal and inland drainages. It would therefore be prudent to 'disinfect' boats and equipment moving between different aquatic environments and this warrants further consideration by biosecurity agencies.

Conclusions

The five known host native fish species, *N. erebi*, *M. ambigua*, *M. peelii*, *L. unicolor* and *B. bidyanus* appear very susceptible to infection by *A. invadans*. Given the invasive nature of EUS, it can be expected to spread to other parts of the MDRS, and the recent confirmation of EUS in farmed and wild fish in the Murray River demonstrates that the disease in not restricted to the Barwon-Darling River. Although no attempt was made to estimate the level of mortality in our study, EUS is known to cause losses of

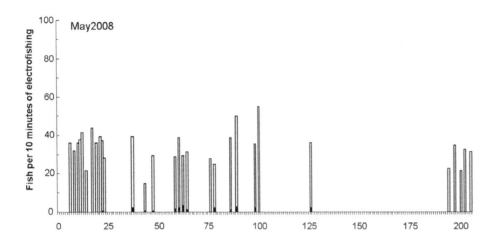

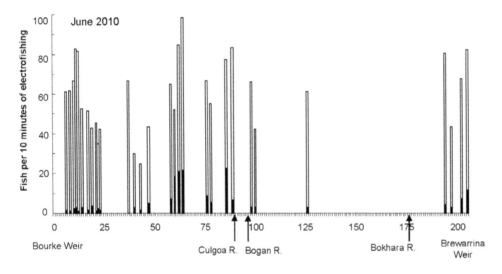

Kilometres upstream of Bourke Weir showing the confluence of main tributaries

Figure 4. Distribution of fish in the Barwon-Darling River between Bourke and Brewarrina weirs with ulcers (black) and without ulcers (grey) in 2008 and 2010.

100% in susceptible species in captivity [29,57]. Oomycete infections cause significant problems in aquaculture and have been implicated in the decline of some wild fish stocks around the world [58]. In *B. bidyanus* culture, the prevalence of EUS can be as high as 90%, with mortality rates >50% in tanks, and winter saprolegniosis can cause total mortality of *B. bidyanus* in earthen ponds [38,59]. The invasive nature of EUS and apparent pathogenicity to at least five endemic species suggest that it poses a significant threat to the fishes of the MDRS warranting surveillance, more vigilant reporting, pathology testing of suspected outbreaks and further research into its potential spread and impact on wild an cultured fish. This will assist in the development of sound biosecurity and fisheries management actions to combat this emerging disease.

Materials and Methods

Ethics statement

All field studies outlined in this paper were authorised under a scientific research permit (permit No: P01/0059) issued by the NSW Department of Primary Industries under section 37 of the Fisheries Management Act 1994. This permit authorises the collection of fish in all waters of NSW. The river sites sampled were not privately owned or protected and no endangered or protected species were involved in this study. All fish collection was carried out in an ethical manner and any fish euthanased were done so in accordance with the Australian code of practice for the care and use of animals for scientific purposes (2004) and NSW Primary Industries (Fisheries) Animal Research Authority 98/14.

Fish and water sampling

Fish were sampled from 30 locations in the Barwon-Darling River over approximately 200 km of river channel between the town weirs of Bourke (30°05'12.77"S, 145°53'39.41"E) and Brewarrina (29°57'27.63"S, 146°51'24.73"E) in north-western NSW (Figure 1). At each location, fish were immobilised using a total of 1080 seconds ('power-on') of electric fishing (boat-mounted 7.5 kW Smith-Root Model GPP 7.5 H.L^{-1}). All fish were collected using a 3 mm mesh dip net and placed in a live-well before being identified, measured (fork length (FL) for fork-tailed

Table 1. Number of ulcerated fish from the assemblage of species collected from 30 locations in the Barwon-Darling River between Bourke and Brewarrina weirs in May 2008 and June 2010.

Common name	Scientific name	May 2008	June 2010
Olive perchlet	*Ambassis agassizii* (Steindachner, 1866)	-	-
Silver perch	*Bidyanus bidyanus*	-	-
goldfish	*Carassius auratus*	-	2 (<1%)
Un-specked hardyhead	*Craterocephalus stercusmuscarum fulvus* (Ivantsoff, Crowley & Allen, 1987)	-	-
Carp	*Cyprinus carpio*	7 (3%)	7 (<1%)
Mosquito fish	*Gambusia holbrooki* (Girard, 1859)	-	-
Carp gudgeon	*Hypseleotris* spp.	-	-
Spangled perch	*Leiopotherapon unicolor*	3 (8%)	**50 (21%)**
Golden perch	*Macquaria ambigua*	3 (2%)	**40 (8%)**
Murray cod	*Maccullochella peelii*	-	**4 (5%)**
Murray-Darling rainbowfish	*Melanotaenia fluviatilis* (Castelnau, 1878)	-	-
Bony herring	*Nematalosa erebi*	**28 (<1%)**	**216 (16%)**
Hyrtl's tandan	*Neosiluris hyrtlii* (Steindachner, 1867)	-	-
Australian smelt	*Retropinna semoni* (Weber, 1895)	-	-
Total number individuals ulcerated		41 (2%)	319 (10%)
Total number of species ulcerated		4 (29%)	6 (32%)

Data in parentheses are the proportion of sampled individuals of each species that were ulcerated. Bold type identifies species in which EUS was confirmed using histopathology. No data means that the species was caught but no specimens were ulcerated.

species and total length (TL) for others) and assessed for disease or abnormalities. The total catch and the number of individuals and species with lesions or ulcers were compared to data obtained using equivalent methods and fishing effort from the same 30 locations in 2008 (obtained from the Department of Industry & Investment NSW, Freshwater Fish Research Database). Dissolved oxygen, pH, temperature and conductivity were recorded at the time of sampling using a Horiba U-10 water quality meter near the surface (<1 m) for each location in 2010.

Histopathology

Diagnostic tests for EUS were carried out on a sub-sample of ulcerated fish (haphazardly selected). After capture, these fish were immediately sealed in bags (one fish per bag) and placed in an ice slurry to induce euthanasia and to preserve specimens until a fixative could be added. Specimens (n = 23) sent for laboratory examination were *L. unicolor* (n = 2), *N. erebi* (n = 7), *M. ambigua* (9), *M. peelii* (n = 1), carp *Cyprinus carpio* (Linnaeus, 1758) (n = 2) and goldfish *Carassius auratus*, (L., 1758) (n = 2).

We used a case definition currently accepted by OIE to confirm the presence of EUS by histopathology. This involved identifying mycotic granulomas in histological sections, with further isolation of *A. invadans* from internal tissues in a subset of cases (Level II diagnosis: [2,6,20]. Within 24 to 48 hours, necropsy examination and sample fixing was carried out on all submitted fish. Samples of skin, underlying muscle tissue and major internal organs were removed and fixed in 10% neutral buffered formalin, and processed for histological evaluation in a standard manner. Slides were stained with Haematoxylin and Eosin (H&E) and Gomoris Hexamine Silver (GMS: [60]). Additionally, samples of tissues underlying ulcers were fixed in 95% ethanol prior to molecular examination.

Nucleic acid detection of A. invadans

DNA extraction from ulcerated lesion of examined fish and EUS PCR was performed as described by Buller *et al.* [61] and the manufacturers recommendations for DNAzol. Briefly, 25–50 mg tissue (about 5 mm) was homogenised in 700 µL of DNAzol reagent (DNAzol® Genomic DNA Isolation Reagent, Molecular Research Centre Inc., Cat. No. DN 127). The homogenate was allowed to stand at room temperature for five to 10 minutes and then centrifuged for 10 minutes at $16,060 \times g$. The supernatant was transferred to a fresh tube and the DNA was precipitated by adding 400 µL of 100% ethanol and mixing by inversion. This was allowed to stand at room temperature for 1 minute and then centrifuged for five minutes at $3421 \times g$. The supernatant was removed and the pellet was washed twice with 600 µL of 75% ethanol by inverting the tube three to six times to re-suspend the DNA then centrifuging for three minutes at $3421 \times g$ to collect the DNA. The remaining ethanol was removed by pipette and the DNA was air dried for five to 15 seconds. The DNA pellet was dissolved by 8 mM NaOH (pellet <2 mm diameter: 50 to100 µL, pellet >2 mm diameter: 150 µL) and then stored at 4°C for immediate use or −20°C for long term storage.

A specific PCR for the direct detection of *A. invadans* in tissue, that targets a 554 bp region of the internal transcribed spacer (ITS) regions [61,62], was undertaken in a 25 µL reaction including: 12.5 µL of Promega PCR Master Mix (Promega Cat. No. M7502), 0.5 µL (800 nM) of each primer (AIFP10, ATTA-CACTATCTCACTCCGC and AIFP 14, CTGACTCA-CACTCGGCTAGC), 2.0 µL of template DNA and purified water. Amplification was then performed in a 96-place thermal cycler (Corbett Research, Sydney, Australia) using the following conditions: one cycle of denaturation at 94°C for five minutes followed by 35 cycles of denaturation at 94°C for one minute, annealing at 55°C for 30 seconds, extension at 72°C for 30 seconds and a final extension of one cycle at 72°C for five

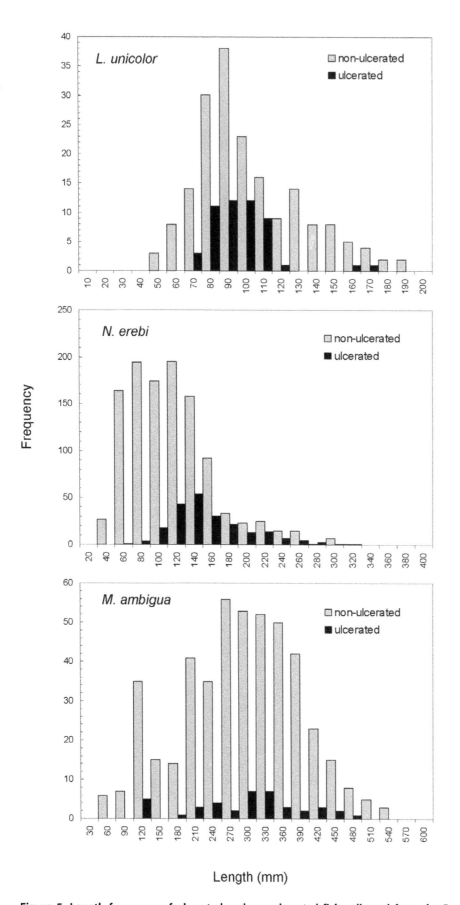

Figure 5. Length frequency of ulcerated and non-ulcerated fish collected from the Barwon-Darling River between Bourke and Brewarrina weirs in June 2010.

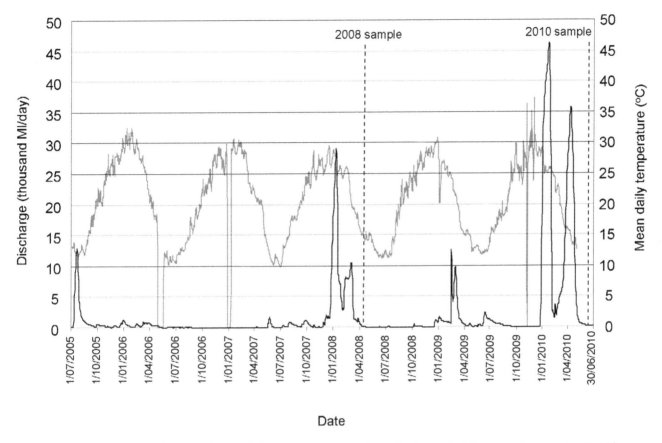

Figure 6. Daily discharge (black line) and mean daily water temperature (grey line) recorded at the Bourke gauge (425003) between July 2005 and July 2010; dotted lines show when the 2008 and 2010 fish samples were collected. Source: NSW Office of Water.

minutes. PCR results were assessed by electrophoresis in 2% agarose gels stained with ethidium bromide. A sample was considered positive when a 554 bp product was produced corresponding to the *A. invadans* positive control.

Acknowledgments

We thank T. Fowler and B. Rampano for assistance with the collection of fish and D. Ballagh for assistance with preparing and fixing samples. Various staff from the Elizabeth Macarthur Agricultural Institute assisted with histopathology and Vanessa Saunders assisted in all PCR analyses.

Thanks to J. Humphrey, B. Ingram, T. Bradley and T. Hawkesford for records and information on EUS. The authors would like to thank N. Otway, A. Boulton and anonymous reviewers whose comments improved this manuscript.

Author Contributions

Conceived and designed the experiments: CAB. Performed the experiments: CAB MG LG SH IM. Analyzed the data: CAB MG LG SH IM. Wrote the paper: CAB SR MG LG SH IM RC.

References

1. Johnson PTJ, Paull SH (2011) The ecology and emergence of diseases in fresh waters. Freshwater Biology 56: 638–657.
2. Baldock FC, Blazer V, Callinan R, Hatai K, Karunasagar I, et al. (2005) Outcomes of a short expert consultation on epizootic ulcerative syndrome (EUS): Re-examination of causal factors, case definition and nomenclature. In: Walker P, Laster R, Bondad-Reantaso MG, eds. Diseases in Asian Aquaculture V. Manila, Philippines: Fish Health Section, Asian Fisheries Society. pp 555–585.
3. Callinan RB, Paclibare JO, Bondad-Reantaso MG, Chin JC, Gogolewski RP (1995) *Aphanomyces* species associated with epizootic ulcerative syndrome (EUS) in the Philippines and red spot disease (RSD) in Australia: Preliminary comparative studies. Diseases of Aquatic Organisms 21: 233–238.
4. Lilley JH, Roberts RJ (1997) Pathogenicity and culture studies comparing the *Aphanomyces* involved in epizootic ulcerative syndrome (EUS) with other similar fungi. Journal of Fish Diseases 20: 135–144.
5. Egusa S, Masuda N (1971) A new fungal disease of *Plecoglossus altivelis*. Fish Pathology 6: 41–46.
6. Lilley JH, Callinan RB, Chinabut S, Kanchanakhan S, MacRae IH, et al. (1998) Epizootic ulcerative syndrome (EUS) technical handbook. Aquatic Animal Health Research Institute, Bangkok. 88 p.

7. Lilley JH, Hart D, Richards RH, Roberts RJ, Cerenius L, et al. (1997) Pan-Asian spread of single fungal clone results in large scale fish kills. Veterinary Record 140: 653–654.
8. Callinan RB, Chinabut S, Kanchanakhan S, Lilley JH, Phillips MJ (1997) Epizootic ulcerative syndrome (EUS) of fishes in Pakistan. A report of the findings of a mission to Pakistann, 9–19 March 1997. Prepared by collaboration between ACIAR, AAHRI, NACA, ODA, NAWFisheries and Stirling University, UK.
9. Sanaullah M, Hjeltnes B, Ahmed ATA (2001) The relationship of some environmental factors and the Epizootic Ulceration Syndrome outbreaks in Beel Mahmoodpur, Faridpur, Bangladesh. Asian Fisheries Science 14: 301–315.
10. Blazer VS, Lilley JH, Schill WB, Kiryu Y, Densmore CL, et al. (2002) *Aphanomyces invadans* in Atlantic menhaden along the east coast of the United States. Journal of Aquatic Animal Health 14: 1–10.
11. Blazer VS, Vogelbein WK, Densmore CL, May EB, Lilley JH, et al. (1999) *Aphanomyces* as a cause of ulcerative skin lesions of menhaden from Chesapeake Bay tributaries. Journal of Aquatic Animal Health 11: 340–349.
12. Choongo K, Hang'ombe B, Samui KL, Syachaba M, Phiri H, et al. (2009) Environmental and climatic factors associated with epizootic ulcerative syndrome (EUS) in fish from the Zambezi floodplains, Zambia. Bulletin of Environmental Contamination and Toxicology 83: 474–478.

13. AFFA (2000) AQUAVETPLAN. Agriculture, Fisheries and Forestry - Australia, Canberra.

14. MacKenzie RA, Hall WTK (1976) Dermal ulceration of mullet (*Mugil cephalus*). Australian Veterinary Journal 52: 230–231.

15. Rogers LJ, Burke JB (1981) Seasonal variation in the prevalence of 'red spot' disease in estuarine fish with particular reference to the sea mullet, *Mugil cephalus* L. Journal of Fish Diseases 4: 297–307.

16. Virgona JL (1992) Environmental factors influencing the prevalence of cutaneous ulcerative disease (red spot) in the sea mullet, *Mugil cephalus* L., in the Clarence River, New South Wales, Australia. Journal of Fish Diseases 15: 363–387.

17. Callinan RB, Sammut J, Fraser GC (2005) Dermatitis, branchitis and mortality in empire gudgeon *Hypseleotris compressa* exposed naturally to runoff from acid sulphate soils. Diseases of Aquatic Organisms 63: 247–253.

18. Callinan RB, Fraser GC, Virgona JL (1989) Pathology of red spot disease in sea mullet, *Mugil cephalus* L., from eastern Australia. Journal of Fish Diseases 12: 467–479.

19. Callinan RB, Rowland SJ (1995) Diseases of silver perch. In: Rowland SJ, Bryant C, eds. Silver Perch Culture Proceedings of Silver Perch Workshops, Grafton & Narrandera, NSW, Australia, April 1994. Sandy Bay: Austasia Aquaculture. pp 67–76.

20. OIE (2003) Manual of diagnostic tests for aquatic animals. Office International des Epizooties, Paris, France.

21. ANZECC (2000) Australian and New Zealand guidelines for fresh and marine water quality. Volume 1, The guidelines. Australian and New Zealand Environment and Conservation Council, Agriculture and Resource Management Council of Australia and New Zealand, Canberra.

22. Ingram BA, Gavine FM, Lawson P (2005) Fish health management guidelines for farmed Murray cod. Fisheries Victoria, Research Report Series No. 32, Snobs Creek.

23. Hatai K, Egusa S, Takahashi S, Ooe K (1977) Study on the pathogenic fungus of mycotic granulomatosis. I: Isolation and pathogenicity of the fungus from cultured-ayu infected with the disease. Fish Pathology 12: 129–133.

24. Ahmed M, Rab MA (1995) Factors affecting outbreaks of epizootic ulcerative syndrome in farmed fish in Bangladesh. Journal of Fish Diseases 18: 263–272.

25. Ahmed GU, Hoque MA (1999) Mycotic involvement in epizootic ulcerative syndrome of freshwater fishes of Bangladesh: A histopathological study. Asian Fisheries Science 12: 381–390.

26. Fraser GC, Callinan RB, Calder LM (1992) *Aphanomyces* species associated with red spot disease: an ulcerative disease of estuarine fish from eastern Australia. Journal of Fish Diseases 15: 173–181.

27. Roberts RJ, Willoughby LG, Chinabut S (1993) Mycotic aspects of epizootic ulcerative syndrome (EUS) of Asian fishes. Journal of Fish Diseases 16: 169–183.

28. Sosa ER, Landsberg JH, Stephenson CM, Forstchen AB, Vandersea MW, et al. (2007) *Aphanomyces invadans* and ulcerative mycosis in estuarine and freshwater fish in Florida. Journal of Aquatic Animal Health 19: 14–26.

29. Saylor RK, Miller DL, Vandersea MW, Bevelhimer MS, Schofield PJ, et al. (2010) Epizootic ulcerative syndrome caused by *Aphanomyces invadans* in captive bullseye snakehead *Channa marulius* collected from south Florida, USA. Diseases of Aquatic Organisms 88: 169–175.

30. Hawke JP, Grooters AM, Camus AC (2003) Ulcerative mycosis caused by *Aphanomyces invadans* in channel catfish, black bullhead, and bluegill from southeastern Louisiana. Journal of Aquatic Animal Health 15: 120–127.

31. Vishwanath TS, Mohan CV, Shankar KM (1998) Epizootic Ulcerative Syndrome (EUS), associated with fungal pathogen, in Indian fisheries: histopathology – 'a cause for invasiveness'. Aquaculture 165: 1–9.

32. Khan MH, Marshall L, Thompson KD, Campbell RE, Lilley JH (1998) Susceptibility of five fish species (Nile tilapia, rosy barb, rainbow trout, stickleback and roach) to intramuscular injection with the oomycete fish pathogen, *Aphanomyces invadans*. Bulletin of the European Association of Fish Pathologists (United Kingdom) 18: 192–197.

33. Pathiratne A, Jayasinghe RPPK (2001) Environmental influence on the occurrence of epizootic ulcerative syndrome (EUS) in freshwater fish in the Bellanwila–Attidiya wetlands, Sri Lanka. Journal of Applied Ichthyology 17: 30–34.

34. Hopley D, Smithers S, Parnell K (2008) Thirty years later, should we be more concerned for the ongoing invasion of Mozambique Tilapia in Australia? Science 57: 1359–1368.

35. Wedemeyer GA (1996) Physiology of Fish in Intensive Culture Systems. Melbourne: Chapman and Hall.

36. Hrubec TC, Robertson JL, Smith SA, Tinker MK (1996) The effect of temperature and water quality on antibody response to *Aeromonas salmonicida* in sunshine bass (*Morone chrysops* × *Morone saxatilis*). Veterinary immunology and immunopathology 50: 157–166.

37. Bly JE, Clem LW (1992) Temperature and teleost immune functions. Fish & Shellfish Immunology 2: 159–171.

38. Rowland SJ, Landos M, Callinan RB, Allan GL, Read P, et al. (2007) Development of a health management strategy for the silver perch aquaculture industry. Report to Fisheries Research and Development Corporation on Projects 2000/267 and 2004/089. NSW Department of Primary Industries, Fisheries Final Report Series No. 93. NSW Department of Primary Industries, Cronulla. 219 p.

39. Chinabut S, Roberts RJ, Willoughby GR, Pearson MD (1995) Histopathology of snakehead, Channa striatus (Bloch), experimentally infected with the specific Aphanomyces fungus associated with epizootic ulcerative syndrome (EUS) at different temperatures. Journal of Fish Diseases 18: 41–47.

40. Quiniou SMA, Bigler S, Clem LW, Bly JE (1998) Effects of water temperature on mucous cell distribution in channel catfish epidermis: a factor in winter saprolegniasis. Fish & Shellfish Immunology 8: 1–11.

41. Puckridge JT, Walker KF, Langdon JS, Daley C, Beakes GW (1989) Mycotic dermatitis in a freshwater gizzard shad, the bony bream, *Nematalosa erebi* (Günther), in the River Murray, South Australia. Journal of Fish Diseases 12: 205–221.

42. Rowland S (1995) High density pond culture of silver perch, *Bidyanus bidyanus*. Asian fisheries science 8: 73–79.

43. Sammut J, Melville MD, Callinan RB, Fraser GC (1995) Estuarine acidification: impacts on aquatic biota of draining acid sulphate soils. Australian Geographical Studies 33: 89–100.

44. Sammut J, White I, Melville D (1996) Acidification of an estuarine tributary in eastern Australia due to drainage of acid sulphate soils. Marine and Freshwater Research 47: 669–684.

45. Hall KC, Baldwin DS, Rees GN, Richardson AJ (2006) Distribution of inland wetlands with sulfidic sediments in the Murray-Darling Basin, Australia. Science of The Total Environment 370: 235–244.

46. Bowling LC, Baker PD (1996) Major cyanobacteria bloom in the Barwon-Darling River, Australia, in 1991, and underlying limnological conditions. Marine and Freshwater Research 47: 643–657.

47. Callinan RB (1988) Diseases of Australian native fishes. In: Bryden DI, ed. Fish Diseases Refresher Course for Veterinarians Proceedings 106, May 1988. Sydney: University of Sydney. pp 459–472.

48. Paperna I (1991) Diseases caused by parasites in the aquaculture of warm water fish. Annual Review of Fish Diseases 1: 155–194.

49. Rowland S, Tully P (2004) Hatchery quality assurance program for Murray cod (*Maccullochella peelii peelii*), golden perch (*Macquaria ambigua*) and silver perch (*Bidyanus bidyanus*). NSW Department of Primary Industries, Sydney.

50. Charalambidou I, Santamaria L (2005) Field evidence for the potential of waterbirds as dispersers of aquatic organisms. Wetlands 25: 252–258.

51. Willumsen B (1998) Birds and wild fish as potential vectors of *Yersinia ruckeri*. Journal of Fish Diseases 12: 275–277.

52. Taylor PW (1992) Fish-eating birds as potential vectors of *Edwardsiella ictaluri*. Journal of Aquatic Animal Health 4: 240–243.

53. Koel TM, Kerans BL, Barras SC, Hanson KC, Wood JS (2010) Avian piscivores as vectors for *Myxobolus cerebralis* in the Greater Yellowstone Ecosystem. Transactions of the American Fisheries Society 139: 976–988.

54. Meyers TU, Scala J, Simmons E (1970) Modes of transmission of whirling disease of trout. Nature 227: 622–623.

55. Johnson LE, Ricciardi A, Carlton JT (2001) Overland dispersal of aquatic invasive species: a risk assessment of transient recreational boating. Ecological Applications 11: 1789–1799.

56. Gates KK, Guy CS, Zale AV (2008) Adherence of *Myxobolus cerebralis* to waders: implications for disease dissemination. North American Journal of Fisheries Management 28: 1453–1458.

57. Pradhan P, Mohan C, Shankar K, Kumar B (2008) Infection Experiments with *Aphanomyces invadans* in Advanced Fingerlings of Four Different Carp Species. In: Bondad-Reantaso M, Mohan C, Crumlish M, Subasinghe R, eds. Diseases in Asian Aquaculture VI. Manila: Fish Health Section, Asian Fisheries Society. pp 105–114.

58. van West P (2006) *Saprolegnia parasitica*, an oomycete pathogen with a fishy appetite: new challenges for an old problem. Mycologist 20: 99–104.

59. Callinan R, Rowland S, Jiang L, Mifsud C (1999) Outbreaks of epizootic ulcerative syndrome (EUS) in farmed silver perch Bidyanus bidyanus in New South Wales, Australia; 26 April–2 May 1999; Sydney. World Aquaculture Society, Baton Rouge. 127 p.

60. Brown GG (1978) Micro-organisms, fungi, virus inclusion bodies, and parasites. An introduction to histotechnology 1st edn. New York: Appleton-Century-Crofts.

61. Buller N (2004) Molecular diagnostic tests to detect epizootic ulcerative syndrome (*Aphanomyces invadans*) and crayfish plague (*Aphanomyces astaci*). Fisheries Research and Development Corporation, Canberra. 99 p.

62. Buller N (2007) Further research and laboratory trials for diagnostic tests for the detection of *Aphanomyces invadans* and *A. astaci* (Crayfish plague). Fisheries Research and Development Corporation, Canberra. 130 p.

19

Glossiness and Perishable Food Quality: Visual Freshness Judgment of Fish Eyes Based on Luminance Distribution

Takuma Murakoshi[1], Tomohiro Masuda[1], Ken Utsumi[1], Kazuo Tsubota[2], Yuji Wada[1]*

1 Food Function Division, National Food Research Institute, National Agriculture and Food Research Organization, Ibaraki, Japan, **2** Department of Ophthalmology, School of Medicine, Keio University, Tokyo, Japan

Abstract

Background: Previous studies have reported the effects of statistics of luminance distribution on visual freshness perception using pictures which included the degradation process of food samples. However, these studies did not examine the effect of individual differences between the same kinds of food. Here we elucidate whether luminance distribution would continue to have a significant effect on visual freshness perception even if visual stimuli included individual differences in addition to the degradation process of foods.

Methodology/principal findings: We took pictures of the degradation of three fishes over 3.29 hours in a controlled environment, then cropped square patches of their eyes from the original images as visual stimuli. Eleven participants performed paired comparison tests judging the visual freshness of the fish eyes at three points of degradation. Perceived freshness scores (PFS) were calculated using the Bradley-Terry Model for each image. The ANOVA revealed that the PFS for each fish decreased as the degradation time increased; however, the differences in the PFS between individual fish was larger for the shorter degradation time, and smaller for the longer degradation time. A multiple linear regression analysis was conducted in order to determine the relative importance of the statistics of luminance distribution of the stimulus images in predicting PFS. The results show that standard deviation and skewness in luminance distribution have a significant influence on PFS.

Conclusions/significance: These results show that even if foodstuffs contain individual differences, visual freshness perception and changes in luminance distribution correlate with degradation time.

Editor: Ayse Pinar Saygin, University of California San Diego, United States of America

Funding: This work was supported in part by grant-in-aids for Young Scientists (B) from the Japan Society for Promotion of Science 23730718 and Grant-in-Aid for Scientific Research (B) from the Japan Society for Promotion of Science: 22300072, awarded to YW. The funders had no role in study design, data collection and analysis, decision to publish, or preparation of the manuscript.

Competing Interests: The authors have declared that no competing interests exist.

* E-mail: yujiwd@affrc.go.jp

Introduction

Every day, consumers make choices about food quality, choosing from among many samples in the market. Freshness is one of the most important factors in fish quality [1,2]. A number of measures, including biochemical, chemical, physical and microbiological techniques, have been developed to measure freshness in food [3,4]. However, consumers usually cannot use these measures for freshness in the marketplace, but must rely on sensory cues such as appearance, texture, sound, taste and smell. Previous research has indicated that the results of these sensory assessments of freshness are highly correlated with freshness scores from chemical parameters, and provide a simple and low cost method of judging the freshness of food [5,6]. In particular, visual properties can include very rich information for food freshness perception [7–10]. For instance, Péneau et al. [9] found that shininess on the surface of food contributes to the perceived freshness of strawberries and carrots.

In the field of vision science, material perception studies focus on the analysis of visual cues that may underlie our ability to discriminate between the different properties of an object. For example, material perception has been studied for stucco-like surfaces [11] and Lambertian surfaces [12]. Motoyoshi et al. [11] revealed that glossiness or the material perception of visual objects varied with image statistics on the surface of objects. In their experiments, the appearance of a visual object was perceived to be glossier as the skewness of the luminance histogram increased.

These statistics of luminance distribution in images are also determining factors for the perceived freshness of food [7,8,10]. Wada and colleagues [7,8,10] found a correlation between perceived freshness and the values of luminance distributions and spatial frequency in images of individual fresh foodstuffs (cabbages and strawberries). However, stimuli used in these studies were patches from photographs of one cabbage leaf or one strawberry, and did not include individual differences among the same foods. For example, when consumers buy an apple in a grocery store, they must choose from many individual apples with various optical differences. Consumers perceive the differences among them and choose the one which looks the best. Of course, the luminance distribution of each image of an individual item would also include these differences. Thus, previous research could not show whether image statistics such as luminance distribution can predict the degradation of food among samples including items with individual differences. If image statistics were important factors in the perception of the freshness of foodstuffs with individual differences, this study might provide a powerful contribution to the development of a simple and low cost freshness estimation system using image statistics. In this study, we investigated whether or not statistics in luminance distribution are among the determining factors in the visual freshness perception of fresh foods even when individual differences are included in stimulus images. We further investigated to determine which parameters were involved and how they affected the perception of freshness. In order to simulate the daily observation and choice of fresh foodstuffs, we used the paired comparison method to measure perceived freshness. The paired comparison method has been used for measuring food preference [13] and visual preference for products [14], and is useful in the determination of preferences. This method allows us to measure freshness perception in comparisons of fresh foodstuffs in a manner that is close to a typical purchase situation. In this study, we chose fish eyes as the stimuli, because the glossiness of a fish's eye plays an important role in assessing the freshness of the fish. After death, fish become dry and wrinkled due to loss of surface moisture, and this initially occurs in the eye. In relation to the glossiness of the fish eye and the freshness of the fish, it has long been known that there is a strong correlation between the degree of fish freshness and the eye fluid refractive index (RI) value of the fish [15,16]. The eye fluid of a fresh fish is bright and transparent. This brightness is lost with time due to drying. Therefore, the light refraction properties of the eye fluid can be used as a quality criterion to assess the freshness of fish [15]. However, refractometers are necessary to measure RI whereas consumers evaluate the glossiness of a fish's eye without machines when they purchase food. In addition, in a previous study on human dry-eye patients, Goto et al. [17] showed that tears contribute not only to ocular surface wetness but also to the extent of light reflection. This finding suggests that the intensity of corneal light reflection reflects tear volume and ocular surface wetness. Thus, we can assume that fish eyes lose wetness and light reflection as the degradation time increases, and that this should be accompanied by a change in the luminance distribution of the fish eye image.

Materials and Methods

Ethics statement

We used three fishes (horse mackerel; *Trachurus japonicus*) that we randomly selected from a local market on April 21, 2011. The research followed the tenets of the Declaration of Helsinki. Written informed consent was obtained after a complete explanation of the study. The study was approved by the institutional ethics committee of the National Food Research Institute.

Participants

Eleven volunteers participated in the experiment (mean age = 31.45 *SD* = 8.19). All of the participants reported normal or corrected-to-normal visual acuity, normal color vision, and no history of neurological problems. No experts on cooking, trading, fish farming, or the sensory evaluation of food were included. We conducted no specific training for participants.

Apparatus

The visual stimuli were presented on a 22-in CRT monitor (Iiyama HM204DA) using ViSaGe (Cambridge Research Systems Co. Ltd.).

Sample

We used three fishes (horse mackerel; *Trachurus japonicus*) that we randomly selected from a local market on April 21, 2011. The photographs used in the experiments were taken on the date of purchase and the day after.

Stimulus Images

The images used in the experiment were taken in a dark room in which the humidity and the temperature were kept at about 23% and 29.0°C, respectively. A digital camera (Nikon D3) was set up using a tripod in a box designed for taking photographs (D' CUBE J; 116×100×100 cm). Illumination was achieved with two floor lamps with a color temperature of 5400K. We took 4256×2832 pixel photos automatically every 2.5 min for 197.5 min (3.29 hours). As stimulus images, we used 128×128 pixel (4.7×4.7 degrees of arc) patches of the eyes of the three individual fishes from the photographs of the freshness degradation process taken at 0, 1.63 and 3.29 hours (see Fig. 1). The purpose of this selection was to investigate whether observers would perceive freshness as a negative function of degradation time. Table 1 shows the statistics of luminance distribution for nine stimulus images (three fishes at three degradation times). These images involve not only the difference between the degradation times, but also individual differences including different positions relative to the camera and illumination.

Procedure

The participants' heads were fixed to a chin rest about 57 cm from the screen. Participants binocularly observed the presented stimuli in a dark room after a dark adaptation period of 10 min. Two stimulus images, which were positioned side-by-side 7 degrees apart in visual angle, were presented on the screen. Participants were required to report which of the two fish in the stimulus images they perceived to be fresher by pressing one of two keys, for a total of 720 trials (comparison of each of the 9 images with each other, yielding 9×8 = 72 comparisons×10 times). For each of the 72 pairs of eye presentations, each of the eyes was presented on the right 50% of the time and on the left the other 50%.

Analysis of data

For each participant, we calculated the perceived freshness score (PFS) for each image using the Bradley-Terry model [18] for comparison on the uni-dimensional scaling. The PFS were analyzed by analysis of variance (ANOVA) with within-subject factors of individual differences between each fish and degradation time. $P < .05$ was considered statistically significant. In order to

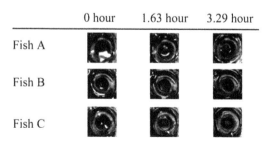

Figure 1. Stimulus images used in the experiment. Stimulus images were patches of 128×128 pixels (4.7×4.7 degrees of arc) of the eyes of three individual fishes from photographs of the freshness degradation process taken at 0, 1.63 and 3.29 hours.

Table 1. Statistics of luminance distribution in each stimulus image.

	Fish A			Fish B			Fish C		
Time (hour)	0	1.63	3.29	0	1.63	3.29	0	1.63	3.29
Average (cd/m²)	66.24	62.73	67.03	60.93	65.72	73.35	47.44	54.88	53.21
SD (cd/m²)	64.69	51.83	49.66	53.01	49.83	49.28	42.47	40.46	41.97
Skewness	1.49	1.44	1.50	1.21	0.87	0.80	1.67	1.06	0.97
Kurtosis	4.46	4.61	5.38	3.83	2.99	2.99	5.41	3.32	3.10

clarify the relationship between the statistics of luminance distribution (average luminance, luminance standard deviation, luminance skewness, luminance kurtosis) and the PFS, a multiple linear regression analysis was also conducted.

Results

Yardstick of perceived freshness scores

The PFS were calculated using the Bradley-Terry Model for each image from the frequency with which one fish was perceived to be fresher than another. Figure 2 shows the yardstick on which one-dimensional lines of PFS for each image are plotted according to their scores. A χ^2 test revealed agreement between participants' responses for perceived freshness ($\chi^2 = 297.11$, $p<.01$). As shown in Figure 2, although each individual fish has different PFS values, an image with a shorter degradation time was perceived as fresher than one with a longer degradation time within each individual fish. In addition, the differences between the images were large when PFSs were high, whereas they were small when the scores were low.

ANOVA of perceived freshness scores

Figure 3 shows the PFS for each stimulus image as a function of degradation time. The PFS for each fish decreased as the degradation time increased; however, the slopes of these scores differed greatly between each fish. As in Figure 2, the differences between the PFS of each fish were larger at the shorter degradation time, and smaller at the longer degradation time.

The ANOVA identified degradation time as having a significant effect on PFS [$F(2, 40) = 36.03$, $p<.01$]. The effect of individual differences was also significant [$F(2, 40) = 16.04$, $p<.01$]. Further, interaction was observed between the effects of degradation time and individual differences [$F(4, 40) = 5.27$, $p<.01$]. The simple

main effects for degradation time and individual differences were examined using a post hoc test, identifying the simple main effects for degradation time in all fishes [$F(2,9) = 22.20$, $p<.01$ for fish A, $F(2,9) = 12.91$, $p<.01$ for fish B, $F(2,9) = 12.36$, $p<.01$ for fish C], and individual differences at 0 and 1.63 hours degradation time [$F(2,9) = 20.69$, $p<.01$ for 0 hr, $F(2,9) = 6.06$, $p<.05$ for 1.63 hr, $F(2,9) = 1.60$, *n.s.*, for 3.29 hr].

Multiple linear regression analysis on statistics of luminance distribution and perceived freshness scores

Multiple linear regression analysis was conducted in order to determine the relative importance of the luminance distribution statistics of the stimulus images in predicting PFS. In identifying the significant variables accounting for PFS, it was found that standard deviation and skewness of luminance distribution had a significant influence on PFS. The adjusted R^2 of this model is .74, which indicates that 74% of variation in PFS was explained by these two dimensions. The significant F-ratio ($F = 8.43$, $p<.05$) indicates that the results of the regression model were unlikely to have occurred by chance. Thus, the goodness-of-fit of the model is satisfactory. Only standard deviation dimensions significantly and positively influenced PFS. Based on the beta coefficient of each independent variable, it is possible to assess the impact of each variable on PFS. As shown in Figure 4, the standard deviation was an important determinant of PFS; it had the highest standardized coefficient value, .79. Figure 5 shows the relation between the scores predicted using this model, and actual PFS. Actual PFS results were distributed near the line that indicates predicted score, meaning that the model in which standard deviation and skewness of the luminance distribution had a significant influence on PFS predicted actual PFS quite precisely.

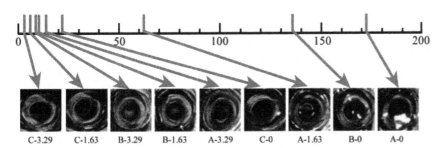

Figure 2. Yardstick which lines up the perceived freshness scores of each image according to their scores on a one-dimensional line. Each vertical gray line on the yardstick indicates the perceived freshness score (PFS) of each image. Distance from the left edge of the scale to each vertical gray line depicts the score size of each image: an image indicated by a vertical gray line positioned nearer the right side was perceived as fresher than that nearer the left side. Labels under the images indicate the individual identification index and degradation time of each image.

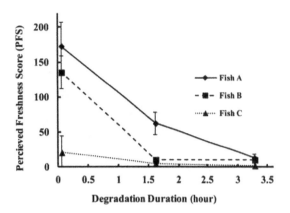

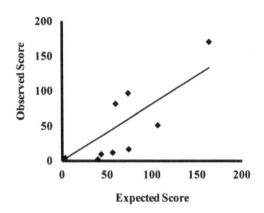

Figure 3. Perceived freshness score (PFS) in each stimulus image as a function of degradation time. The x-axis indicates the degradation time (hour) and the y-axis indicates PFS. Each line represents a function of degradation time of each individual fish. The vertical bars indicate standard error (SE).

Figure 5. Predicted score from the regression model vs. actual perceived freshness score.

Discussion

We investigated whether luminance distribution would continue to have a significant effect on visual freshness perception even under the condition where visual stimuli included not only the degradation of a food, but also individual differences between foods. The results indicate that even if foodstuffs contain individual differences, the visual freshness perception and changes in luminance distribution still correlate with degradation time.

Results of ANOVA with factors of individual differences and degradation time on PFS revealed that when people assess the freshness of fresh fish from its visual appearance, perceived freshness varies greatly between individual fishes. On the other hand, the effect of degradation over time is consistent within each fish; the PFS of each fish decreased as degradation time increased. These findings suggest that freshness perception from fish eyes is affected by degradation time, and also individual differences between each fish.

The result of multiple linear regression analyses of PFS with statistics of luminance distribution as independent variables shows that PFS are predictable from these statistics in an image. The regression equation implies that the change in the perception of freshness over degradation time in a single fresh foodstuff can be estimated from the variation of luminance distribution in accordance with our previous studies [7,8,10]. Moreover, our

findings showed that even if some foodstuffs contain individual differences, degradation in the freshness of those foodstuffs can be predicted by their luminance distribution. Multiple linear regression analysis revealed that perceived freshness can be predicted by standard deviation and skewness of luminance distribution; in particular, fresh food is perceived as fresher as standard deviation in its image becomes higher.

The high-adjusted $R^2 (= .74)$ of our model indicates that standard deviation and skewness of luminance distribution in an image have an important role in the perception of freshness in food. These statistics may be related to the wetness of the eye. It is indicated that the wetness of the eye affects the luminance distribution in images of the human eye [17], and the light refraction properties of eye fluids in fish change with drying [3]. Fish eyes became dryer with a longer degradation time due to the low humidity (about 23%) in our experiment, so that the luminance distributions of the images of the fish eyes changed according to degradation time. Thus, it can be suggested that changes in standard deviation and skewness of luminance distribution may correlate with wetness on the surface of the eye. This correlation might enable us to perceive the freshness of fish from its image with our visual systems using this information.

Here it should be noted that a skew in luminance distribution does not exactly equate to glossiness. Some recent studies have suggested that the strong correlation between glossiness and histogram is violated if the extremes of distribution do not correspond to the locations of specular highlights of the visual objects [19,20]. Since the same spatial correspondence may not apply to translucent objects such as fish eyes, there is the possibility that the correlation between luminance distribution change and the perceived freshness of a fish eye might involve not only the surface change of eyes, but also changes in volumetric light-transport properties such as scattering or absorption.

In addition, there were individual differences in the images of the fish in our experiment due to photographic conditions such as the relative position of each fish to the lighting and camera. Previous studies on the relationship between corneal light reflection and ocular surface wetness in humans suggest that the position of the light source, object, and observer may be an important factor that possibly affects the measurement of reflection [17]. Furthermore, the human visual system allows multiple images to be obtained simultaneously or sequentially through binocular disparity and sequentially motion parallax, and such multiple images enhance glossiness perception [21]. Since the effect of motion parallax on glossiness is enhanced by head motion

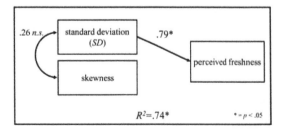

Figure 4. The regression model, which is most appropriate for accounting for PFS computed by multiple linear regression analysis. Two luminance distribution dimensions (standard deviation and skewness) have a significant influence on PFS ($R^2 = .74$, $F = 8.43$, $p<.05$). Only standard deviation significantly and positively influenced PFS (beta = .79, $p<.05$).

[22], the relationship between the light source, object, and viewpoint cannot be ignored for quality perception. Further research which strictly controls artifacts such as the photographic conditions is necessary in order to elucidate the details of the effect of individual differences between foodstuffs on visual freshness perception.

To conclude, we found that humans perceive the freshness of food as a function of luminance distribution in an image, and this perception correlates with degradation over time even when individual differences between foodstuffs are included. The present findings are potentially useful not only towards determining the mechanism of visual freshness perception, but also towards developing a new technique for the nondestructive evaluation of the freshness of fish or any other fresh food. This approach to revealing the function between human perceptions and optical parameters might allow us to establish a foundation upon which the functions between human perceptions and biochemical parameters can be objectively measured.

Recent studies [23–25] using fMRI adaptation and visual object agnosia have suggested that texture and color activate different regions in the human cortex and that glossiness does not depend exclusively upon processing in the same constellation of regions. In addition, Nishio et al. found that particular cortical areas in macaques possess selectivity for glossiness [26]. These findings imply that visual properties such as shape, color, texture and glossiness are separately processed in the brain. Another study using fMRI suggested that the ventral cortex around the fusiform gyrus is related to categorization of materials in humans [27]. The current study, which implies that visual properties such as glossiness might be useful cues for food quality, provides another plausible line of reasoning for the evolutionary advantage of the visual systems to extract glossiness.

Acknowledgments

We are most grateful to the participants. We would like to thank ELCS (English Language Consultation Services) for examination of the manuscript.

Author Contributions

Conceived and designed the experiments: YW T. Murakoshi KT. Performed the experiments: T. Murakoshi T. Masuda. Analyzed the data: T. Murakoshi KU. Contributed reagents/materials/analysis tools: T. Murakoshi T. Masuda. Wrote the paper: T. Murakoshi YW.

References

1. Halbrendt C, Wang Q, Fraiz C, O'Dierno L (1995) Marketing problems and opportunities in Mid-Atlantic seafood retailing. Am J of Agricl Econ 77: 1313–1318.
2. Lebiedzińska A, Kostrzewa A, Ryśkiewicz J, Żbikowski R, Szefer P (2006) Preferences, consumption and choice factors of fish and seafood among university students. Pol J Food Nutr Sci 15/56: 91–96.
3. Huss H (1988) Fresh fish - Quality and quality change. FAO Fisheries Series no29 Food and Agriculture Organization of the United Nations. 132 p.
4. Ólafsdóttir G, Martinsdóttir E, Oehlenschäger J, Dalgaard P, Jensen B et al. (1997) Methods to evaluate fish freshness in research and industry. Trends Food Sci Technol 8: 258–265.
5. Baixas-Nogueras S, Bover-Cid S, Veciana-Nogués T, Nunes ML, Vidal-Carou MC (2003) Development of a quality index method to evaluate freshness in Mediterranean Hake (Merluccius merluccius). J Food Sci 68: 1067–1071.
6. Ben-gigirey B, Vieites Baptista De Sousa JM, Villa TG, Barros-velazquez J (1999) Chemical changes and visual appearance of albacore tuna as related to frozen storage. J Food Sci 64: 20–24.
7. Arce-Lopera C, Masuda T, Kimura A, Wada Y, Okajima K (2013) Luminance distribution as a determinant for visual freshness perception: Evidence from image analysis of a cabbage leaf. Food Qual Prefer 27: 202–207.
8. Arce-Lopera C, Masuda T, Kimura A, Wada Y, Okajima K (2012) Luminance distribution modifies the perceived freshness of strawberries. Iperception 3: 338–355.
9. Péneau S, Brockhoff PB, Escher F, Nuessli J (2007) A comprehensive approach to evaluate the freshness of strawberries and carrots. Postharvest Biol Technol 45: 20–29.
10. Wada Y, Arce-Lopera C, Masuda T, Kimura A, Dan I et al. (2010) Influence of luminance distribution on the appetizingly fresh appearance of cabbage. Appetite 54: 363–368.
11. Motoyoshi I, Nishida S, Sharan L, Adelson EH (2007) Image statistics and the perception of surface qualities. Nature 447: 206–209.
12. Wijntjes MWA, Pont SC (2010) Illusory gloss on Lambertian surfaces. J Vis 10 (9): 13, 1–12.
13. Wagner MW, Green KF, Manley MB (1965) Paired comparison method for measurement of sugar preference in squirrel monkeys. Science 148: 1473–1474.
14. Courcoux P, Smenou M (1997) Preference data analysis using a paired comparison model. Food Qual Prefer 8: 353–358.
15. Gokoglu N, Yerlikaya P (2004) Use of eye fluid refractive index in sardine (Sardina pilchardus) as a freshness indicator. Eur Food Res and Technol 218: 295–297.
16. Yapar A, Yetim H (1998) Determination of anchovy freshness by refractive index of eye fluid. Food Res Int 31: 693–695.
17. Goto E, Dogru M, Sato EA, Matsumoto Y, Takano Y, et al. (2011) The sparkle of the eye: The impact of ocular surface wetness on corneal light reflection. Am J Opthalmol 151: 691–696.e1.
18. Bradley RA, Terry ME (1952) Rank analysis of incomplete block designs I. The Method of paired comparisons. Biometrika 39: 324–45.
19. Anderson B, Kim J (2009) Image statistics do not explain the perception of gloss and lightness. J Vis 9(11): 10, 1–17.
20. Kim J, Anderson B (2010) Image statistics and the perception of surface gloss and lightness. J Vis 10(9): 3, 1–17.
21. Sakano Y, Ando H (2010) Effects of head motion and stereo viewing on perceived glossiness. J. Vis 10(9): 15, 1–14.
22. Tani Y, Araki K, Nagai T, Koida K, Nakauchi S, et al. (2013) Enhancement of glossiness perception by retinal-image motion: Additional effect of head-yoked motion parallax, PLoS One 8(1): e54549.
23. Cavna-Pratesi C, Kentridge RW, Heywood CA, Milner AD (2010) Separate processing of texture and form in the ventral stream: Evidence from fMRI and visual agnosia. Cereb Cortex 20(2): 433–446.
24. Cavina-Pratesi C, Kentridge RW, Heywood CA, Milner AD (2010) Separate channels for processing form, texture, and color: Evidence from fMRI adaptation and visual object agnosia. Cereb Cortex 20(10): 2319–2332.
25. Kentridge RW, Thomson R, Heywood CA (2012) Glossiness perception can be mediated independently of cortical processing of colour or texture. Cortex 48: 1244–1246.
26. Nishio A, Goda N, Komatsu H (2012) Neural selectivity and representation of gloss in the monkey inferior temporal cortex. J Neurosci 32: 10780–10793.
27. Hiramatsu C, Goda N, Komatsu H (2011) Transformation from image-based to perceptual representation of materials along the human ventral visual pathway. Neuroimage 57(2): 482–94.

PERMISSIONS

All chapters in this book were first published in PLOS ONE, by The Public Library of Science; hereby published with permission under the Creative Commons Attribution License or equivalent. Every chapter published in this book has been scrutinized by our experts. Their significance has been extensively debated. The topics covered herein carry significant findings which will fuel the growth of the discipline. They may even be implemented as practical applications or may be referred to as a beginning point for another development.

The contributors of this book come from diverse backgrounds, making this book a truly international effort. This book will bring forth new frontiers with its revolutionizing research information and detailed analysis of the nascent developments around the world.

We would like to thank all the contributing authors for lending their expertise to make the book truly unique. They have played a crucial role in the development of this book. Without their invaluable contributions this book wouldn't have been possible. They have made vital efforts to compile up to date information on the varied aspects of this subject to make this book a valuable addition to the collection of many professionals and students.

This book was conceptualized with the vision of imparting up-to-date information and advanced data in this field. To ensure the same, a matchless editorial board was set up. Every individual on the board went through rigorous rounds of assessment to prove their worth. After which they invested a large part of their time researching and compiling the most relevant data for our readers.

The editorial board has been involved in producing this book since its inception. They have spent rigorous hours researching and exploring the diverse topics which have resulted in the successful publishing of this book. They have passed on their knowledge of decades through this book. To expedite this challenging task, the publisher supported the team at every step. A small team of assistant editors was also appointed to further simplify the editing procedure and attain best results for the readers.

Apart from the editorial board, the designing team has also invested a significant amount of their time in understanding the subject and creating the most relevant covers. They scrutinized every image to scout for the most suitable representation of the subject and create an appropriate cover for the book.

The publishing team has been an ardent support to the editorial, designing and production team. Their endless efforts to recruit the best for this project, has resulted in the accomplishment of this book. They are a veteran in the field of academics and their pool of knowledge is as vast as their experience in printing. Their expertise and guidance has proved useful at every step. Their uncompromising quality standards have made this book an exceptional effort. Their encouragement from time to time has been an inspiration for everyone.

The publisher and the editorial board hope that this book will prove to be a valuable piece of knowledge for researchers, students, practitioners and scholars across the globe.

LIST OF CONTRIBUTORS

Sophie Depiereux, Mélanie Liagre, Lorraine Danis and Patrick Kestemont
Unit of Research in Environmental and Evolutionary Biology (URBE-NARILIS), Laboratory of Ecophysiology and Ecotoxicology, University of Namur, Namur, Belgium

Bertrand De Meulder and Eric Depiereux
Unit of Research in Molecular Biology (URBM-NARILIS), University of Namur, Namur, Belgium

Helmut Segner
Centre for Fish and Wildlife Health, Vetsuisse Faculty, University of Bern, Bern, Switzerland

Jun Hong Xia, Xiao Ping He, Zhi Yi Bai and Gen Hua Yue
Molecular Population Genetics Group, Temasek Life Sciences Laboratory, National University of Singapore, Singapore, Republic of Singapore

Demetra Andreou
Centre for Conservation Ecology and Environmental Change, School of Applied Sciences, Bournemouth University, Fern Barrow, Poole, Dorset, United Kingdom
Cardiff School of Biosciences, Biomedical Building, Museum Avenue, Cardiff, United Kingdom

Kristen D. Arkush
Argonne Way, Forestville, California, United States of America

Jean-Franc¸ois Guégan
Maladies Infectieuses et Vecteurs : E´cologie, Génétique, E´volution et Contrôle, Institut de Recherche pour le Développement, Centre National de la Recherche Scientifique, Universities of Montpellier 1 and 2, Montpellier, France
French School of Public Health, Interdisciplinary Centre on Climate Change, Biodiversity and Infectious Diseases, Montpellier, France

Rodolphe E. Gozlan
Centre for Conservation Ecology and Environmental Change, School of Applied Sciences, Bournemouth University, Fern Barrow, Poole, Dorset, United Kingdom

Michael H. H. Price
Department of Biology, University of Victoria, Victoria, Canada
Raincoast Conservation Foundation, Sidney, Canada

Stan L. Proboszcz and Craig Orr
Watershed Watch Salmon Society, Coquitlam, Canada

Rick D. Routledge
Department of Statistics and Actuarial Science, Simon Fraser University, Burnaby, Canada

Allen S. Gottesfeld
Skeena Fisheries Commission, Hazelton, Canada

John D. Reynolds
Earth to Ocean Research Group, Department of Biology, Simon Fraser University, Burnaby, Canada

Shyh-Chi Chen
Department of Biology, Queen's University, Kingston, Ontario, Canada

R. Meldrum Robertson and Craig W. Hawryshyn
Department of Biology, Queen's University, Kingston, Ontario, Canada
Centre for Neuroscience Studies, Queen's University, Kingston, Ontario, Canada

Kevin Alan Glover, Anne Grete Eide Sørvik, Egil Karlsbakk and Øystein Skaala
Institute of Marine Research, Bergen, Norway

Zhiwei Zhang
Jiangsu Institute of Marine Fisheries, NanTong City, P. R. China

Mark D. Aurit
Tampa Bay Water, Clearwater, Florida, United States of America

Robert O. Peterson
Water Resources Bureau, Southwest Florida Water Management District, Brooksville, Florida, United States of America

Justine I. Blanford
Department of Geography, GeoVISTA Center and Dutton e-Education Institute, Penn State University, University Park, Pennsylvania, United States of America

Gen Hua Yue, Jun Hong Xia, Peng Liu, Feng Liu, Fei Sun and Grace Lin
Molecular Population Genetics Group, Temasek Life Sciences Laboratory, 1 Research Link, National University of Singapore, Singapore, Singapore

Andrea Di Cesare, Carla Vignaroli, Sonia Pasquaroli, Sara Tota, Paolo Paroncini and Francesca Biavasco
Department of Life and Environmental Sciences, Polytechnic University of Marche, Ancona, Italy

Gian Marco Luna
Institute of Marine Sciences, National Research Council, Venezia, Italy

Hsiao-Che Kuo, Young- Mao Chen and Tzong-Yueh Chen
Laboratory of Molecular Genetics, Institute of Biotechnology, National Cheng Kung University, Tainan, Taiwan
Research Center of Ocean Environment and Technology, National Cheng Kung University, Tainan, Taiwan
Agriculture Biotechnology Research Center, National Cheng Kung University, Tainan, Taiwan

Ting-Yu Wang, Peng-Peng Chen and Tieh-Jung Tsai
Laboratory of Molecular Genetics, Institute of Biotechnology, National Cheng Kung University, Tainan, Taiwan

Hao-Hsuan Hsu
Laboratory of Molecular Genetics, Institute of Biotechnology, National Cheng Kung University, Tainan, Taiwan
Agriculture Biotechnology Research Center, National Cheng Kung University, Tainan, Taiwan

Szu-Hsien Lee
Institute of Nanotechnology and Microsystems Engineering, National Cheng Kung University, Tainan, Taiwan
Department of Engineering Science, National Cheng Kung University, Tainan, Taiwan

Chien-Kai Wang
Division of Environmental Health and Occupational Medicine, National Health Research Institutes, Zhunan, Miaoli, Taiwan

Hsiao-Tung Ku
Research Division I, Taiwan Institute of Economic Research, Taipei, Taiwan
Office for Energy Strategy Development, National Science Council, Taipei, Taiwan

Gwo-Bin Lee
Institute of Nanotechnology and Microsystems Engineering, National Cheng Kung University, Tainan, Taiwan
Department of Engineering Science, National Cheng Kung University, Tainan, Taiwan
Department of Power Mechanical Engineering, National Tsing Hua University, Hsinchu, Taiwan

Ekrem Misimi, John Reidar Mathiassen and Ulf Erikson
SINTEF Fisheries and Aquaculture, Trondheim, Norway

Svein Martinsen
Nekton AS, Smøla, Norway

Maya L. Groner, Crawford W. Revie and Ruth Cox
Centre for Veterinary Epidemiological Research, Department of Health Management, Atlantic Veterinary College, University of Prince Edward Island, Charlottetown, Prince Edward Island, Canada

George Gettinby
Department of Mathematics and Statistics, University of Strathclyde, Glasgow, Scotland, United Kingdom

Marit Stormoen
Centre of Epidemiology and Biostatistics, Norwegian School of Veterinary Science, Oslo, Norway

Inge Geurden, Mukundh N. Balasubramanian, Johan W. Schrama, Sadasivam J. Kaushik, Stéphane Panserat and Françoise Médale
INRA, UR1067 NUMEA Nutrition, Métabolisme et Aquaculture, Aquapôle INRA, Saint Pée-sur-Nivelle, France

Peter Borchert
INRA, UR1067 NUMEA Nutrition, Métabolisme et Aquaculture, Aquapôle INRA, Saint Pée-sur-Nivelle, France
Aquaculture and Fisheries Group, Wageningen Institute of Animal Sciences (WIAS), Wageningen University, Wageningen, The Netherlands

Mathilde Dupont-Nivet and Edwige Quillet
INRA, UMR1313 GABI Génétique Animale et Biologie Intégrative, Jouy-en-Josas, France

Diogo Teruo Hashimoto
Centro de Aquicultura, Universidade Estadual Paulista, (UNESP), Campus de Jaboticabal, Jaboticabal, SP, Brazil

José Augusto Senhorini
Centro de Pesquisa e Gestão de Recursos Pesqueiros Continentais, Instituto Chico Mendes de Conservação da Biodiversidade, Pirassununga, SP, Brazil

Fausto Foresti
Departamento de Morfologia, Instituto de Biociências, Universidade Estadual Paulista, (UNESP), Campus de Botucatu, Botucatu, SP, Brazil

Paulino Martínez
Departamento de Genética, Universidad de Santiago de Compostela, Facultad de Veterinaria, Lugo, Spain

Fábio Porto-Foresti
Departamento de Ciências Biológicas, Faculdade de Ciências, Universidade Estadual Paulista, (UNESP), Campus de Bauru, Bauru, SP, Brazil

Angel Avadí
Université Montpellier 2- Sciences et Techniques, Montpellier, France
Institut de Recherche pour le Développement (IRD), UMR212 EME IFREMER/IRD/UM2, Séte, France

Pierre Fréon
Institut de Recherche pour le Développement (IRD), UMR212 EME IFREMER/IRD/UM2, Séte, France

Jorge Tam
Instituto del Mar del Perú (IMARPE), Callao, Peru

Lars Holten-Andersen
Department of Veterinary Disease Biology, Faculty of Life Sciences, University of Copenhagen, Frederiksberg C, Denmark
Division of Veterinary Diagnostics and Research, National Veterinary Institute, Technical University of Denmark, Copenhagen, Denmark

Inger Dalsgaard
Division of Veterinary Diagnostics and Research, National Veterinary Institute, Technical University of Denmark, Copenhagen, Denmark

Kurt Buchmann
Department of Veterinary Disease Biology, Faculty of Life Sciences, University of Copenhagen, Frederiksberg C, Denmark

Craig A. Boys
New South Wales Department of Primary Industries, Port Stephens Fisheries Institute, Nelson Bay, New South Wales, Australia

Stuart J. Rowland
New South Wales Department of Primary Industries, Grafton Fisheries Centre, Grafton, New South Wales, Australia

Melinda Gabor, Ian B. Marsh and Steven Hum
State Veterinary Diagnostic Laboratory, NSW Department of Primary Industries, Elizabeth Macarthur Agricultural Institute, Menangle, Narellan, New South Wales, Australia

Les Gabor
State Veterinary Diagnostic Laboratory, NSW Department of Primary Industries, Elizabeth Macarthur Agricultural Institute, Menangle, Narellan, New South Wales, Australia
Pre Clinical Safety, Novartis Animal Health Australia, Yarrandoo, Kemps Creek, New South Wales, Australia

Richard B. Callinan
Faculty of Veterinary Science, University of Sydney, Camden, New South Wales, Australia

Takuma Murakoshi, Tomohiro Masuda, Ken Utsumi and Yuji Wada
Food Function Division, National Food Research Institute, National Agriculture and Food Research Organization, Ibaraki, Japan

Kazuo Tsubota
Department of Ophthalmology, School of Medicine, Keio University, Tokyo, Japan

Index

Printed in the USA
CPSIA information can be obtained
at www.ICGtesting.com
JSHW051445221024
72173JS00006B/1585